Advance Praise for

Green Remodeling

For 20 years, David Johnston has been a leader in combining environmental design and
excellence in residential construction. Now he has put this deep experience into *Green Remodeling*,
for the rest of us to draw upon. His on-the-ground experience, through the good and the difficult,
makes for entertaining reading, while providing highly useful and accessible information.
Not only does it address the latest thinking about green building, it's one of the best
guides to building in general. David, as usual, doesn't pull any punches.

— Bill Reed, AIA, LEED® AP, Natural

D0730859

At a time when every American is feeling helpless about affecti
to make a difference, along comes *Green Remodeling* — one of the most empowering, comprehensive
how-to books ever! Yes, we *can* change the world for the better by the way we remodel our home.

— Honorable Claudine Schneider, former US congresswoman, and author of
energy efficiency and renewable energy legislation

Imagine having factually accurate *and* emotionally calming info at your finger tips when you remodel
your home. Now, add energy-conscious, durable, healthy, and environmentally-responsible. Sound
implausible? In *Green Remodeling*, David Johnston uses rich language, anecdotes and anec-don'ts to help
homeowners translate their needs and desires into refreshed spaces that are gorgeous, great, and green.

— Helen English, Executive Director, Sustainable Buildings Industry Council

The only solution for the condition of our planet today is for each of us to do all we can to
preserve and enhance the natural capital on which the future of our children depends. The air we
breathe, the water we drink, and the integrity of the ecosystems on which all life and thus all economic
activity depends is vital for a sustainable future. *Green Remodeling* is a tool you can use today to
make your contribution to ensuring a secure future for all the world's children.

— Hunter Lovins, president of Natural Capitalism, Inc., and author of *Natural Capitalism*

As co-host of the integral sustainability domain, David is one of the leading thinkers on how to transform our consumer culture to a more sustainable world. Everyone has a role to play in this transition and *Green Remodeling* is a great place to start. No matter what your project may involve, there are more environmentally sustainable ways to build and this book takes you step by step toward a better future for all....

— KEN WILBER, author of *A Theory of Everything*

David Johnston is such a good writer, readers will readily absorb the wealth of information offered in *Green Remodeling*.

— KATHERINE SALANT, nationally syndicated newspaper and online columnist, and author of *The Brand New House Book*

Remodelers are the unsung heroes of the building industry — adding functionality and beauty to extend the life of existing buildings. Their work is much trickier than new construction because the rooms they start with are rarely square or plumb, and are often full of surprises. *Green Remodeling* shows how this inherently resource-efficient industry can become even more environmentally conscious, and it does so in a way that is both accessible and comprehensive. This book is a real treasure!

— NADAV MALIN, editor, *Environmental Building News* and *BuildingGreen Suite*

Green Remodeling

Green
Remodeling

Changing the World One Room at a Time

DAVID JOHNSTON
and KIM MASTER

NEW SOCIETY PUBLISHERS

Cataloguing in Publication Data

A catalog record for this publication is available from the National Library of Canada.

Cover design by Diane McIntosh. Image Anthony Saint James, Photodisc Green.
Interior design by Jeremy Drought.
All illustrations, except where otherwise noted, by Jill Haras and Kim Master.
Printed in Canada by Transcontinental Printing. Second printing April, 2005.

New Society Publishers acknowledges the support of the Government of Canada through the Book Publishing Industry Development Program (BPIDP) for our publishing activities.

Paperback ISBN: 0-86571-498-3

Inquiries regarding requests to reprint all or part *Green Remodeling: Changing the World One Room at a Time* should be addressed to New Society Publishers at the address below.

To order directly from the publishers, please add $4.50 shipping to the price of the first copy, and $1.00 for each additional copy (plus GST in Canada). Send check or money order to:

New Society Publishers
P.O. Box 189, Gabriola Island, BC V0R 1X0, Canada
1 (800) 567-6772

New Society Publishers' mission is to publish books that contribute in fundamental ways to building an ecologically sustainable and just society, and to do so with the least possible impact on the environment, in a manner that models this vision. We are committed to doing this not just through education, but through action. We are acting on our commitment to the world's remaining ancient forests by phasing out our paper supply from ancient forests worldwide. This book is one step towards ending global deforestation and climate change. It is printed on acid-free paper that is 100% old growth forest-free (100% post-consumer recycled), processed chlorine free, and printed with vegetable based, low VOC inks. For further information, or to browse our full list of books and purchase securely, visit our website at www.newsociety.com

NEW SOCIETY PUBLISHERS   www.newsociety.com

This book is dedicated to John Steiner and Margo King who have been a constant source of inspiration: one person can change the world. Without their support, evocation of and belief in a hopeful future, this book might never have been written.

BOOKS FOR WISER LIVING FROM MOTHER EARTH NEWS

Today, more than ever before, our society is seeking ways to live more conscientiously. To help bring you the very best inspiration and information about greener, more sustainable lifestyles, New Society Publishers has joined forces with *Mother Earth News*. For more than 30 years, Mother Earth has been North America's "Original Guide to Living Wisely," creating books and magazines for people with a passion for self-reliance and a desire to live in harmony with nature. Across the countryside and in our cities, New Society Publishers and *Mother Earth News* are leading the way to a wiser, more sustainable world.

Contents

Part I

Getting Started

PART II

Changing the World One Room at a Time

PART III

Everything You Need to Know (and more)

Acknowledgments

I<small>T TAKES A VILLAGE TO WRITE A BOOK LIKE THIS.</small> There are many people on whose shoulders I stand to have this work in print. Thanks to opportunities so many people have given me, my life has been enriched with experiences that have figuratively built the foundation of this book.

To my clients at Lightworks Construction who taught me how to be a professional remodeler, I am forever grateful. My mentors in the green building industry are Bill Reed, Bob Berkabile, Ray Anderson, Gail Lindsey, Bill Browning, Helen English, Fred Morse, Greg Franta, and so many others who have made a sustainable future their life's work.

To the "GEMS" group — Jim Turner, Richard Perl, John Steiner, Alisa Gravitz, Paul Ray, Ana Mika, Michael Ostrolenk and our circle of change agents: the world is a better place because each of you have such passion and commitment to healing the planet.

To my patrons at What's Working — Mike Weil and Elizabeth Vasetka from the city of Boulder; Wendy Sommer, Meri Soll, Ann Ludwig, and Karen Kho at the Alameda County Waste Management Authority; Lynn Simon at Global Green; and all the remodelers who I have had the honor of teaching about green remodeling: I am grateful for your patience and support.

Special thanks to Pete Nichols for his dedication and persistence as we researched this book. To my collaborators who have reviewed and contributed to this text — Marc Richmond, George Watt, Mercedes Corbel, John Brockett, David Lupberger, Kate Leger, Claudine Schneider, Sandy Butterfield, John Shurtz, Noah Lieb, Nancy and Larry Master: your guidance and insight have been invaluable.

Most of all I appreciate my collaboration with Kim Master, a brilliant and tenacious soul; her dedication and commitment made this book possible.

Foreword

by Marc Richmond

FOURTEEN YEARS AGO, I FELT TIRED OF BUILDING; I had been building for five years and needed to expand my horizons. So I left the industry to acquire additional education in graduate school, to pursue a career in a completely different sector — the environmental field — that would, I hoped, prove more meaningful to me and more valuable to others.

While researching the greatest causes of environmental damage, I read a statistic that changed my life forever: 40 percent of all the world's energy and material resources were used in only one industrial sector — the building industry. I was shocked. I had seen all of the dramatic photographs of air, water, and land pollution, as well as images of natural resources being used at an unprecedented rate, but I didn't understand that my previous work caused environmental ills greater than any other industry. I also felt removed from nature because I have always lived in cities or suburbs, oblivious to strip mines, oil fields, clear-cut forests, steel mills, refineries, and cement kilns. Like others, I was overwhelmed by environmental information and confused about how to make a dent in reducing the problem.

However, I did have a connection to buildings through my work experience, and like most other people, I lived in a building and worked in a building. I didn't need to go elsewhere to help the environment and follow my dream! I needed to wake up, look at where I already was, and fix things from the inside out. I was about to begin my new career in a new, little-known field called green building. To this day, that building industry statistic still inspires me more than anything else to continue to educate people about the benefits of green building.

Knowledge is power. We can't measure people's ideas in their heads or feelings in their hearts, but we can measure what people do. If people have the education and training to understand their choices in life and how they can exercise them, then they will make the appropriate change for themselves and for society in general. I hope that you already have had or will shortly have an awakening of sorts as I did years ago, and that the education you get from this book will drive you to take action as well. Picking up this book is already a piece of that action. Reading it and then using its valuable information will be the measurable piece.

I met David Johnston about ten years ago when he interviewed me for a job opening with an exciting new green building products company. He was a tough interviewer, knew what kind of person he was seeking for this job, and told me directly that it wasn't going to be me. He did seem to like me, though, and asked me to visit him if I ever came to town. I stopped by a few months later and immediately found him to be a man of great character, vision, experience, curiosity, warmth, and balance. I have had the honor of coming to know David over the past ten years as a mentor, a brother, a collaborator, and a confidant. He is someone whom I, and many others who know him, respect as an experienced professional and as a deeply thoughtful human who holds a great passion for people and the planet. He has taught me the quality of being impatient in fighting for the need to educate people on how they could do things better for themselves and their world. He has also taught me to be patient with the world and the slow and nonlinear process of change. I truly admire his manner of always respecting people and trying to understand their perspective. This characteristic is most evident in this book, as he thoughtfully and comprehensively offers an educational pathway to empower readers to make simple, positive changes in their lives that affect them directly as well as all of the rest of us on this planet.

David's previous book, *Building Green in a Black and White World*, was aimed at professional builders — to educate and inspire them to build green for you, their customer. This book is aimed at you, the consumer and the true implementer of the movement. The subtitle, *Changing the World, One Room at a Time*, is truly appropriate because the book is intended to inspire you with the new-found knowledge that small room changes are indeed significant decisions, with ramifications not only in your home but also in your locality and in our global environment.

Whether you intend to hire a professional or do it yourself, this book offers hundreds of useful, doable ideas and steps. As one of the country's most knowledgeable educators of green building, David has assembled all of his experience in this book. He shows that you have many choices in the types of products and services you can buy; that you can make definite choices in your remodeling work, allowing you to express your personal values as you build a healthy, affordable, and durable home.

Most of us feel helpless in a massive global economy that seems to be run by powerful, faceless corporations moving forever forward. In fact, we have a great deal of control, but we rarely exercise our power. We are the consumer of all products and services — those faceless corporations cannot move one inch without our daily purchases. We owe it to ourselves to become more educated consumers and ask for what we want. What do we want? We want our home to be our castle. To build such a castle, we want a good design and quality installation of products that work, that are durable, that are low maintenance, that save us money on utilities and maintenance, and that are healthy for our families. What we are seeking is a concept called green building. We also deserve to know the answers to questions regarding how a product is made, what type of pollution its manufacture creates outdoors and inside our homes, and what the real cost of this product will be over the long term. Many of us have exercised this power, as evidenced by the plethora of products now available versus ten years ago, and in the change in typical manufacturing processes. Today, most carpet manufacturers are actively battling to "out green" their competition by offering products with high recycled content, low emissions, plant-based fibers, and recyclability. The forest products industry has made great strides in implementing sustainable harvesting practices and in making engineered products that use half as much fiber as solid wood while being stronger, straighter, and more durable. Paint and adhesive manufacturers have moved from offering mainly highly volatile, dangerous products to offering numerous products that are completely solvent free.

I hope you will see the knowledge and power within this book. The action to make it happen lies with you. Please read this book carefully, consider its recommendations seriously, and allow yourself — and all of us — to benefit from your new education.

Marc Richmond
Austin Energy Green Building Program
Austin, Texas, 2003

Introduction

My entire childhood was spent in a suburb of Chicago. As far as I knew, June Cleaver was how everyone's mom behaved. I could ride my bike anywhere and be safe. Most of the houses looked alike and many of the people acted the same way. School was interesting and Little League was what boys did in the summer. But no matter what the planned activities were, I spent most of my time in the woods and prairies. I lived outdoors, catching insects, taming raccoons, and fishing in the local streams and ponds. I reveled in the way nature changed throughout the year from spring green to summer lush to fall leaves to winter silence.

Wooded areas are made for building "forts" and treehouses. Or so I believed when I was 13. The highlight of high school was building a three-story treehouse in a giant elm tree deep in the woods; two friends and I worked on it during summers after football practice and on cold fall afternoons. The gratification from building that tree house fulfilled me as nothing else did. My friends and I planned and improvised with materials neighbors had thrown away. Then we built like mad until the designs in our heads stood in physical reality before us. After a long weekend of work, I'd look back and feel great about what we had accomplished. Best of all, we could get away from the grown-ups at will! Thirty feet up and hidden in a dense forest, they couldn't find us, and if they did, they couldn't get to us.

When I arrived at the University of Colorado in 1968, I thought studying engineering would be the key to continuing my building experience. I bent my mind around calculus and physics and statistics and structures, but it was too abstract. I missed the practical experience of building; I missed nailing boards together. So in the summers I worked on construction sites, where my real learning took place.

College and I parted ways after three and a half years so that I could return full-time to building houses. I thought my treehouse and summer building projects would be all the experience I needed to become a carpenter in the "real" world. Wrong! I had no idea there were such dues to be paid! Builders that I met were renegades; young bucks like me were minions to be ordered into servitude. "Boy, see that pile of two by fours over there? They need to be over here. Go move 'em and see me when you're done." After two long hours of hauling boards, I heard, "Boy, what are you doing? Those two by fours are in the way! Move 'em back to where they were, pronto!"

College days started to look sweeter and sweeter.

My savior was a great man named Bob Prentice. He was a third generation Colorado carpenter whose granddaddy had built bars in Cripple Creek during the gold rush days. Bob was a compassionate perfectionist. He taught me a lesson I had never learned in my twenty-something years: there is only one way to do things — right! If it went up, it was plumb. If it went across, it was level. When boards came together, they were square. Period.

I am grateful to Bob to this day because while the rest of the world was saying, "Good enough is good enough," Bob challenged me to build it right. What a concept! What if that philosophy of doing it right were applied to the rest of my life? What would that look like? My entire life I had been searching for black and white rules to follow that would always be true — Bob gave me the first chapter in the rule book.

I followed in Bob's bar-building family tradition and moved to Mississippi to build three bars in a college town. I jumped all over the chance to design and build commercial spaces without being a licensed architect or contractor. Each of the three buildings was unique. We used recycled stained glass, turned balusters, and wrought iron salvaged from torn-down antebellum mansions. Design was now in my blood. Drawing pictures and creating three-dimensional constructions, from paper to completion, was thrilling. I wanted more.

A client retained me to design and build a geodesic dome country retreat. I had never built one, but I knew that the inventor, Buckminster Fuller, fondly known as Bucky, was teaching at Southern Illinois University. The design school at the university taught classes in constructing geodesic domes, so I went to the source for the information I needed. When I arrived in Carbondale, Illinois, in 1974, I knew my life would be different from that point forward. In one geodesic dome that had been converted into a workshop, I saw an electric vehicle being built on a VW chassis. It was one of the first hybrid designs to get over sixty miles to the gallon. In another shop, carcasses of wind machines were being rebuilt using contemporary electrical technology. Site-built solar panels were everywhere.

I was home! I started school a month later. The Design Science program required that you create your own major so I invented what I called "Environmental Systems Design."

Life took on a new reference point for me: there was life before Bucky, and life after Bucky. From day one, my thinking was challenged and changed: we were taught to think in systems. Everything was connected, like nature. Nothing existed separately from anything else. A shift in one system affected the equilibrium of everything else. Any object, place, or thought was made up of subsystems and was a part of a metasystem, or larger context. All systems were created equally and followed the same rules. Smaller systems and larger systems dynamically interacted constantly. Nothing had meaning in isolation. It made so much sense!

Systems thinking became part of my daily life. In order to study buildings, I had to learn about urban planning and forest ecology. And how could I understand heating and cooling systems without a solid background in human comfort conditions and climatology? (The term "passive solar" had not been coined yet. It was called "applied climatology.")

Passive solar.

The building's structure (or an element of it) is designed to allow natural thermal energy flows such as radiation, conduction, and convection generated by the sun to create heat. The home relies solely or primarily on non-mechanical means of heating.

So many interrelationships became obvious to me. A building was an integral part of its surrounding ecosystem. If it respected the ecosystem and interacted with the natural flows, everything functioned better. When it was designed to capture solar heat in winter or prevailing breezes in summer it was more comfortable. Either it fit systemically or it was an irritant to the larger system. Buildings that interrupted animal habitats or changed water courses or microclimatic conditions had larger systemic impacts than just at the site itself. A building also had significant impact on the occupants, for good or ill. I learned that drafting an architectural wonder was unacceptable unless its existence was environmentally justified.

I graduated with my degree in Environmental Systems Design. Each word held a special significance for me. Designing environmental systems was a delightful challenge. Solar heating, wind-generated electrical energy, hybrid electric vehicles, and structures that danced with the elements all stimulated my thinking, my imagination, and my soul.

The horizons of my thinking said that systems were the basis of understanding everything. With systems thinking we could redevelop cities, eliminate environmental pollution, even cure cancer. I faced a design opportunity to take what works today and redesign the rest. It was so logical! I also developed a new appreciation for architects. I assumed architects understood good design because they had gone through similar training; architects had to synthesize an amazing amount of information to be able to design buildings that worked on many levels.

After graduation, my optimism led me to believe that it was just a matter of time until all design would be based on environmental systems. I wanted to be on the cutting edge of this exciting time; I was sure a new day was dawning that would shape how we built buildings from now on. Environmentally responsive, solar powered, human engineered, aesthetically integrated — the home of the twenty-first century had arrived 25 years ahead of schedule! Environmentalists had predicted that we would hit our limits to growth in the year 2000, but with this new vision we were going to make it! We had the tools, and the world would obviously understand its dilemma and transform itself for the sake of the future. Once again, I believed I had a clear view of the rules of the game.

Many years and many miles later, I realized how woefully inaccurate my naïve assumptions had been. What I had thought was inevitable — a world based on systems thinking, renewable energy, and human engineering — never truly came to pass. Solar had a short life in the late '70s and early '80s. Geodesic domes made great radar enclosures for the defense department, but they never made it to the suburbs. Bucky died in 1983, loved but not understood. His clear guiding principles seemed to get lost in the oil embargo and the subsequent recession.

Trying to recapture the enthusiasm of my early treehouse days, I designed homes, taught college, consulted for the government, and ran a construction industry trade association. But something was missing — something about looking over my shoulder at the end of the day and feeling gratified at seeing the frame of a new structure that wasn't there days ago. Building was still in my blood.

In 1983, I started Lightworks Construction in Washington, D.C. I wanted this company to combine the solar work I had done and the construction I loved. I was going to design and build solar homes that would embody all the principles I learned in school with the latest state-of-the-art passive solar design. I had been doing business by the book, following all the black-and-white rules of starting a construction company. But I realized I didn't want a black-and-white company — I wanted a "green" construction company! I didn't want to build typical houses that looked like all the others on the block. I wanted to build houses that incorporated all I had learned about systems, energy efficiency, and quality construction that would last a hundred years. My hard work, marketing, and friendly customer relations paid off when Lightworks became the contractor of the year in Washington, D.C. and was named one of the top 50 contractors in the country by *Remodeling Magazine*. Building green in a black and white world paid off in spades!

My days as an official contractor ended ten years ago when I moved to Boulder, Colorado and began What's Working, an environmental construction consulting firm. I have had the pleasure of working with dozens of policy makers, developers, builders, architects, and homeowners to create thousands of healthy, environmentally friendly homes and buildings around the country. Still, I call myself a "recovering contractor" because I just can't leave my own house alone. Not only have I built a new office (the inspiration for this book), but I have replaced, repaired, upgraded, or added every major appliance in my home, not to mention the windows, doors, siding, roofing, insulation, the entire basement apartment, and a new solar system.

Green remodeling is the beginning of creating a lifestyle that is more in tune with the natural rhythms and flows of the planet. With all of the ecosystems on the planet in rapid and severe decline, the extinction of 27,000 species per year, and the onset of global climate change, we can no longer wait for "them" to fix it all. There is no "them" out there. *We* are the ones we have been waiting for to make the world a better place for our children and grandchildren. Each decision we make today regarding energy use, building materials, and water consumption will have a long-lasting impact on future generations. We have a responsibility today to make the wisest decisions possible to address the myriad environmental issues we face.

My work is based on the belief that each one of us makes a difference. Collectively, Americans spend $160 billion each year on remodeling. If just a fraction of this money was focused on greener construction, we could restore a significant portion of the world's ecosystems while immediately creating a healthier indoor environment for our families. Just think in terms of changing the world, one room at time.

David Johnson
Boulder, CO
July 2004

How to Use this Book

IT IS OUR INTENTION THAT THIS BOOK BE THE ONLY ONE you will need to start on the adventure of remodeling your home. It is divided into several sections so that you can readily find the information that is pertinent to your specific project. The first chapter is a journal of my own remodeling projects, complete with stories and details not covered elsewhere. But primarily the book is a "how-to" resource, including how to finance your remodeling project (Chapter 2); how to work with an architect, how to love your contractor, and how to stay sane during the remodeling process (Chapter 3); and how to understand building science basics (Chapter 4).

Given that buildings have such an enormous impact on the global environment, we have also developed a brief synthesis of the major environmental issues facing our planet (Chapter 5), particularly as they affect the choices you must make when remodeling. It is difficult to find a single source of environmental facts and figures that tell the real story instead of simply giving the media spin on our planetary conditions. This chapter is meant to stand alone as a reference, providing you with a basis to find out more about what is happening around the world — but it also contains information that relates specifically to the green building process.

The remaining chapters form a resource of green building materials and techniques — why you should use them, and how they provide benefits to your family and your checkbook. You can use Chapter 6 to look up individual rooms that you want to renovate, and then find information on stages of construction and categories of products in the following chapters (Chapters 7–19), to guide your remodeling team in selecting the green products that are most appropriate to your project.

The Resources section provides information for further research on a variety of subjects. The glossary will help you find terms commonly used in the green building world that may be unfamiliar to you, and which may help you to communicate clearly with your architect and contractor. In Appendix 2, you will find conversion tables that will allow you to cross-reference between US measurements and metric equivalents.

The most important thing to us is to provide you with a smorgasbord of options on green building. You don't have to do everything! Pick and choose among the hundreds of options for the design features and green building products that serve your best interests. There is no such thing as the right or wrong set of products. This book was designed to give you as much information as you need in order to make informed decisions that will lead to your pleasure and comfort for years to come, when you are living in your remodeled home.

Part I

Getting Started

CHAPTER ONE

The Story of My Remodeling Project

GREEN REMODELING IS NEW TO MOST PEOPLE, so I want to share my experiences remodeling my 1973 vintage home and building a new office addition. They represent two entirely different approaches: first, working within the confines of an existing house with incremental improvements; and second, putting a second-floor addition over a detached garage that was designed to be as green as I could make it. My intention is to introduce you to some new building materials in the hope of easing their integration into mainstream construction, and to make green building standard building practice for you. What is unique about green remodeling is that each green material holds importance to your home; in a sense, each material has a story. You end up telling people about the retaining walls that were reclaimed from your dentist's yard or the structural beams that were compressed together from fast-growing aspen trees.

In my home renovations, I did many things that you will probably never consider. There is no need, unless you have a compelling desire, as I did, to do everything. Building green is a thinking process, not a contest to see how many green things you can incorporate into your home. Do what you can within your budget and motivation. By taking control of the project myself, I was able to get many products directly from the manufacturer. I also intervened with the trade contractors and instructed them on how to use the products. I was able to select my tradesmen by their willingness to try something different. In a typical remodeling project, you are somewhat at the mercy of a remodeler's relationship with subcontractors. In any case, here is my story; I hope it encourages you to create your own!

My 1973 Vintage Home

When I moved to Boulder, Colorado in 1993, I bought a "fixer upper" with a view. The house was built in 1973 and was rented for eight years before I bought it. Like

Fig. 1.1: My home in the hills. Credit: Kim Master.

any rental property, it showed the results of hard use and little maintenance. It needed everything, especially in the kitchen. All I could think about was getting rid of the avocado refrigerator and the lemon yellow countertops. I wanted to transform the kitchen from the '70s look-of-the-day to something modern and functional.

Several months later we were having our first dinner party. The house looked great, the table was set, and there were candles everywhere sparkling in the windows that overlooked the lights of Boulder. I didn't think the house could look any better, except with a new coat of paint. Just as friends were starting to arrive, our first November winter windstorm blew in. The windows were closed — yet all the candles blew out in the living room! The windows leaked air so badly that we had to wear jackets while we entertained. I realized that the kitchen renovations would have to wait; first, the single pane windows had to go.

Educated about window technology from my remodeling days, I called a local manufacturer that made Heat Mirror™ windows. These are often called "super windows" because they have a film of plastic suspended between two panes of glass that make them super energy-efficient — they're almost twice as effective as conventional windows. I started in the living room where most of the room was glass, by spring I had replaced all the fixed glass in the house. On the south side I wanted to enhance the passive solar gain, to heat my home naturally with the sun, so I installed high solar heat gain coefficient (SHGC) glass that lets lots of sunlight through the glass. On the east side I put in low SHGC windows that minimize sunlight through the windows, reducing overheating on hot summer mornings. The windows worked like a charm, making the house more comfortable and reducing our heating bill significantly.

Now I was ready to take on the avocado kitchen. Well — not quite. I came home one night to find water all over the basement floor. The water heater had rusted out and leaked. I talked to plumbers who suggested low-end models that would "serve the purpose," but I wanted better. I finally found a water heater that had twice the insulation around the tank. It included a valve that prevented hot water from rising up into colder pipes and returning to the tank as cold water, a phenomenon known as "thermosyphoning." Luckily, the model was also the "California code builder model," so the company manufactured a great many of them for builders to implement easily and inexpensively into their projects. My energy-efficient water heater ended up costing only slightly more than the low-end model the plumber suggested, due to transportation expenses from California. But more importantly, my monthly energy bills have been noticeably lower.

The upgraded windows and water heater were just the beginning of the improvements I've made to my home. I rebuilt the boiler and made it more efficient. I insulated all the hot water pipes and the crawlspace. I replaced the '70s washing machine when it died with a front-loading unit that used half as much water per load. I replaced burned-out light bulbs with compact fluorescent bulbs that are still burning ten years later. I repainted with low volatile organic compound (low-VOC) paint that contains fewer toxins that conventional paint. And I replaced the avocado refrigerator that finally died (thank God!) with the most energy efficient model on the market that uses 50 percent less electricity and comes in a lovely off-white color.

Even if it wasn't broken, I tried to replace old technology with new technology. Now, with the push of a button, my Metlund "on-demand hot water pump" circulates water in the pipes until it reaches a predetermined hot temperature at my bathroom faucet. I no longer waste water waiting for the shower to heat up; rather, the pump shuts off and the water is instantly hot.

Photovoltaic (PV) panels that generate electricity from the sun have been my best and biggest upgrade. Living in the mountains, we are at the end of the pipeline for everything, including electricity, water, gas, and cable. If anything went wrong somewhere below us we lost our service. At least once a month, especially in big snowstorms or thunderstorms, our power would go down, stopping my home-based business in its tracks. So in 1999, when the State of Colorado and the US Department of Energy had a rebate program for PV systems, I jumped on the opportunity. Interest rates had come down and I was able to refinance the house, allowing me to take enough money out of equity to pay for the PV system and simultaneously save money on my monthly payment.

The 1.2-kilowatt system is enough to power my office. In an emergency, the system also powers the hot water baseboard heating, lights, and the refrigerator. I

Solar Saves the Day

One day several months after installing my photovoltaic system, I was working frantically on a final report for a contract in Washington, D.C. The report had to be in that day by close of business, D.C. time — period. I was furiously writing, sending faxes and e-mails, and finalizing the report. At 3:00 p.m. mountain time, I hit send to deliver the report by e-mail to D.C. Only when I went into the kitchen to get a drink did I realized that all the digital clocks were flashing and that the power had been out for six hours! My computer never even blinked. The money I would have lost if I hadn't gotten the report in was more than the cost of the PV system. As far as I was concerned, the system paid for itself in three months.

don't believe in payback periods for energy equipment, but I do believe in never being powerless again.

Being powerless is not my only pet peeve — I also dislike the idea of my house on fire. We live in a mountain desert that is prone to drought and fire. The summer of 2002 was especially dry and it seemed as if the entire state was on fire. In every direction the smoke-filled sky looked like the Los Angeles basin. Just a quarter of a mile south of us a fire broke out. Fortunately a fire plane was just taking off after refilling its tanks in a nearby reservoir and it was able to put out the fire. A few weeks later another fire started just north of us and burned a couple of neighbors' houses. I panicked. I worked out the details to put one inch of rigid insulation over the existing plywood siding and under new James Hardie fiber-cement siding. That served two purposes: it increased the insulation all around the house and it fireproofed my home, reducing my home insurance rates by 15 percent. Needless to say, I slept better for the rest of the summer.

All this green retrofit work I did in the house was within the "footprint" of the existing foundation with very little cosmetic change (yes, I still have the canary yellow formica in the kitchen). I worked on mini projects one at a time, but the accumulation added up to reducing energy bills by 50 percent and water consumption by 30 percent. Yet these savings are just a small part of the gratification; peace of mind is priceless. The PV system has kicked in many times, allowing me to be productive in my office during down times. Fireproof roofing and siding are major contributors to my sense of safety. Reduced water usage protects the limited neighborhood water system. In general, greening my vintage home has been a fulfilling way to connect with my environment and to support my community.

My 2003 Office Addition

The following section details my office addition process, from planning and deconstruction through interior work and computer networking. I have written it in a journal format to give you a better sense of renovation time frames and chronology, as well as to prepare you for the feelings of elatedness and stress that tend to accompany specific remodeling stages. In sidebars, I've highlighted the green features in my office. This format reflects the way the green features were constantly on my mind, motivating the renovation process.

Planning

July 1, 2002

We are finally planning our office addition! When my wife and I first considered buying the house ten years ago, we climbed up on the garage roof and fantasized about working there. The view is breathtaking! And now it is really going to happen!

This is going to be the "world headquarters" of What's Working, my environmental construction consulting company, so it needs be the best example of what is working. I want to build a signature piece that will reflect what my company is all about — a prototype green building. I want unique architecture that will blend into my mountain neighborhood; a playful space in which to work and entertain; a room big enough for presentations and meetings. Additionally, it has to be energy efficient, it has to use all resources as efficiently as possible, and it has to have clean air without the toxic chemicals that are released from so many building products today.

Consider multipurpose spaces.
Rooms with multiple functions are more adaptable to changing needs. Think ahead when designing spaces — it will save you the time, energy, resources, and money involved with renovating again!

Fig. 1.2: View from garage.
Credit: Kim Master.

Consider "green" financing.

Contact a local professional who is familiar with financing green, energy efficient renovations. For further information, go to <www.DreamSourceFinancial.com>

I ruminate over what I want in my space and how it will work as a multi-purpose building. I need office space now, but what about the future? Will my company outgrow this building as I have outgrown the office in the house? What if I accept an assignment in another country (a fantasy that I hold dearly, especially after attending the World Summit on Sustainable Development in Johannesburg, South Africa, in 2002)? Could we rent out the house and convert the office into an apartment for our return visits?

At the same time, I have a limited budget to work with. My mortgage broker, who originally financed the house, calculates that refinancing my home in light of low interest rates and my home's appreciated value is equal to $100 per square foot of construction costs. This is enough to make my dream office a reality!

September 15, 2002

The plans are complete enough to start calling contractors for bids. I trained many of the remodelers in Boulder and figure I can get one of them to work with me on a project.

September 25, 2002

I am shocked when I get the contractors' figures back — $200 per square foot, twice my budget! The specifications freak out even those contractors I trained through the Boulder Green Remodeling program because they have never used materials such as structural insulated panels. The difference between my aspirations to "walk my talk" and reality have never been more dramatic.

I decide that it would be too expensive to use a contractor so I bite the bullet and choose to be my own general contractor. I haven't built anything significant for over a decade. Will I find good people to get the work done? Am I crazy? Despite the disconcerting voices in my head, I feel committed. Sometimes the voices of internal terror are just there to push us to reach higher and try new things.

October 23, 2002

My architect, George Watt, helps me flesh out my plans. After going through iterations of floor plans, elevations, and finish treatments, we arrive at a final design that accomplishes all of our objectives. My motivations for using green products are twofold. First, I train builders, architects and remodelers how to build green. I have hands-on experience with many of the products I talk about, but not with all of them. Now is my chance to gain practical experience with even more of the materials, rather than simply relying on manufacturers' literature or word of mouth for information. Second, I developed the Boulder Green Points program and counsel many builders on how to comply with the program's energy efficient, resourceful, and healthy material requirements. Often I hear, "The products are too expensive," "I can't get my subs to use them," or "I can't find the products anywhere." Using green products in my own office addition will help me address these issues firsthand.

Focus on what's important to you.

Practicing what I preach — including energy conservation, saving old-growth trees, and healthy indoor air — is important to me, but different people will obviously have different concerns. Whether you feel passionate about reducing your dependence on foreign oil, saving old-growth forests, or creating a healthier environment for your partner and children, you can focus the time and money you spend on your remodel according to your specific motivations.

Deconstruction

January 1, 2003

We rent a U-Haul to empty out the garage in preparation for building. We recycle some of the stuff accumulated over two lifetimes stacked in the garage and put the rest in storage. It feels great to lighten the load — a symbolic beginning to getting rid of the old to make space for the new.

January 2, 2003

The 30-year-old siding is T-111 cedar plywood and the roof is cedar shakes — tinder waiting for a match. They reflect an architectural style of the '70s that today are a liability due to the high likelihood of fire in the mountains. Each fall I am grateful to have made it through another dry summer. So, as the roof of the garage is taken apart piece by piece to make way for the second floor we save the

Fig. 1.3: Garage before construction.
Credit: David Johnston.

One Person's Trash is Another's Treasure

The roof of the garage was deconstructed and donated to a non-profit reseller. One of the challenges I set out for myself was to keep the waste from deconstruction and the addition to a minimum. I have served on the board of directors for a local non-profit organization, The Center for Resource Conservation, for several years. In 1995, the group founded a business called ReSource 2000 (R2K) to sell used building materials. Builders and remodelers donate these building products from deconstructed buildings or leftover materials. Sometimes when a window order is messed up and the windows are returned, the window supplier donates them to R2K. I have supported the business and helped it to grow to a nearly $500,000 per year enterprise. The combination of being on the board and keeping my trash out of landfills inspired me to deconstruct whatever building materials I could for reuse. The roof was the beginning.

shingles as tinder for the wood stove in the winter. The least I can do is use the wood for its best purposes at this late stage of its life.

We deconstruct the plywood roof sheathing in a similar fashion and resell it at ReSource 2000 (a recycled building materials outlet) since it's still in great shape. (One forgets that 30 years ago they made plywood from big enough trees that there were almost no knots on the faces of the structural grade plywood. Today there are "footballs" — cutouts where the knots used to be — all over the faces of plywood.) We then remove the trusses one at a time and load them onto the truck. By the end of the day we are down to concrete walls and one wood frame wall. It is like the garage got a crewcut and is ready for boot camp.

Job Site

January 3, 2003

Fig. 1.4: Roof deconstruction. Credit: David Johnston.

Before you start digging, you never think there will be so much dirt! However, rather than trucking the dirt off site and having to import new dirt for the driveway, we have built a retaining wall for it onsite. The wall is made of old railroad ties graciously donated by my dentist and

Approximately 200 Railroad Ties were Reused for Retaining Walls

I love my dentist. She is not only a master of her trade, but she is interesting and fun. When I'm at her office, we always talk about trips, adventures, life in the abstract and in the pragmatic. (It doesn't get more pragmatic than a root canal.) She and her husband were about to remodel their house and asked if I would help them make it as green as possible.

I visited their lovely home, met with their architect, and we agreed on plans for an extensive addition/remodel that involved considerable deconstruction. I talked them into using ReSource 2000 (R2K). However, there were almost 100 old railroad ties retaining their back yard that would be excavated for their addition; I said I wanted first dibs on the ties. They agreed and R2K deconstructed the house and delivered the ties to my yard. There they sat for about six months; I knew I would figure out how to put them to good use. Eventually, they became retaining walls to support the new driveway.

her husband, who dug them up after renovating their own home. This structure also allows us to extend the driveway around the garage all the way up to the house — eliminating the need to hike up 50 steps to get to the house. Finally, we cover the red mud that has plagued us throughout the construction with 30 tons of recycled concrete. (Colorado red mud is much like concrete itself until it gets wet and then it is like walking through silly putty.)

When Denver International Airport (DIA) opened, Denver officials had the conundrum of what to do with all the infrastructure of the old Stapleton airport, including all the concrete from the old runways. As it turned out, one of the largest concrete recycling facilities in the country was built to take the runways and turn them into gravel that could either be used in new concrete or used for other purposes, like my driveway.

January 5, 2003

Building on a hillside requires extensive underground water drainage. I decide to use drain tile all around the building and in the retaining walls. It collects to central areas where it penetrates the retaining wall and out to the hillside. All of the gutters go into that drainage system so the area around the building stays dry.

Foundation

January 7, 2003

I have always been a closet pack rat. Well, perhaps not too "closet," since I'm admittedly building this addition to store ten years of old papers. Not only do I

Ensure quality construction, especially in the beginning of your renovations.

Any error in the foundation is compounded as you get higher up in the framing process, culminating in serious roof framing dilemmas. When building, there are certain things that always hold true:

- **Square means square.** That's ninety degrees; not eighty-nine, not ninety one.
- **Plumb is plumb, or perfectly perpendicular to the ground.**
- **Level is level.**

All these rules must be adhered to from the beginning or you will have ugly consequences later. For example, consider how a building starts from the ground up. That may seem obvious but it has major implications. If the foundation is out of square, then the rest of the building is difficult to square. Especially when working with structural insulated panels for framing, there is no margin of error for the foundation. The panels should come in pieces that fit together like a jigsaw puzzle; if the pieces don't fit, it's a building nightmare.

Fig. 1.5 (above right): It is important that the foundations are square, level and plumb. These men are making sure the concrete floor is level. Credit: David Johnston.

keep files on all past jobs, I probably still have copies of the first green building articles in magazines ten years ago. The same holds true for building materials. I use framing lumber I have been stacking in the back of the garage for years to reframe the new east wall that will serve as part of the foundation for the second floor. We save some of the plywood from the roof to sheath the new garage walls. In some cases we go to ReSource 2000, the local reused building materials lumberyard, to get framing material. This saves money and reuses materials that might have gone to waste otherwise.

January 13, 2003

We put a lot of faith in our concrete subcontractor for the foundation. I find a company that had been in Boulder for years; everyone in the building trades knew their company and loved working with them. Their reputation, combined with the relationship I am able to establish with the owner, sells me on doing business with them. To make a very long story short, Marvin Clyncke's company pours the foundation within a sixteenth of an inch of the required dimension, making it perfectly level for the structural insulated panel (SIP) floor. It is close to a miracle to see so many laborers working with big concrete forms and a massive concrete pumping truck 50 feet away, with concrete pipes 40 feet in the air pumping high volumes of concrete. Neighbors are flocking to the end of the driveway to check it out. I know if something goes wrong I won't hear the end of it....

Establish good relationships with contractors.
In the remodeling industry, relationships can determine the quality of your project. See Chapter 3 for more information about establishing good relations with your contractors.

Reuse onsite resources.
Reusing stone from a local resource saves the cost and energy of buying new patio materials.

Exterior Work

January 15, 2003

Each time the earth moving equipment unearths flat stones, I have the laborers pick them out of the pile and stack them for later reuse. We have reclaimed stone from all over the property that will become the flagstone patios around the house and the new office.

Framing

January 20, 2003

One of the reasons I am so committed to green building is to protect big trees. I love the big things on the planet including big animals and big trees. They stimulate my imagination to envision what the planet used to look like before the industrial revolution started reducing the absolute numbers of all of the above. The loss of large life forms on earth reduces humans to a lesser species by letting us believe that we can conquer nature. I believe that we are more human by preserving and protecting the ancient things. How can we remember our place in the larger ecosystem if all that is left is second growth forests and tigers in a zoo?

I've been in the construction industry long enough to know what a vice-grip the forest products industry has on wood production. Their approach is to cut the last of the big stuff as fast as possible before regulations get in their way. Once a forest is clearcut, it becomes a plantation, perhaps for corn. The ecosystem is fundamentally changed. The plants and berries, and animals that live on them, are gone. The very soil bacteria that support the forest ecosystem changes. It takes hundreds of years for an ecosystem to return to a stable state after it has been clearcut. The forest products industry likes to brag that there are more trees in the

Fig. 1.6: Reusing stone from a local resource saves the cost and energy of buying new patio materials. Credit: Kim Master.

Structural Insulated Panels (SIPs) Used for Floors, Walls, and Roofs.

SIPs are one of the building products of the future. SIPs are framing materials made by sandwiching expanded polystyrene (Styrofoam) insulation between two pieces of engineered wood (OSB) that are in turn made of small pieces of wood compressed together. Of all the foams, expanded polystyrene is the least toxic to the environment. It uses steam or pentane to expand the foam pellets rather than gases that eat the ozone, such as chlorofluorocarbons (CFCs) or hydrofluorocarbons (HCFCs).

SIPs are energy efficient because they have little wood that spans from the inside of the structure to the outside, a feature known thermal bridging. Without this unwanted transfer of heat through the wall, the SIPs create a structure that is better insulated and therefore more affordable to heat and cool than conventional framing. Making your home more efficient is one of the best investments you can make. I have been searching for the crystal ball that will tell me what the price of energy will be in five years: no one has a definite answer, but the outlook is not good. Natural gas prices increased in Colorado 75 percent in 2003 alone. I want my office to be affordable to heat and comfortable to work in 10–20 years from now when energy prices are high, so I'm building my office 40–50 percent better than the local energy code. If you are doing a major remodel, you have probably decided to stay where you are awhile. Building to the highest standards possible, in part by using efficient materials like structural insulated panels, will ensure that your comfort will be affordable in the future.

US than there were when the pilgrims landed but they are in tree farms. In reality, we have far fewer forests and animals than there were then.

One of the most upsetting things I've learned about forests versus tree farms is that the forest products industry managers shoot bears to protect their "agriculture." Last year in Oregon alone, big forest producers shot 110 bears and cubs and uncounted elk that were damaging the trees on the edge of their tree farms. I want to do my best to preserve what is left of our indigenous heritage.

And that's why I have chosen to use many of the engineered wood products and Forest Stewardship Council (FSC-certified lumber on the office instead of traditional wood framing. Structural Insulated Panels (SIPs) are the first step.

January 20, 2003

We've been moving dirt for weeks. It looks like a war zone, with foxholes everywhere.

While we're digging I am sending design plans to the SIP manufacturer in British Columbia who will precut the panels. I chose this SIP company because they "dry build" the entire structure inside their warehouse to make sure all the panels perfectly fit my office's complex architectural form. One of the down-sides of working with SIPs is that when they don't fit together, it is a real pain — you end up cutting the panels with a chain saw and Styrofoam dust flies everywhere, resulting in a less-than-perfect fit in the end.

To complicate matters, the panel installers only have one week to comp-lete my job, or a little less than a month after the day we broke ground. It puts immense pressure on the architect, the panel salesman, and me to work out all the details in such a short time.

January 21, 2003

There is considerable drama around the arrival of the panels. We scramble around to make sure the top plates of the walls are perfect, that the steel beam over the garage door is secure, and that miscellaneous objects are out of the way. The road to the house is filled with big trucks and heavy equipment day after day, most charging by the hour. The crane that would unload the truck and move the panels to the driveway is $150 per hour. I feel pressure to get the job done quickly.

As the SIP truck arrives, I am amazed to think that there is an entire building stacked on the bed. The neighbors are already starting to collect on the next door lawn and film crews from the city of Boulder and the SIP manufacturer are set up to videotape the entire production. There is a crackle of excitement in the air.

The panels are all marked according to their order of assembly: floor-1, wall-6, roof-3. The crew segregates each set by the order they will be placed, and the crane operator keeps that boom moving for two hours without missing a beat. The first floor panel is in place before we know it. Floor-2, floor-3… they all fit perfectly! There isn't half an inch to spare in 35 feet. The first anxiety attack is over. The crew screws down the entire perimeter of the floor panels and lays down the bottom two-by-eight plates for the wall panels to stand upon. In two hours I have a new floor for the office and there is a ceiling on the garage for the first time in a month. Just as I was getting used to the "topless" garage experience…

Fig. 1.7: This temporary panel madness will transform into roof and walls. Credit: David Johnston.

Fig. 1.8: SIP walls can be erected faster than frame construction, saving labor costs. Credit: David Johnston.

January 22, 2003

This morning the wall erection goes even faster because the panels are smaller and it only takes two guys to install each one. By the end of the day we have walls — it is just amazing!

January 23, 2003

The large, open architecture of the addition requires many large beams both for support and for decorative purposes. The "prow" is able to overhang the driveway because of the large 7" × 14" beam that supports the weight of the structure. A similar 7" × 14" beam is required for the beam over the driveway to hold up the deck. Any beam that size, or for that matter down to 2" × 10" lumber, comes from old growth trees. To avoid cutting any old growth trees, we are using a variety of engineered wood called Parallam, or parallel laminated lumber. Parallam is made from cellulose that is stripped from aspen trees and then reassembled into wood beams. The final result is stronger than pine or fir beams cut from old trees. It is also stunning — with stain it adds a bold architectural accent.

January 24, 2003

Today we install the beams that hold up the roof panels. I am excited to get these beams up, but I'm also feeling stressed. Building is always a complex series of activities that must follow a critical path — something has to be in place before you can take the next step. This is particularly true with structural work. Excavation must be complete before concrete can be formed and poured. The concrete has to cure before you can put a load on top of it. All of the utilities must be installed before you can backfill the dirt against the concrete foundation. The structural wood has to be in place before you can put a roof on it — and so forth.

At 7:00 A.M. I go outside and count the beams to be sure they are all there. I measure each one and compare it to the plans. When I get to the major roof beam, the longest beam and the beam all the other beams attach to, I am horrified to realize that it is two feet too short! I feel like screaming and telling everyone to go home. It is the first thing to go in and it is the one that was custom manufactured for our job. I had asked my carpenter to double-check all the beams a week ago and I *assumed* that he did. He didn't. It's like the old joke, "Sorry boss, I cut it twice and it's still too short." I measured it again and it was still two feet too short.

Problem: The panel crew has today and tomorrow to finish the job before they leave the state. The beam took two weeks to special order. The roof panels are the most complicated part of the job and only the panel crew has the special tools to do the assembly. It looks like the project is going to stop here for who knows how long.

Desperate, I know one person who can save my project: Jeff Booms from *County Line Lumber*. Between begging, praising his prowess at the impossible, and bribery, he finds the only beam this side of Idaho that fits my specifications. He delivers it just in time — the SIP crew doesn't even know it was missing.

Whew! I know now the angels are on my side!

The beam is lifted off the delivery truck and into place at the very top of the roof peak. The carpenters spend the rest of the day cutting and lifting the big beams into place. By dusk we are ready for the final roof panels.

Roofing

January 25, 2003

Schedule extra time to do the job well.

Be aware that when working with a variety of trade contractors there will be times when you are beholden to other people's schedules. It can sometimes make the time allotted for your job unrealistic — and unattainable. Scheduling too much in too small a time frame will inevitably stress out you (and everyone around you)! Someone on the crew will likely make a mistake under pressure and the whole process (including fixing rash mistakes) will take longer than if you had allowed for extra time to do the job well.

Fig. 1.9: Seeing the roof beams lifted into place seems magical. Now I remember why I love building! Credit: David Johnston.

The clouds are rolling, the wind is kicking up, and it is starting to feel like January in the mountains. The first two roof panels go in without much difficulty and fit perfectly. Then, almost out of nowhere, the wind takes off and starts blowing dust devils all over the job site. We lift the next panel with a worker rope-attached to the bottom. Three of us hold on to the rope with two carpenters on the roof to catch the panel. With the panel in mid air, it is like a kite — and we are the tail. It blows out horizontally and almost lifts us off the ground. The crane is swaying and the panel is out of control. It slams into the building and crushes one of the edges. At that point the crane operator makes a unilateral decision and lowers it back down to the ground. We are all shaking from the effort.

By now it is late afternoon. The chilling wind continues to beat on us. The panel crew chief looks at the sky and says he is going to pull his guys off the job.

> **Ice and Water Shield™ used on entire roof.**
>
> Most people misunderstand their roof and believe that it is the roofing material, or shingles, that keeps the underlying structure dry. That is only partially true. The real purpose of shingles is to protect the underlying roofing membrane from sunlight degradation. Ice and Water Shield™ is a quality roofing membrane that will protect the roof for 50 years.

Fig. 1.10: The roofing membrane (typically made of felt paper), keeps water out of the roof, the white flashing keeps moisture out of the roof valley, and 50-year composition shingles protect the roofing membrane from sunlight degradation. Credit: David Johnston.

I ask him if he will be back tomorrow, but apparently they are going to their next job so we would have to finish it ourselves. It's an extremely tense moment; everyone is exhausted, cold, and upset. There are two panels to go and it is getting darker by the minute.

After a few moments, some not-so-subtle extortion on my part, and intervention by the crane operator, the crew chief agrees to finish the last two panels. By now it is pitch black outside, so I pull out high intensity halogen work lights, stand up on the existing roof and shine the lights for the carpenters to set the next two panels. About an hour after dark the last panel is screwed down and the roof is intact. Although it was a trying, stressful ordeal, I am under roof after only four days.

January 26, 2003

Oriented strand board (OSB) is not the most water-resistant material; therefore, I want to make sure that the roof sheathing is totally protected. When OSB gets wet it can expand and lose its structural capacity. Typically, an elastomeric membrane like Ice and Water Shield™ is used only on valleys (where two roof planes come together) or on overhangs where ice dams may build up and allow water to get underneath the shingles in freeze-thaw cycles. But since structural insulated panels can be damaged by water, I need to cover the entire roof with the membrane. Well, we are under a roof and the Ice and Water Shield™ is on!

January 27, 2003

It's snowing! In construction this is not a good thing, but at least the roof is protected. Not only has it dropped 40 degrees in temperature, but it looks like it's going to continue to snow. The panels saved us at least a month of framing and enclosure time — possibly even more time because we were able to build the structure before snow ground the project to a halt. As it is, the building is enclosed, waterproof, and insulated. With a heater, the carpenters and electricians can work inside in short sleeves while the winter snows are swirling outside.

Electrical

February 1, 2003

The weather has slowed our momentum and the carpenters are so cold that their productivity seems like a crawl. Even though the walls and roof are up, the windows don't arrive for another few weeks. All we can do is cover the openings with plastic to try to keep out the cold winter winds. After everything moving so fast for the last month, it is now hard to see any progress.

Fig 1.11: After we had waterproofed the roof, it snowed for a month. Credit: David Johnston.

I want to be able to operate lights from the house, have the computer network connect both places, maintain phone lines from one to the other, and so on. The buildings are about 40 feet apart and all of the utilities start at the house. We have to dig a four foot deep

Think ahead when planning the electrical infrastructure. The primary challenge when configuring electrical infrastructures is to determine how much electricity you will need in the coming years, and to decide where the switches and outlets need to be. The key is to think in terms of future needs rather than current conditions.

trench between the two buildings to get the gas line deep enough. The electrical line has to be one foot higher than the gas. Phone and CAT5 computer cable (CAT5 can carry phone as well as computer network), all run through the same trench. The last thing I want to do is dig the trench up again, especially after I make it a stone walkway! I put all the wiring in conduit and over-wire everything. I need two CAT5 cables for now, so I ran four, just to be confident I will have enough connectivity capability in the future. I put one size larger electrical line than I need in case the load in the office grows with new high-tech toys. The gas line is sufficient for twice the floor space, not even taking into account the energy efficiency of the building.

Work out contractual details in advance.

The job of the general contractor is to bring experience to the project. Part of what makes a job run smoothly is the contractor's relationships with his trade contractors. When serving as your own general contractor, it is vital to develop the working ground rules in advance of starting the work. When will they start? How long will it take them to finish? What happens when they uncover unforeseeable impediments? How much is their time and material work? The more detailed the contractual understanding, the less opportunity there will be for unpleasant negotiations at the end of their job.

Compact fluorescent bulbs used for exterior lighting.

Compact fluorescent bulbs are all over my house, both inside and outside. I was told not to use them outside because they might not work in cold weather. Well, never one to follow directions well, I put them in my outside lamps anyway, and five or six years later they have worked every time. This is one thing we can all do to reduce the demand on utilities. One compact fluorescent bulb can save an amount of energy over its lifetime equivalent to driving a car from New York to Los Angeles. I must have 40 of them around both buildings so I guess I have a great road trip coming up!

February 2, 2003

I am beginning to think the electricians are clueless as they run wire in the SIPs. When I sent architectural plans to the SIP manufacturer, we took exacting care to have them run holes (called "chases") in the Styrofoam for the electricians to run the wire through. It should be a relatively simple process to wire the outlets and switches.

February 12, 2003

Most of the electrical work happened while I was traveling for work. While I was gone, lack of experience and lack of supervision created a nightmare. The apprentice electrician started chopping away at the SIPs to run his wires, drilling holes everywhere. Now it looks like giant rats have eaten away the bottom 18 inches of the walls; all the wire is hanging down into the garage. The electrician has not used any of the chases that were formed into the walls for him. Just now he handed me a bill for $12,000 and he is not even finished roughing-in. He is over budget by 100 percent and the quality of his work is terrible. How did I get myself into this madness? I guess it serves me right for believing in a company named "Wisdom" electrical. The only wisdom in that transaction was how well he robbed me.

Exterior Siding

March 1, 2003

The weather has finally broken and we can move back outside. The first thing I want to do it to protect the oriented strand board (OSB), or the outside layer of the SIPs. I don't need a housewrap like Tyvek™ because the OSB doesn't leak air

Exterior wall moisture protection.

Use two-foot overhangs to protect siding

To protect the walls from water running off the roof, I design the roof with two-foot overhangs. This keeps the walls cooler in summer, protects the office from water, and, just as importantly, I like the look.

Wrap siding in 30-pound felt paper

Felt paper protects the oriented strand board (on the outside of the structural insulated panel frame) from moisture.

Flash all windows with Flex Wrap™

Flex Wrap™ is a special material used to "flash" a window — to divert water away from the window and prevent future problems associated with excess moisture.

and the panels were caulked meticulously when they came together. Rather, I need to keep moisture away from the not-so-waterproof OSB. The roof overhangs in the original design protect the OSB siding to some extent, but in addition I wrap the walls in 30 pound felt paper, similar to how you would protect a roof. We tack the felt on at the bottom of the exterior wall and work up, just as a fish's scales are layered so that water runs off them. Extra precaution is taken at the corners and penetrations that are more vulnerable to water seeping through them. For example, windows are "flashed," or waterproofed with Flex Wrap™ that diverts water away. Felt paper then goes *over* the window-flashing wrap so all the drainage is to the outside (if the felt paper were installed under the window flashing, water could leak behind the window).

Figure 1.12: Roof overhangs can be sized to suit the solar conditions in your region. Credit: Kim Master.

March 15, 2003

I want to stucco the building but I don't like synthetic stucco. Real stucco, or two layers of cement hand-troweled over wire mesh, turns out to be too expensive for my budget. I hire a siding company to install cement-based James Hardie Panel™ which is a fire-resistant siding made from fiber cement that comes in 4' × 8' sheets with a stucco pattern. Working with cement board takes a particular set of skills and tools. With the wrong saw blades, clouds of dust are created every time you make a cut. You also have to cut precisely because it is harder to work with than wood. The first contractor I hired said he knew how to work with the material but quickly

Protect wood on all sides to prevent warping.

Trim is one way to turn unsightly siding seams into an attractive architectural feature. Exterior trim is usually put up first and painted later, leaving the rear surface to absorb water. "Back priming," or painting all surfaces before it is installed, reduces the wood's ability to absorb water, thereby reducing its potential to warp or twist.

Fig 1.13: Hail the size of golf balls in Boulder, CO. Credit: Kim Master.

Consider using hail- and fire-resistant 50-year composition shingles.

Green roofing is a real dilemma for me. In order of importance I want my roof to be functional, aesthetically pleasing, and green. Why is "green" the last priority? Because there are few green options! Aside from expensive green roofing materials like faux-cedar shakes made from recycled garden hoses, most roofing is generally energy-intensive to produce and releases toxic contaminants when it's manufactured, installed, and landfilled. So it's best to find the longest-lasting roofing available so you don't have to waste energy, resources, and money to replace the roof. Like most green building materials, "green" roofing implies "durable" roofing.

proved otherwise. I replaced him with another crew that only worked with cement siding products.

March 17, 2003

We put two-by-fours over each of the seams in the James Hardie Panel™ siding, creating a wood trim pattern that echoes the overall building architecture. The top surface of each two-by-four is beveled like a windowsill so water will run off and away from the building. All the wood trim is "back primed."

Roofing

March 18, 2003

As evidenced by the month-long hiatus in my addition construction, Colorado often has severe weather. Not just snow, but wind over 125 miles per hour (hurricane strength is 75 mph), and hail bigger than golf balls at least once a year.

Given this weather, Colorado is the #1 roof replacement state for insurance companies in the country. Underwriter's Laboratory (UL) now has a hail rating for roofing products, and most roofing fails the test for Colorado, including cement tiles, clay tiles, and lightweight fiber-cement tiles. Fifty-year-rated composition asphalt/fiberglass shingles prove to have the best durability for the best price. These shingles resist hail and the uplift of strong winds; moreover, they are easily removed in case I decide to replace them with solar panels.

Interior Trim

March 20, 2003

For highlight trim, I install Forest Stewardship Council (FSC)-certified ipê, a tropical hardwood from South America. When oiled, ipê looks like a cross between teak and walnut — it's beautiful as baseboards and, accents around the window and door casing, and as edging around the old reclaimed doors that I will use as desks and cabinets.

Exterior Finish

March 25, 2003

The architectural crowning glory is the deck and trellis on the south side of the office. It has the best view on the entire property. It also is the first thing you see as you drive up the driveway and acts like a picture frame announcing there is something special going on inside. It also serves to shade the south glass in the summer to keep the building cool.

It all hangs from three 6" × 6" beams on each front corner and is supported by a 7" × 14" Parallam beam. Decks are unusual in that the framing structure is also finish trim so this had to look great. It was the last day of having the crane in the driveway so we carefully raised the 6" × 6" beams into place and then hung the large beam from them. Once the beam was in place we could remove all the bracing that kept the posts plumb. After that, we placed the joists, decking, and rafters of the trellis, creating the "craftsman" look we wanted to achieve.

Windows

April 7, 2003

Window technology has advanced dramatically in the last 20 years. Today, average windows are twice as efficient as those in the '80s. That still isn't good enough for me. I want the addition to be as bright and energy-efficient as possible, therefore high-tech windows are my best option. I choose low emissivity (low-E) windows with specific solar heat gain coefficients (SHGC). I use low-SHGC window for the east and west windows that tend to be the hottest, so that I can enjoy the view overlooking the plains without cooking in the summer. I use high-SHGC on the south facing glass to maximize solar heat gain in the winter and keep my office

Green alternatives for interior trim.

Oriented Strand Board (OSB)

OSB is made from small pieces of wood that are pressed and glued together. It is a stable product that doesn't require lumber taken from old growth trees.

Highlight trim is Forest Stewardship Council (FSC)-certified ipê

FSC certification ensures the wood is managed and harvested sustainably, without damaging forest ecosystems.

Bona Kemi MEGA™, water-borne acrylic wood finish

Wood finishes are one of the most toxic finishes you can use in the home, affecting the respiratory, nervous, and immune systems. The so-called "Swedish floor finish" has been illegal in Sweden for over 20 years — don't let your floor finisher talk you into it. MEGA™ by Bona Kemi is a safer, durable finish for wood trim. It is a water-based urethane that has very little odor and dries to a satin sheen. MEGA™ protects the wood and highlights deep grain textures.

Fig. 1.14: The deck and trellis. Credit: Kim Master.

"Green" Wood for Your Exterior Trim

The decking material of choice is a composite of wood fiber and recycled plastic. It is my solution to the existential angst in the grocery line when I'm asked, "Paper or plastic?" Do I kill trees for a bag or do I use petrochemicals to carry my groceries to the kitchen? I still can't answer that question, but I used the bags—recycled along with wood fiber—for my decking.

Alkaline Copper Quaternary (ACQ) deck framing

Pressure-treated lumber using chromated copper arsenate (CCA) has been used for "ground contact" applications for years to keep wood from rotting or being eaten by termites. The chromium is hexavalent chromium, a known carcinogen that caused Pacific Gas and Electric (PG&E) to pay out one of the largest class action settlements in history (made famous in the movie *Erin Brockovich*). Arsenate or arsenic is used for rat poison. We have built decks and playgrounds for decades out of this material. It is so toxic that when it is burned, a tablespoon of the ash is potent enough to kill a cow. Finally, the EPA and the makers of CCA have come to an agreement to phase out CCA with several new replacements. Natural Select™ is another safe alternative wood treatment that uses alar, the coating used to protect apples from insects.

FSC-certified lumber used for trellis and three-by-twelve wood steps

FSC certification ensures that the wood comes from a sustainably managed forest.

Exterior trim is made from oriented strand board (OSB)

Wherever I can, I look for alternatives to old-growth wood. So for exterior trim I used exterior-grade OSB that is pre-finished at the factory to be waterproof. Without the pre-finish, moisture can seep into the OSB and damage the integrity of the structure.

Fig. 1.15: Low-E windows help keep us comfortable when it's 105 degrees outside.

warm; the sun is low enough in the sky that the overhangs that block unwanted sun in the summer will not obstruct this winter light.

I install Heat Mirror™ for all the fixed glass (see sidebar), or glass windows that do not move or open. The down side of heat mirror is that it requires a metal spacer between the panes of glass because of the tension of the plastic film; in turn, the spacer tends to lose heat through conduction. To counter the transfer of heat through the metal, I buried the spacer in the wood trim.

Interior Finishes

April 15, 2003

In an effort to avoid toxic paints, I use Kelly-Moore's Eco-Spec™ product line — a non volatile organic compound (VOC) paint that, unlike conventional paints, does not release toxic chemicals into the indoor air. It works as well as conventional paint for about the same price with no "new house smell" when it is applied.

Heating, Ventilation, and Air Conditioning (HVAC)

May 15, 2003

My heating options are limited by the open space design and the use of SIPs. Forced air won't work well because of minimal access for running ductwork. The office is so energy-efficient that it would be hard to find a small enough furnace to be efficient. Many contractors who looked at the job said to just use radiant electric resistance heaters. To me it was like telling me to drink hemlock! Electrical resistance heating is the least efficient source of heating next to having your own mini nuclear reactor for your heat source.

I decide to use radiant hot water in the floor — an easy decision but a challenge to execute. Radiant floors are created by laying tubing in serpentine coils with lightweight concrete (gypcrete) poured over them, and the concrete is then covered with wood, carpet, or tile. The problem is that I don't have enough room from the floor height to the bottom the doors and windows to pour the gypcrete. I also don't want to use baseboard radiators because filing cabinets and desks would cover many of the walls and reduce the heaters effectiveness. So I have found a product called Warm Board™ made of 0.875-inch-thick high-density OSB, which is specially grooved for the tubing to run through. I lay the Warm Board™ on top of the floor and fill in

High-tech windows.

All windows are low-emissivity (low-E), tuned by orientation

Low-E windows have a film "sputtered" onto the inside surfaces of a double paned window, which reflects heat out in the summer and into the space in the winter, increasing the insulation value of the window. Windows with specific solar heat gain coefficients further determine how much heat can enter through the windows.

Fixed glass is Heat Mirror™

This is a special type of low-E glazing made by suspending a thin plastic film like shrink wrap between two panes of glass with low-E coatings on all interior surfaces. The window is then filled with the inert gas, argon, to reduce the movement inside the window that leads to increased heat loss.

Climate-specific design.

Energy in your home has to be designed as a system. Whether or not you think about it, your home is always interacting with the environment. Hot sunny days create one response from your home's cooling system; cold snowy days create a totally different response from your heating system. When you consciously question your environment (When does the sun rise? What rooms does the sun shine into and when? In what direction does the wind blow in different seasons?), you create design requirements that are more efficient than automated systems because they are in tune with your specific environment.

the rest of the floor with two layers of 7/16-inch-thick OSB sheathing. It leaves just enough room to cover the floor with carpet and allows the doors to open. It also provides another layer of subfloor so the floor now is able to support 100 people dancing Greek style!

Water Heating

May 20, 2003

For a heat source, I first considered a tankless water heater that hangs on a wall and is plumbed like a typical water heater but doesn't store water in a tank. Unfortunately, they are expensive and aren't designed for space heating. As an alternative, I found an efficient A.O. Smith water heater with extra thick foam insulation around the tank and high efficiency combustion — it works great!

June 2003

All the contractors are gone; the rest is up to me. It's a mixed blessing. On one hand, all the disruption, noise, trucks, and muddy boots in the house are gone. On the other, it's all up to me to finish. I have been looking forward to this stage because I love to do finish work. The down side is that I travel all over the country training others in green building, so I'm gone much of the time. The process slows down dramatically.

Flooring

Early August 2003

My wife and I install the InterfaceFlor™ carpet tiles ourselves —which takes just a few hours since they incorporate a peel and stick process for attaching to the subfloor. Very satisfying! This is a brand new product on the market for residential applications. Not only is it 100 percent recycled content, but the 19-inch square tiles also come in many colors and patterns. The tiles also allow for flexibility in terms of how I want to use the space at a later time. In the future, the space may become an apartment, so I will have to run electrical lines and plumbing through the floor. Now I will be able to pick up only the tiles I need to get the work done. I will also change the flooring in both the bathroom and kitchen-to-be. The tiles allow me to remove just the pieces I don't want and replace them with, say, ceramic tile.

Fig. 1.16: InterfaceFlor recycled-content carpet tiles over radiant heat. Credit: Kim Master.

Natural Cooling

Late August

Thanks to natural cooling, the building works like a charm. The western sun is typically the hottest, and we are blessed to be on an east slope so the mountain blocks the late summer

afternoon "furnace" sunlight. Large pine trees (that we took great care to design around) also provide plenty of shade. I open and close the awning and casement windows (placed strategically on each side of the building) as the wind patterns shift. Casement windows open like doors, hinged on the side so you can use them as wind scoops. Awning windows are hinged on the top and provide ventilation even when it's raining. When the wind isn't blowing, I use a centrally located ceiling fan for air movement. When it has been over 105 degrees Fahrenheit outside, the highest temperature in the addition was 84 degrees. No need for air conditioning with the building doing its job so well!

Wireless Networking

September

It is critical to keep the connection between the house and the new office, but I have wires everywhere in my life and I'm tired of tripping over all of them. So I install a wireless network, which is a real boon to civilization. I can now sit on the deck, barefoot, and check my e-mail on my laptop or I can work from the house and access the computers in the office. My commute has tripled from the den to the new addition fifty feet away, but the delight of working in the new space far outweighs the inconvenience of the additional travel!

Fig. 1.17: Ceiling fan provides natural cooling while the skylight behind it provides natural light. Credit: Kim Master.

Fig. 1.18: The new "world headquarters" of What's Working, September 2003. Credit: Kim Master.

Remodeling: An Emotional Rollercoaster

I love my new office, but I have to remind people that remodeling is a wild ride. It seems like it should be straightforward, but in reality, it is an emotional experience from start to finish. There is the excitement of the design process, the shock of the first cost estimates, the thrill of the initial construction, the frustration with the contractors, the elation of seeing the structure in form, the impatience of wanting it to be finished. All are part of the process. The more prepared you can be for the experience, the better you will handle the ride.

CHAPTER TWO

Affordable Green Dream

G REEN BUILDING IS FRAUGHT WITH MYTHOLOGY: "It is too expensive;" "I can't get the products where I live so why bother?" "I don't want to live in some weird strawbale house;" "Solar is ugly;" "Nobody I know has ever done it;" "I'm not an old hippie and don't want to live that way." Other people assume falsely that green building is too good to be true — that it is all greenwashing and a marketing ploy on the part of the industry to dupe us once again.

Yes, some building elements do cost more. But many cost less! More importantly, green building is the way buildings will be built in the next decade and this is just the tip of the iceberg. When we describe what we do at What's Working, we tell people that we consult in the growing field of applied common sense. Most of what green building brings to the construction industry is just a more systematic way of thinking about buildings. Looking at the big picture rather than just at the bricks and mortar adds a new perspective to how all the pieces fit together and the consequences, both near and far, of the decisions we make at the design stage of a project and the products we use to build it.

Green building need not be too expensive. When it is part of the initial process of setting goals for the remodeling project it becomes matter of fact — you, your architect, and your builder just make it happen. Many builders have found that the real cost is in the learning curve, not in the implementation of the building process. The products are becoming more available and more affordable all the time as major manufacturers develop new lines to meet the "green" demand. Paints are a classic case in point; paints that are low in volatile organic compounds (VOCs) are now featured in the product lines of all the major manufacturers. Keep in mind that green building doesn't have to look any different from conventional buildings; it is how it's built, not what it looks like. And most significantly, people from coast to coast and across the midlands are building and remodeling green.

> *Green building doesn't have to look any different from conventional building; it's how it's built, not what it looks like.*
>
> **David Johnston, What's Working, Boulder, CO**

Green remodeling actually makes and saves money. And this is not just long-term energy saving costs; the cost to implement green features ("first costs") is often less than remodeling by conventional standards. In the long term, green renovations increase the resale value of your home. There are also financing options available for people who remodel with energy efficient features that can save hundreds of dollars. No matter how you look at it, green remodeling is a smart, moneymaking investment.

Be careful not to get caught in the "payback" trap for energy conservation features. Payback is an illusion for many reasons. If you save $2.50 per month on your energy bill will you even notice it? What is the payback of a night on the town? We don't think about payback for anything but energy conservation or solar products: the only things that do have a payback. There are two better ways of thinking about the energy you use. The first is to look at your mortgage payments plus your monthly energy bills as one collective cost. Most often, the increase in monthly payments for energy upgrades is less than the savings on your utility bills, so it is money in your pocket. Perhaps more important is how much will it cost you in five years to heat and cool your house; just look at the increase in energy costs over the last five years. Installing energy conserving products such as insulation or solar panels is a cheap insurance policy against rapidly rising costs for fossil fuels.

Another way to look at the cost of green is what it is worth to you to reduce the possibility that your children will develop asthma or other respiratory problems. How valuable is the health of your family? There are many subtle savings and preventative measures that can't be put into a bottom line cost/benefit analysis. How much do you value your Saturdays for family time rather than refinishing the deck? Is comfort on cold winter nights — thanks to energy efficient windows and increased insulation — important to you? All of these issues go into the "value proposition" of making green decisions.

For clarification, we've broken down how green remodeling saves money into five categories: general design strategies; landscaping; energy systems; products and materials; and job site considerations. Then we explain how you can fund your renovations in a way that maximizes the financial benefits of added energy-efficient features. Finally, we demonstrate the advantages of making investments in your dream — green.

How Green Remodeling Saves Money

Alex Wilson is president of BuildingGreen, Inc., the publisher (based in Brattle-boro, Vermont) of *Environmental Building News (EBN)* and the *GreenSpec* directory of green building products. Alex has been the executive editor of EBN since the newsletter's founding in 1992 and is co-editor of GreenSpec. We value his concise, clear, and informative writing, which helps demystify green building concepts. The following section was adapted, with permission, from his article, "Building Green on a Budget," published in *Environmental Building News*, Volume 8, Number 5. It has been slightly altered so that it is most applicable for residential remodeling projects.

General Design Strategies

- Remodeling existing homes instead of building new saves significant quantities of materials and energy, in turn benefiting the environment. Project costs are reduced and there may be significant savings in time and money associated with not needing extensive regulatory review and approvals.

- Good design is the best way to make your project affordable. It is much less expensive to make changes on paper than after construction has started. A few extra hours of your architect's time can save hundreds if not thousands of dollars in construction changes.

- Integrated building design often results in first-cost savings, especially in the case of energy conservation (see Energy Systems below), but also in other areas of design. For example, including contractors in discussions with the architect and engineer can help identify ways to streamline the process and save on materials. Involving a landscape architect early may reduce the need for new costly plantings because he or she can protect existing plants from construction harm.

- Smaller renovations require fewer resources during construction, disturb less land during site work, and use less energy during operation.

- Paying attention to solar orientation by locating more windows on south-facing walls can reduce energy costs by 10 to 40 percent right off the bat. Using your home to respond to the natural energy flows of your microclimate means you reduce your need (and cost) for expensive mechanical systems.

- Minimize cooling loads (and costs) by reducing window area on east and west facades and, where there are windows, by incorporating shade plants to reduce the heat buildup.
- Leave floor slab exposed. Eliminating carpet will avoid mold and other biological pollutants, the environmental impacts of manufacturing the carpeting, and the cost of the carpet. Consider texturing and pigmenting concrete: it is beautiful and less expensive than carpet alternatives.
- Optimal Value Engineering (OVE) and advanced framing reduce waste without compromising structural performance. Some builders have found that they can reduce the framing materials by 20 percent and increase energy savings at the same time. You reduce overall material use, thereby benefiting natural resources and saving you money.
- Open layouts help distribute daylight, reduce ducting requirements for conditioned air distribution, simplify reconfiguration of space, and reduce material use.
- Optimize building dimensions to reduce cutoff waste. Anytime you reduce waste you save resources and money by buying less material, reducing onsite labor (for measuring and cutting), and paying less for disposal.

Landscaping

- Indigenous landscaping like prairies, woodlands, and desert gardens support wildlife and biodiversity better than conventional turf. Native landscaping does not require irrigation or chemical treatments, generally making native plantings less expensive to maintain. Moreover, native plantings reclaim weekend time for the family by eliminating the need to mow the lawn.
- In areas with low annual precipitation and areas that are prone to drought, provide xeriscaping (dry-adapted plantings) to obviate the need for irrigation systems and more expensive plantings.

Energy Systems

- An integrated design makes it possible to pay for increased energy conservation measures through savings in heating, ventilation, and air conditioning (HVAC) equipment. For example, a tight, well-insulated building envelope with high-performance glazing and shading strategies may enable you to downsize or eliminate conventional heating and/or cooling equipment. We have worked on many homes in the harsh Colorado climate that use only a typical water heater to heat the entire home.

- Using a water heater to satisfy heating loads saves first costs for an unnecessary furnace.

- Specifying specific glazings (windows) for different window orientations usually does not cost anything more than conventional windows. This allows the south windows to help heat your home and reduces heat gain from east and west windows that would otherwise drive up air conditioning costs. It almost always results in future savings from passive solar heating and cooling — reducing the need for mechanical equipment.

- Using ceiling fans increases air flow and occupant comfort in homes without mechanical cooling. This significantly reduces equipment costs.

- Keeping ducts away from exterior walls will improve energy performance and save money because less ducting is required.

- Reducing outdoor lighting by using motion sensors will save energy and reduce light pollution.

Products and Materials

- Salvaged materials can often be obtained at lower prices than new, virgin materials, depending on labor costs. This benefits the environment because fewer resources are used, thereby also saving landfill space. Salvaged materials include lumber, millwork, windows, cabinets, some plumbing fixtures, and hardware.

- Use more durable materials that require less maintenance. Brick facades may cost more up front but you never have to paint or repair your siding.

- Recycled content composite decking will outlast wood by two to three times, never requires paint or finishes, and will not burn or splinter bare feet in the summer.

Fig. 2.1: How much does green cost: Green remodeling can be less expensive when you consider options in the context of the entire house. What you save in one area can be applied to another area.

KEY:

ø = no cost

¢ = low cost

$ = more costly

$ - energy-efficient washer & dryer

$ - light-colored, reflective roofing

$ - whole-house fan

$ - l.c. recessed lighting

$ - ridge vent

$ - radient foil stapled to rafters

$ - PV panels

$ - check & seal ducts with mastic

$ - ceiling fan

$ - energy-efficient refrigerator

¢ - removable shade cloth

¢ - deciduous trees on south & west aspects of house

¢ - trellis with deciduous vines

¢ - permeable paving

¢ - low voltage landscape lights

¢ - portable fan

¢ - low-flow toilet & showerhead

ø - set water heater to 120˚

¢ - compact flourescent floodlights

$ - low-E skylight & screen which opens

$ – R-38 attic insulation

¢ - drape or blind all sun-facing windows

¢ – awning to shade window

- Using structural materials as finish materials eliminates a costly and resource-dependent building component. For example, use exposed beams or concrete floor slabs.
- Downsize the supply pipe diameter with water-conserving fixtures (such as low-flow showerheads and toilets) to deliver hot water faster, reduce standby losses from hot water pipes, and reduce water waste. Smaller diameter pipes are less expensive.

Job Site

- Protect existing trees during remodeling. It may cost a bit more, but these costs are easily recouped by having to spend less on plantings when renovations are finished. Large trees also boost property value, and the shade and cooling effect they provide allows homeowners to downsize air conditioning equipment.
- Recycling job site waste avoids expensive landfill disposal costs.

Source: "Building Green on a Budget," *Environmental Building News,* 8(5): 11 to 14.

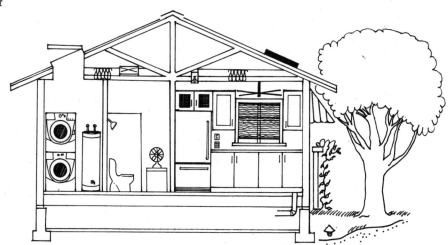

Credit: Jill Haras & Kim Master

Although this list is far from complete, it demonstrates just a few ways green remodeling is more affordable on the first cost side. The most important aspect is thinking in terms of building integration, or the house as a system. An over-arching framework makes it possible to incorporate many strategies that, taken alone, might cost you more.

The following section on financing your renovation was written by our good friend and financial expert, Steven Schueth. Steven is President and Chief Marketing Officer of First Affirmative Financial Network, LLC. He lives in Boulder, Colorado, but his influence as a leader in socially responsible invest-ments is recognized internationally.

Financing Your Renovation

There is a growing awareness and a deep yearning to understand how "greening up" a home can create a healthier living environment and save money. Financing healthy home upgrades and improvements can be easy, depending on your income, the equity you have in your home, and your credit scores.

If you have money in the bank or in an investment portfolio that, if liquidated, won't generate substantial capital gains taxes, the simplest way to finance your renovations might be to write a check. Once your renovations are complete, you can refinance the property at a higher market valuation, take out some or all of the additional equity invested, and put it back in the bank or into the stock market.

If you have good credit and a long-term low interest rate credit card and don't need extensive renovations (i.e., under $20,000), you might want to use your card (and maybe even get frequent flyer miles). Once the work is complete, you should refinance and pay off your credit card(s).

A local bank will often finance renovations. This should be a relatively painless process if you have substantial equity in your home; the bank will simply extend you a home equity loan. If you have good credit, solid, verifiable income, and/or substantial investment assets, the bank may be willing to lend you money for renovations as a personal loan, secured by your personal guarantee.

Construction-to-permanent one-time closing loan packages make a lot of sense if current interest rates are lower than your existing permanent loan. Essentially, you apply for a mortgage based on the "future completion value" of your home as determined by an appraisal. Loan amounts range up to 95 percent of the future completion value and closing costs are rolled into the construction loan. Your existing mortgage is paid off, and additional payments are made to the contractor as the renovation work proceeds. Upon completion, the lender extends a permanent loan to you based on current market interest rates with no extra cost. Overall, closing costs are lower, especially when compared to paying for two separate loan transactions. This approach works best for higher cost renovations, say $50,000 or more. Only a few select lenders offer this type of product. You are not likely to find it at your local bank.

Another unique loan package that you won't find at local banks is generically called an Energy Efficient Mortgage (EEM). The Federal Fannie Mae, a financial products and services provider, is piloting a green mortgage product in a few states. A certified energy rater will analyze future energy savings you can expect from your renovation and calculate the present value of those savings. That dollar amount can be considered additional equity in your home, thus increasing the amount of money you can borrow, or reducing the amount of cash you need for a down payment. For some borrowers, the additional equity could mean hundreds of dollars saved annually by reducing or possibly eliminating private mortgage insurance.

Energy efficiency upgrades can help increase the value of your home and provide a resale advantage over other properties in your neighborhood. According to a study published in *The Appraisal Journal* in October 1999, the selling price of homes increased by $20.73 for every $1 decrease in energy bills. Think about it: if you could save $350 in utility costs annually, your home might increase in value by more than $7,000.

You can profit by investing in efficiency measures regardless of how long you live in your home. If the reduction in your monthly energy bills exceeds the after-tax mortgage interest paid to finance the energy efficiency upgrades, you will enjoy positive cash flow as long as you live in that home. And if energy prices spike as expected, energy efficiency may be one of the best investments you will ever make.

A healthy, energy efficient home often means you can borrow more or finance a higher percentage of the property value. Or you could use all of your energy savings to pay down your mortgage faster. If you make one extra payment per year, it is possible to pay off a 30-year loan in closer to 20 years — and save thousands of dollars in interest!

These are just a few of the financing options available to you; many solutions must be customized. Contact a local professional who is familiar with financing green, energy efficient renovations for help with these critical decisions.

The Mystery of FICO Scores Revealed

A majority of lenders use a credit score developed by Fair Isaac & Co., known as a FICO score, to determine the likelihood that credit users will pay their bills. People with high FICO scores are likely to repay loans and credit cards more consistently than people with low FICO scores. FICO scores are remarkably predictive, which is why lenders rely on them so heavily for credit decisions.

As a group, consumers in the 700–749 score range, for example, have a delinquency rate of five percent, as illustrated below. This means that for every 100 borrowers in this range, lenders expect that approximately five will default on a loan, file for bankruptcy, or fall 90 days past due on at least one credit account in the next two years. Most lenders consider consumers in this score range as very low risk.

Delinquency Rates by FICO Score

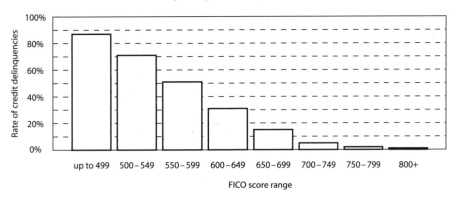

Source: Steve Schueth and Teresa Lopez, personal conversation.
Principals of DreamSource Financial LLC. Boulder and Rigway, CO; Santa Fe, NM.

This chart demonstrates the delinquency rate (or credit risk) associated with selected ranges of FICO scores. In this illustration, the delinquency rate is the percentage of borrowers who reach 90 days past due or worse (such as bankruptcy or account charge-off) on any credit account over a two-year period.

FICO scores are calculated based on information drawn from your credit history. While knowing your actual score is a good start, understanding the key factors affecting your FICO score is much more important. For more information and direction on how you can increase or maintain your FICO score over time, go to <www.myfico.com>.

Choose the Options You Can Afford

Our renovations are limited by what we can afford to finance, therefore we hope this chapter's overview of cost-effective green building options combined with expert financing tips will enable you to consider creative green remodeling options. Green remodeling is not about opening your wallet; it's about opening your mind!

CHAPTER 3

Working with Building Professionals

MY CONSTRUCTION COMPANY IN WASHINGTON, D.C. was a design/build firm that specialized in remodeling. Lightworks Construction was intended to be a solar construction company, but by the mid-eighties solar had faded and we became a full service company specializing in large upper-end additions. We also built a "million" basement remodels and quite a few kitchens. By designing what we built, we found that we could streamline the process and avoid many of the pitfalls that occur when the design-construction process is linear.

For a typical remodeling job, the homeowner works with an architect who draws up a set of plans or blueprints. The architect then hands the plans off to an engineer, an interior designer, and a landscape architect who each contribute their respective insights. The plans are then put out to bid to three to ten remodeling contractors who dissect the plans into various trades for bid prices, including carpentry, roofing, insulation, plumbing, and electrical. The contractor puts his "head scratching" time into the plans and comes up with a bid. This whole procedure occurs in a vacuum with no interaction between the trades. When the bids come in, the homeowner looks at all the apples and oranges proposals with numbers everywhere and has to make an existential decision on which stranger to cohabit with for the next three to nine months. It can be a bit like throwing darts at a dart board to come up with a decision.

All too often the worst of all possible outcomes emerges: the homeowner chooses the lowest bid! Would you choose the cheapest heart surgeon? Probably not. Unfortunately, the conventional bidding process wastes time and costs more due to change orders and hidden costs. All too often, it doesn't accomplish the intention of building a great team to manifest your dream home, and is eventually brought to a close in a court of law — not what anyone ever intended.

We seek spaces that satisfy not only the basic requirements of size and function, but also our beliefs and philosophy. You, your architect, and your builders will define those beliefs in tangible terms that can be made real with wood, steel, glass, and stone. In other words, you have dreams for your home, your architect informs those dreams, and your builders materialize the informed dreams. However, the process is rarely this linear and straight-forward. For example, those who build, including carpenters, masons, electricians, and plumbers, are an incredible resource for ideas. Builders can significantly improve the look, feel, and efficiency of your home, especially when their ideas are considered early in the design process. Moreover, by including the builders from the start, you can create a team with a common understanding of the history of the project. This proves extremely helpful when a construction problem needs to be resolved along the way.

George Watt, Architect, Boulder, Colorado

A better outcome occurs when remodeling professionals collaborate as a design team from the first sales meeting. My architect, myself (as contractor), and often an interior designer would meet with our prospective client and brainstorm solutions to their design needs. The interaction was often stimulating and the clients loved it. Not only did these meetings help clarify the dreams of the customers, but all of us were able to ask questions at the beginning to further clarify the desires of the prospective client. It was much easier to get the client's ideas on paper with the designers present up front.

Once we got the job, we would all work iteratively on the design — costing different design solutions along the way, improving the design based on past experience — and come up with a better overall solution than any one of us would have developed on our own. By knowing the cost of construction we usually met the client's budget the first time out.

Today, the team approach, or collaborative design, is still an anomaly. The remodeling process is fraught with difficult decisions, conflicting information, and challenges right up to the very end. After years of remodeling we learned to prioritize client-architect-contractor relations. Ask yourselves, "Can we work with this remodeler? Does the architect share our values and worldview of what is important? Can we have deep discussions about our dreams and feel heard? Will the construction company protect our twenty thousand dollar landscaping? How will we solve problems and come to mutual understanding?" If you treat your relationships in the home remodeling process with the same care and consideration that you treat your friends and neighbors, the outcome is often much better than any contract can ensure.

Given the collaborative nature of remodeling a home, we thought it would be most informative to let you hear from the architects, contractors, and homeowners themselves. Together, our stories will help you find and work with the best professionals to remodel your dream home.

Working with an Architect

One of the most important relationships you will establish when remodeling is with your architect. He or she will capture your dreams and ideas and translate them into a language that the contractors can understand. We can't emphasize the importance of this process enough. Your architect helps you see what all the talk means on paper so that you can visualize the project before it starts and avoid disappointment later. It is a lot easier and less expensive to erase a line than to tear down a wall. For some, hiring an architect may seem like an unnecessary expense because you already have a design or plan in mind. However, even if you know the outcome you desire, an architect is like a lawyer who will help you write the brief that will lead to that outcome more smoothly than if you attempt a remodel design on your own. Hire an architect — your dream home is worth it!

Expert Advice from an Architect: Mercedes Corbell

We could think of no better person to explain the nuances of working with an architect during the remodeling process than our good friend and talented green architect, Mercedes Corbell. Mercedes Corbell Design and Architecture is located in Oakland, California, and her green designs can be viewed online at <www.mercedescorbelldesign.com>.

What kind of projects need an architect?

That depends on how one defines "need." For some projects, depending on their size and the requirements of the local jurisdiction, an architect is required to prepare the plans. For other projects an architect is not required by the local building department to prepare the plans. Basically, if there are design decisions to be made and drawn, you should strongly consider hiring an architect. Architects are the most broadly trained of the design professionals, and they look at a building in terms of its aesthetic qualities, the functionality of its spatial layout, and relationship between rooms, the building and planning code implications, construction feasibility (though contractors may dispute the architect's expertise here), and the compatibility of a design idea with the existing architecture. Investing dollars in the construction of a remodeling project without investing the careful attention that an architect brings to a design problem seems like a big risk to take. That said, some projects are simple in terms of construction and design — these projects could benefit from a simple set of permit drawings, which can often be produced by the contractor.

What do architects do in a remodeling project?

An architect designs buildings, and this overall activity encompasses a series of smaller activities. An architect is also the legal agent of the owner as well as a "neutral arbiter" of disputes that might arise between the owner and the builder. The architect does not decide the result of the dispute but is the first mediator. The architect traditionally has been the lead in the process and one of three in the triangle of architect-owner-builder. In this capacity, the architect acts not only as a musician in the orchestra but also as the conductor, coordinating the work of consultants into the design, assisting the owner in choosing a contractor and in getting construction prices, applying for the building permit, and visiting the job site during construction to see that the results align with the design intent.

An architect can be involved once the owner has decided to remodel or add onto their house. The first tasks start during the "programming and predesign" phase and include establishing project constraints, drawing up a measured set of drawings of the existing building (called "as built" drawings), and identifying the owner's list of requirements and requests. This is the time to establish your list of green goals and expectations.

The second phase is referred to as the "schematic design" phase. This is akin to a brainstorming process. Your architect produces a series of schemes that the owner chooses to pursue. Actually, the accepted scheme often results from the combination of a series of schemes. The chosen design is expanded during the "design development" phase, resulting in more views of the design, material lists, and the preliminary work for consultants like the structural engineer or green consultant. If required by law, the architect then applies for design review or other planning department applications and navigates through that process. Sometimes redesign is required by the planning department. Once the design is well described and developed, the final construction drawings and specifications are prepared during the "construction documents" phase. At that point, it is time for the architect to apply for the building permit and for the contractor(s) to provide their final bid price during the "bidding/ negotiation" phase. Finally, it's time for construction, and the architect is available to visit the site and review the course of the work.

Table 3.1: Architectural Remodeling Phases

Phase I:	Programming and Predesign	• Establish project constraints
		• Sketch measured set of drawings
		• Identify owner's list of requirements and requests
Phase II:	Schematic Design	• Produce series of schemes that the owner chooses to pursue
Phase III:	Design Development	• Chosen design developed
		• More views of design
		• Materials list
		• Preliminary work for consultants
		• Planning Department applications
Phase IV:	Construction Documents	• Final construction drawings and specifications
		• Architect applies for building permit
Phase V:	Bidding and Negotiation	• Contractors provide final bid
Phase VI:	Construction	• Architect visits site and reviews course of work

Source: Mercedes Corbell, Personal Conversation. Mercedes Corbell Design + Architecture, Oakland, CA.

<www.mercedescorbelldesign.com>

What kind of experience do architects have?

Since the job of an architect is technical, artistic, communicative, and regulatory, the education and training of an architect typically contains these elements as well.

The formal education of an architect, whether it takes place during undergraduate or graduate coursework (or both) varies between schools. Some schools emphasize theory and aesthetics, others construction technology, etc. At a minimum an architectural student will study the preparation of architectural drawings, architectural design, architectural history, structural design, and often calculus and physics (though in

> *There is a sense of disconnect in the way that most of us live. It is very unsustainable and fairly toxic to the environment, generally speaking. If you want to live in a house that is less toxic and if you want to leave something for our great grandchildren, build green. Building green is exciting. And it is a challenge too. Anyone can build a house. But can you build a green house?*
>
> **Cate Leger, Architect, Berkeley, California**

architectural practice these are rarely if ever used). Ideally, the architectural student will also study freehand drawing, color, sculpture, the sociology of design, construction methods, environmentally sustainable building practices, design theory, and professional practice.

The architectural profession is regulated by each state and so the architect must meet the requirements of that state. Each state requires that an architect perform a certain number of years working for a licensed architect as an intern.

Following the internship period are the licensing exams. The exams are extensive and cover the areas of architectural, structural, mechanical, building technology, construction administration, site planning and design, architectural layout and design, and professional practice. They are rigorous exams and in some states there is the additional requirement of an oral exam.

Builders often criticize an architect's training because it rarely includes doing construction or spending many hours on a construction site. This is a drawback and it's up to the architect to spend as much time on job sites as possible. It is a great question to ask your architect, "How much actual hands-on construction experience have you had?"

Lastly, an architect is always learning. Given the complexity of the profession there is no way to know it all. Just choosing to grow and learn in the aesthetic area is the work of a lifetime. Some architects continue their growth solely on their own, and others take part in continuing education courses offered by trade groups such as the American Institute of Architects (AIA).

How do I find the right architect for my project?

While interviewing architects, it makes sense to use some of the typical interviewing guidelines: check their references, meet the architect and see examples of their work (often just photos work, but you can visit the buildings as well), get a feeling for the rapport between you and the architect.

When looking at the work of the architect, keep in mind that an architect is the servant of the homeowner to a large extent, and must translate their goals and aesthetic rather than providing their own aesthetic. So if you don't see the style you are looking for in their portfolio, do not decide too quickly that it is not a fit. Instead, look for well executed examples of the style they are working in and for interesting ideas (aesthetically or practically) that show a creative spirit. The other element to look for is the architect's ability to listen to you and to communicate

Questions to Consider When Choosing an Architect

- What do the owner and architect each want to accomplish with the project?
- What would the owner and architect each need from the experience to feel successful?
- Are there any fears about the project?
- Are the owner's goals for their home and the philosophy of design that informs the architect's practice consistent? (e.g., if the owner wants to build their project green, is sustainability a foundation of the architect's practice?) If so, can they go beyond a systems approach to environmental building based on resource conservation to weave in ideas of lifestyle, quality of light and space, and relationship to the landscape?

- How will decisions be made as you progress through the design process?
- Are there others that will round out the design team, including the builder, landscape architects, structural engineers, civil engineers, mechanical engineers, financiers?

In general, is there a fit between the owner and the architect? The answer is part intellectual (what can be understood quantitatively) and part emotional (what can only be felt intuitively).

Source: George Watt, Architect, Boulder, CO

clearly, since so much of the process will involve listening to your ideas and concerns and translating them into the project design.

What is the best way to work with an architect?

Have fun!

Although the architectural design process can be tedious, and it involves large sums of money (both in terms of design fee and construction costs), it can also be fun. After all, you are investing in making your home reflect your life and dreams, and the things and experiences that bring you pleasure. Also, it's a chance to learn about what may be an entirely new area — that of design.

Keep an open mind

Hiring an architect means that you should end up with ideas that are better than the ones you've arrived at on your own. So the first rule of thumb is to let the architect look at your project with fresh eyes; do not dictate how you want the design to turn out. Instead, let him/her know what things you want to accomplish (and this can be very specific: I want a master bedroom of about 14 by 15 feet which extends off of the back of the house). It's fine to let him or her know what solutions you've thought of so far, and it could be that it is the best solution. But the architect needs to get their head around the problem before arriving at that conclusion.

> *Design is iterative. It is the architect's task to create a working record of the process of design, influenced by the opportunities and constraints brought to the project over time. It is the owner's responsibility to engage in the discussion inspired by the drawings and models created in order for the architect to evolve the design in the next set of drawings and models.*
>
> **George Watt, Architect, Boulder, CO**

Ideally you'll get at least three design solutions, even if some of them don't contain all the items on your list and if some are completely different than you discussed or expected.

During the design process homeowners typically become clearer about the details of what they care about. At the beginning of the project, it's difficult — if not impossible — for the homeowner to put in writing all the likes and dislikes they experience in their home. It is while reviewing the schematic designs that the architect and the homeowner refine their understanding of each other and of the project requirements. It may not be until one of the schemes does not allow a view of the neighbor's canary island date palm that the owner realizes the importantance of that view, and should be included.

Expose yourself to architectural and interior design

Expand your view of what's possible in home design in terms of room types, spaces, and materials. Go on home tours, check out books from the library on both domestic and international design, subscribe to design magazines, and start to notice buildings around you. The more exposure you have to interiors, buildings, and gardens the easier the design process will be, because you will have a context in which to place the design ideas that your architect is proposing. Most schools in the United States have a dismal record of educating students in visual and aesthetic literacy, and so the gap between the training of an architect and that of a typical homeowner is often great.

Communicate what you like and don't like to the architect

At the start of the project it's important to communicate as best you can your design preferences. The best way, since an architect is a visual person, is to show pictures and tour local buildings together that strike your fancy.

If you don't understanding the drawings, get help from the architect

It's not uncommon for a homeowner to have difficulty understanding the plans, and it's also not uncommon for the architect to forget that this happens. The architect has become so accustomed to speaking the language of lines within an architectural project that he/she forgets that the owner does not speak this language. Add the fact that often one of the most used drawings is a floor plan, which by its nature is abstract (rather than an illustrated perspective drawing), and the opportunity for homeowner confusion is great. Ask the architect for an "acting out" of the idea. Get out the tape measure together and map out where things will go, using masking tape to mark important spots on the floor or wall. Use stakes or sticks at the exterior to map out the extent of an addition. You can also ask for perspective sketches or additional drawings to illustrate an idea.

The other part of this is — pay attention during the design process. All too often an owner will coast along during the design process and wake up during construction deciding that they in fact want something different. Changes made during construction are expensive, and can cause frustration for all parties involved.

Get cost estimates at each design phase

One of the biggest complaints about architects is that they don't meet construction budgets. The fact is, architects are not cost estimators and cannot guarantee what price a contractor will be willing to work within. The architect can refer to both square foot figures and recent projects as a benchmark, but the nuts and bolts of an estimate should be left to those who produce them all the time; namely, a cost estimator or a general contractor. It is both possible and wise to ask a contractor to perform this service from the schematic design phase onward. If you make your selection early enough, your contractor can become part of your design team, estimating costs and recommending practical construction details through the rest of the design process.

In turn, the architect faces the challenge of gently curbing the owner's enthusiasm for things they may not be able to afford. It can be difficult to dissuade a homeowner from an element or space that their heart desires until there are some cold hard facts (the cost estimate) to review and process.

Lastly, a good architect will be naturally excited about design and materials and all that is possible, and can get carried away. The key is that the architect responds to cost information and scales back the plan in an attempt to meet the budget.

How do I manage construction costs?

No matter how much cost estimating is done or how carefully it is done, there will always be changes and unexpected conditions that result in cost increases. There are many factors influencing the design and construction of a project; additionally, remodeling involves the integration of a new project into or attached onto an existing structure. The foibles and intricacies of the old structure are not entirely visible even during the demolition stage. Homeowners change their mind and want a different material. Architects come up with a new idea that is wonderful but costs more. Surprises keep happening. Instead of being totally shocked, the better strategy is to plan to be surprised and set aside at least 15 percent of the estimated construction cost as a contingency amount. This means that if your budget is, for example, $100,000, then you'll want to actually aim for a construction cost of $85,000 so that you have $15,000 set aside.

What are the architectural design fees?

There are several ways that architects charge for their services. Hourly, fixed fee, a fixed fee based on a percentage of the cost of construction, and/or hourly for some phases with fixed fees for others. Reimbursable expenses are added to the fee (printing costs, etc.). Consultant costs are either included or charged for separately.

What products does the architect produce for this fee?

After all of the design phases are complete, the architect provides the owner with a set of architectural plans and often a written specification of products and materials to be used on the project. These "contract documents" form the basis of the contract between the owner and the builder. The builder's contract with the owner to provide construction services references the architect's name and the date of the plans and specifications. Note that often there are different "editions" of the design, with the first set often being that used to obtain the building permit. *It is crucial to the success of the project that everyone uses the same dated prints and that old versions are thrown away.*

It's tempting for some homeowners to try to save architectural fees during construction by not having the architect visit the site or by diminishing the services during construction. Unfortunately, the end result almost always suffers. The design continues to evolve during construction, with a number of factors influencing the project's design at this stage: unexpected conditions popping up and having design ramifications, code issues that the site inspector interprets differently than the plan checker did during the review process, and new ideas about how to handle something now that the plans are in three dimensions. In general, having the skills of an architect through only part of the process short-changes your dream home; imagine using the skills of a surgeon for only part of the surgery!

Choices, choices, choices!

The number of choices to make during the process can be staggering. It's important to start looking at materials, finishes, appliances, hardware, and so on as early on as possible, so that you aren't faced with a huge list at the end. In fact, the architect typically does at least a preliminary selection of most of these, since they are integral to the design, but at some point the owner must at least review and approve or disapprove the suggestions. Many times the architect leaves specific decisions up to the contractor. Unless you pay attention, you may get substitutions that compromise the integrity of your green design. Another way choices are presented to an owner is as "allowances." The builder has set aside a budget for items that the homeowner must select. Be careful that the allowance reflects the true cost of items you may want. An allowance for $12.00 per yard of carpet may be a builder price but you probably wouldn't want to live with carpet that cheap.

> *There are so many resources that go into new houses and so much of it is built without thinking about it carefully. The house is going to be there for a long time; don't let it be a missed opportunity.*
>
> **Cate Leger, Architect, Berkeley, CA**

How do I collaborate with my architect and builder?

Traditionally an architect worked on the design and the drawings and then provided them to the contractors during the bidding process. The problem with this model is that beautiful buildings were designed in somewhat of a cost vacuum, and all the effort and money and falling in love with the design were wasted once it was discovered that the project exceeded the budget. Working with a contractor during construction for cost estimating is the first step in collaboration, and the other is to have the contractor present during design (whether at part of the architect/client meetings or just with the architect) to suggest efficient construction details, bring up possible conflicts with other parts of the building such as the plumbing system, and in general bring their extensive experience in construction to bear on the project. Typically contractors need to be paid for these services, just as any project consultant is paid. Some contractors are willing to credit some of these fees if they are hired to do the job.

Another benefit to working with a contractor during design is that it gives the homeowner a chance to get to know a contractor before working with one. And working with one contractor during design does not preclude the homeowner from working with another during construction, although it is often best to maintain continuity if the relationship goes smoothly.

In general, an architectural project has so many bits of information to manage that no architect will do it perfectly. It's everyone's job to manage the flow of information and it's advisable for at least one person to prepare meeting notes and distribute them. *At the very least, the homeowner should keep their own dated notes of meetings and decisions and review them with the architect on a regular basis. The same habit should extend to the homeowner's communication with the builder.*

Working with Remodeling Contractors

If anyone knows how to find, hire, and establish a relationship with remodeling contractors, it's Dave Lupberger. We met years ago when we were both designing and remodeling homes in the Washington, D.C. area. He is now an industry consultant, and author of the *Turn-Key System for Remodelers* program and *The Emotional Homeowner*. Through his company, Remodelers Advantage, Dave works with remodeling contractors across the country.

Expert Advice from a Contractor: David Lupberger

Finding the Perfect Remodeler

Here are some simple steps to take to find the right remodeler for your home:

- Drive around your neighborhood. Most remodelers will post signs promoting their services in front of homes they are working on. Knock on the door and speak with the homeowner. I know of no better source of high-quality referrals than a happy homeowner, so the better remodeler will work hard to leave a legacy of satisfied customers. And you'll find that visiting remodeling projects is an excellent source of design ideas.
- Talk to friends, or friends of friends. Be bold! The more people you ask, including colleagues at work, clubs, professional organizations, charities, or service organ-izations, the more names you'll be able to gather. Be sure the people have personal experience with the remodeler they recommend. Six to twelve months after a job has been completed on their home is the best time to ask questions. During that interval their remodelers will have responded to some warranty item claims — an

excellent test of their reliability and professionalism. Many folks are overflowing with information from this once-in-a-lifetime experience and are full of stories they want to share.

- Yellow Pages. You can use the phone book, but are you willing to spend thousands of dollars based on a random lead you get?
- Contact the National Association of Home Builders (NAHB) Remodelers Council at 800-368-5242, ext. 216. The NAHB has published three consumer information brochures in a series: *Remodeling Your Home: How to Find a Professional Remodeler; Understanding Your Remodeling Agreement; and How to Live With Your Remodeling Project*. These can all be purchased from the NAHB bookstore for $3 each. The purpose of NAHB is to promote professionalism and image within the remodeling industry, therefore association members are more reliable than remodelers who are not members.
- Call the local National Association of the Remodeling Industry (NARI), 800-611-NARI (6274) or e-mail info@ nari.org. Ask for the most recent NARI Home Remodeling Guide that will list industry members in your area. Like NAHB, this association can provide you with more dependable, qualified "green" remodelers than the Yellow Pages; in fact, NARI members must attend a training seminar a pass a final exam to qualify as a NARI-certified green builder.
- Get a copy of Green Spec, <www.buildinggreen.com>. If you live in a remote community, you may have difficulty finding a green building remodeler. However, this book details specifications for building green, like using only certified wood, using solar energy, or recycling at least 50 to 80 percent of construction waste. Include in the contract that renovation plans will adhere to these specifications.

Now you know where to begin — *but*, before you sign that remodeling contract, learn the difference between a good remodeler and the telltale signs of a *bad* remodeler!

Table 3.2 Signs of a Good (or Bad) Remodeler

Signs of Good Remodeler...	Signs of a BAD Remodeler...
Returns your first or second calls.	Does not return your second call. They might just be too popular, but imagine how stressful that could be after work on your house begins.
Shows up promptly at your first meeting.	Fails to show up to your first meeting and does not call to reschedule.
A well-organized work site with workers moving like they know what they are doing.	A messy and chaotic work site.
A licensed, bonded, or registered business, when law requires it.*	An unlawful business.
Proof of insurance. The remodeler supplies you with a certificate of insurance indicating the company has sufficient general liability and workers' compensation insurance. The remodeler also has coverage against theft of any materials delivered to the jobsite but not yet installed.	No insurance.
A business address to confirm their permanance. It could be an office, or, as often is the case, a home office.	Works out of the back of a truck.
Does not mind you asking for a credit report, that you can obtain easily from your local banker.	Does not want to give you their name, address, and social security number to secure a credit report.
After contacting a list of suppliers and trade contractors the remodeler has worked with, you determine he or she manages business responsibly, pays trade contractors and suppliers on time, and that the business has a good reputation in the building community.	Poor reputation in the community increasing the risk that you might get a construction lien placed on your property by an unpaid trade contractor or supplier.

Note: *Call the building department in your local jurisdiction to find out the exact city, county, and state requirements and verify the appropriate licensing of your candidates.

Signs of Good Remodeler...	Signs of a BAD Remodeler...
After calling the Better Business Bureau and your local consumer affairs office to check the company for consumer complaints, you determine the remodelers are honest advertisers and sellers with few complaints. If your favorite remodeler has a complaint, it was addressed quickly and to the client's satisfaction — a sign of professionalism.	Complaints recorded by the Better Business Bureau and local consumer affairs office. Keep in mind unreasonable homeowners also exist, so if your favorite candidate has a complaint, ask for information about its resolution.
Committed to fulfilling your desires.	Does not seem to prioritize your concerns.
The homeowners he or she has worked for in the past testify this remodeler provided excellent value and delivered high quality work.	Past homeowners fail to praise the remodeler's craftsmanship.
Provides a contract price you can afford. Remodeling team making an effort to keep prices down by "engineering" certain features in your home in a cost-effective way.	Does not provide price checks and value engineering to help you stay within your preliminary design budget. This suggests a remodeler may not be interested in keeping you within budget.
Provides peace of mind and security.	Makes you stressed.
Progress on design.	Lack of progress on design.
Design-revisions contain the exact changes you asked for, and the revisions come back to you within a time frame the remodeler agrees to.	Shows general inattentiveness and lack of attention to detail.
Communication clear and easy, like when you're talking with a good friend.	Difficult or uncomfortable discussions. Dismissive.
Contractor comments, "This is your house, we do it the way you say."	Contractor comments, "Lady, I've been building 25 years, don't tell me how to build."
When you ask if he is a green builder, contractor replies, "Why yes! If it's not VOC-free, it was not meant to be!"	When you ask if he is a green builder, contractor replies, "What do you mean by that?"

Source: David Lupberger, Personal Conversation. Remodelers Guild. Olney, MD.

Working with Your Emotions

We used to laugh about calling Lightworks a design, build, and family therapy firm because we were in the middle of so many relationship issues that surfaced between couples during remodeling. It really helped develop my interpersonal communication skills. Remodeling is stressful and when it is over, all you remember is that is took longer than you thought but mainly that you love the new space. Nonetheless, understanding before you start that you will have moments of "buyer's regret," cost overruns, friction with some of the workers, more dust in your life than you ever imagined, and times when no one on the job seems to see eye-to-eye. For those times here are some tools to help get you through.

John Brockett is an unusual combination of trained psychologist and builder; he has eased homeowners through the remodeling process for over twenty years. Given John's sense of humor and insight, he is the perfect person to forewarn you of the emotions you may feel as you remodel your home.

Expert Advice from a "recovering" psychologist and builder: John Brockett

Remodeling and Love: There Really Is a Connection
Dating
In this initial stage, you call up some prospective contractors. If they fail to call you back, you call someone else. You meet with them, ask your friends what they think of them, and make sure they are not in trouble with the law or have a history of troubled homeowner relationships. Finally, you narrow down your search to someone you listens to your needs, communicates well, and is committed to fulfilling your needs.

Honeymoon
The beginning of your relationship with your contractor is the "honeymoon period." This is when the other can do nothing wrong and you usually overlook any faults the other may have because you are so enamored. This is natural and inspiring. With a little help the ecstatic feeling can be transferred throughout the project.

Labor of Love
Nothing stays the same, so the "honeymoon" will naturally become a "labor of love" given the chance. Again, your feelings, emotions, and thoughts, when clearly

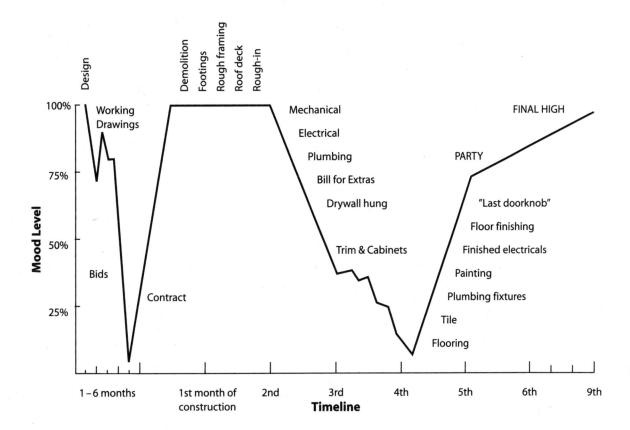

named, will become a natural corrective device when the ship gets off course. Naming them allows for necessary shifts, whereas harboring them is equal to constipation (and that is always uncomfortable for everyone).

Fig. 3.3: Relationship between mood and remodeling. Credit: David Lupberger, redrawn by Jeremy Drought.

Mediation

When trouble arises, it is best not to try to negotiate the really difficult problems of communication with your contractor by yourself. At the first signs of trouble, get some help. Call in your architect. You have hired experts to build your dream home, therefore you should consult an expert to ensure that your participation is in support of that dream in every way possible.

Growing Stronger from the Relationship

Building a new home is synonymous with building a new life. It is an emotional experience and a wonderful opportunity to manage a major undertaking through to completion. With diligence and commitment you will step back and smile, wanting everyone on the planet to see.

" *I have been reading about the California Native Americans and how they built their own structures just out of materials that were available to them. It was a community effort ... such a different kind of process from how we live now, where there are people who build, and then there is everyone else who just goes into these structures and doesn't really know what they are all about. But when I remodeled my home, I learned how it works. When my husband says, "The plug is not working," I can usually figure out the problem, whereas in all the other houses I've lived in it's been a complete mystery. It feels good to understand the building you are living in.*

Susan Jones, Homeowner "

Don't Stress! A Tool for Relaxing During Home Renovations

Like any major undertaking, building a home is a naturally stressful endeavor. From major university studies we know that moving or remodeling your home is in the top five most stressful life events, which include divorce, loss of job, severe illness and death of a loved one. Are you scared yet? Well, if you are, you are absolutely normal! What we call stress is simply the mind and body's response to fear of the unknown. Even really good things in life, like remodeling your dream home, are naturally stressful. However, a little information and a few special tools can go a long way to making your project a pure and lasting joy.

One tool is a centering device, kind of like an Eastern religion "mantra." A mantra is a word or words that when brought to the foreground of your attention will calm and focus your energy and poise. So, here's what to do when your emotions and thoughts are running wild after a construction worker accidentally installed a hot tub where the kitchen should be:

- Control your instinct to pick up a hammer and do something with it other than drive nails!
- Find a pen and three pieces of paper.
- Find a quiet, relaxing place to be by yourself. Try to rid yourself of the day's trials and tribulations, empty your mind, get quiet and relaxed.
- Ask yourself, "What are the three most important aspects of my new home?" They can be actual parts of the physical structure, feelings about having the project, reasons why you are building, relationships with others involved, etc. Don't force it; remain quiet until what comes, comes. Try not to second guess what comes usually comes from a deeper place than normal everyday functioning.
- Write one on each piece of paper as they appear.

- Place the three pieces of paper in front of you. Look for feelings or emotions that arise as you read each one.
- Discard one of the three. Notice what you feel.
- Again, do the same until there remains only one. Notice how you feel, notice what this remaining value means to you, notice that above all else this can represent the most important aspect of your new home.

> *When you start paying attention, you start to realize the impact we have. That is one of the things about this project that has been a life-changing experience for me. I started to ask, "What's in that? Where did it come from? What's its impact?" I began to see the world differently.*
>
> **Susan Jones, Homeowner**

There is a method to this madness and if you use it, some magic can happen. Believe it or not, this value has come to you through your uncon-scious and you can depend on it. If you are willing to maintain that value in every aspect of the building process, the power of that dedication will fill your new home. The gift of this process is to stay true to yourself in the face of adversity.

Helpful Tip: Maintain Positive Energy (Your house will show it!)

Remember, this home is a reflection of you. No one will see it as you do, including your contractor, however competent she may be. The difference between satisfaction and disappointment rests in one's ability to navigate the territory between the inner truth and outer reality. In this gray area, conflict and stress can arise. Either the contractor can be engaged to do her best work or discouraged to their worst. In other words, like the law of attraction, good energy produces more of the same; a stress-filled job site can produce a comedy of errors. This doesn't mean the homeowner must take a Pollyanna approach to the project. Rather, it means that when there is a problem, communication is clear, concise, and free of blame. Trust that the overall larger project will succeed even if certain smaller concessions must be made, and your home will radiate the caring energy that went into it.

Tools for Change

The bottom line for this chapter is to develop healthy relationships with your team. Take the time to find just the right architect, designer, contractor, and their workers. The best insurance policy you can provide for a successful remodeling project is to spend time interviewing and "dating" several of each type of professional. Do your homework. Everyone who has gone through the process of remodeling their home will tell you that this is the most important step. With the right teammates, you can make your remodeling dreams come true.

Working with building professionals can be exciting and fulfilling. We hope we have given you some tools to help you choose your professionals wisely and to remain calm when something doesn't work out exactly as you planned. Remember, this is your bathroom, kitchen, bedroom, or addition we're talking about — and while our goal is to *change* the world, a quarrel with a #$!@-ing electrician is not the *end* of the world.

CHAPTER 4

Building Science Basics: The physics you never wanted to know in high school

BUILDINGS ARE ESSENTIALLY A MANIFESTATION of the basic laws of physics. What holds them up, what keeps them dry, and what makes them comfortable are all just applied physics. Buildings fail when we ignore these laws. My childhood was filled with mops and buckets and backed up sewer drains, but this could have been avoided if our home had been built in accordance with physics. This brief overview will help you understand how your house works and when to be concerned if your contractor or his subcontractors start to ignore the basics.

Heat Movement: Thermodynamics

Energy is basically the "go" of things. Without energy the planet would be at rest, and nothing would ever happen. It takes energy for everything we know to exist. We rarely think about energy because it is the mainly invisible: only the results of energy in action are apparent to us. Energy has two basic laws that determine its behavior and the results we can achieve by using it.

The First Law of Thermodynamics

The first law of thermodynamics says that energy can neither be created nor destroyed, only changed from one form to another. We are familiar with various forms of energy on a daily basis. *Atomic or nuclear energy* is what powers the sun and nuclear power plants, and is the basis of nuclear bombs. Basically the energy that holds atoms together is broken and energy is released, called *radioactivity*.

Electrical energy occurs when electrons are released from an atom and they flow from a higher concentration to a lower concentration. A battery has more electrons on one end than the other, and when a light bulb or a motor connects them, the electrons flow from one end to the other. That is why a light switch works: we connect the concentrated electrons in the power lines to the "ground" and the light bulb is in the middle of the flow, creating resistance — which causes it to glow. When we flip the switch, the flow stops and the bulb goes out.

Chemical energy is released every time we eat. Digestion breaks the chemical bonds in food and releases energy. Petroleum is stored chemical energy. The energy is stored in the bonding of molecules and released when those bonds are broken by refining the oil or by burning it.

Mechanical energy is the energy of anything in motion. The chemical energy in petroleum is burned and released to create the mechanical energy of our cars going from place to place.

Potential energy is stored in differences in altitude. Hydropower is based on gravity pulling water down from higher elevations to lower elevations spinning a generator to produce electricity in the process.

We constantly change energy from one form to another every day.

The Second Law of Thermodynamics

The second law of thermodynamics says that energy can be changed from one form to another, but something is always lost in the process. In other words, there is no such thing as a free lunch when we convert energy — just as when you translate from one language to another, something is lost in the process. When you burn petroleum, you lose some of the chemical potential energy in the form of heat.

Heat is the final or lowest form of energy. At the end of the day, all the energy we use turns into heat that is radiated from the earth to the universe, an infinite heat sink. The more times energy is converted from one form to another, the more heat is lost in the process. The difference between what we started with and what we have to work with in the end is called "entropy." Entropy measures how much is lost in conversion. The higher the entropy, the less efficient the energy conversion process was. The lower the entropy, the more efficient the process. Understanding energy use requires an understanding of how entropy works.

Different forms of energy have different concentrations, or the ability to do more or less work. Uranium has many times more potential for doing work than a hot rock. Electricity is a much higher form of energy than the heater in your car. You can do some things with one that you cannot do with another, so we change it back and forth to do the work we want to accomplish.

Sunlight is the only form of incoming energy we have on the planet. Everything else comes from stored energy like oil, coal, or natural gas. When we use solar energy to heat our homes, there is very low entropy because we are going from the energy source to a direct application, making us comfortable. Very little is lost in the process of sunlight coming through our windows and heating our homes in the winter.

On the other extreme, if we use electricity generated by an oil burning power plant then there is a whole string of high entropy processes that take place. Oil exploration is often in remote places, like Alaska or under water. It takes energy to do the exploring. When it is found, it then takes energy to manufacture massive equipment to extract the oil. The equipment must then be transported to the site, using more energy. It takes more energy to run the pumps to extract the oil. Once the oil is extracted it must be transported to another location for conversion to useable products like diesel oil. It takes a lot of energy to reconfigure the chemical bonds in petroleum to make other products at refineries. The oil is then transported to a power plant that burns the oil to create steam to turn the big generators that create electricity. The electricity is then distributed through power lines that lose energy in the process, until it is converted by a transformer from high voltage to a form that you can use in your house. At the outlet, you can plug in a computer that uses electricity very efficiently and does amazing work, or you can plug in an electric baseboard heater that converts electricity to heat to make you comfortable for a few minutes. Energy, or entropy, is lost at every step of the process. The final usable energy available is called net energy.

Fig. 4.1: Direct passive solar heating is a good example of low entropy energy use, because there's only one step from sunlight to indoor heat. Credit: Reid Yalom, Leger Wanaselja Architecture, San Francisco, CA.

Fig. 4.2: Multiplying the efficiency losses at each stage in the process of delivering electricity to your home shows a high entropy energy process.
Credit: Jill Haras and Kim Master.

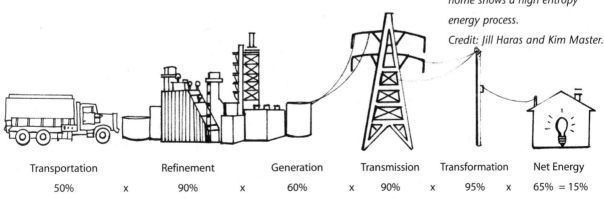

Transportation		Refinement		Generation		Transmission		Transformation		Net Energy
50%	x	90%	x	60%	x	90%	x	95%	x	65% = 15%

Efficiencies multiply. Net energy from a coal or oil-fired power plant is only 15 percent by the time it reaches your home! This is an example of high entropy energy use.

All energy moves from higher concentration to lower concentration. So that means that warm air always moves from hot to cold. The difference is called the temperature differential. The higher the differential the more heat moves from hot to cold. In the winter, your house is warmer than outdoors so heat moves from your house to the universe. The only thing that stops it is your insulation or reducing the *conduction* of heat through the walls and ceilings.

Hot air is also lighter than cold air so hot air rises like a hot air balloon. Often, the second floor is warmer than the first floor or the basement. That is from the process of *convection*, warm air rising and cooler air falling.

Also, warmer objects radiate heat to cooler objects. That is the experience of standing in the sun on a hot day. We feel warmer in the sunlight because of the *radiant* heat from the sun. That is why you feel the heat from a teapot even though your hand may be several inches away.

So heat transfer has three characteristics:

Conduction is the process whereby heat flows *through a material*

Thermal conduction is analogous to electrical currents; if it conducts electricity, it will conduct heat. Insulation slows the rate of conduction and is a better insulator than wood. Wood is a better insulator than metal. Would you rather pick up a hot frying pan with a cast iron handle or a wood handle?

Convection is the heat transfer in a gas (air) or liquid by the circulation of currents

Convection is based on the fact that warm air (or water) rises and cold air falls. A chimney works because of convection. Drafts form at single-glazed windows because the windows cool the air, which gets heavier and moves down the window and across the floor.

Radiation is energy radiated or transmitted as rays or waves (or in the form of particles for the subatomic physicists in the family)

Radiation is how the sun works to heat the earth. On a warm day it is hotter in the direct sun than in the shade, even though the temperature is the same. Warm surfaces radiate toward cold objects.

Our homes use these principles all the time to keep us comfortable — or not. The intention with green building is to use all of the laws of thermodynamics to our best advantage. We can incorporate as much insulation as possible to reduce the conduction of heat to the environment; reduce drafts by sealing the house well and incorporating ventilation where and when we want it, by directing and controlling convection; and we can take advantage of the radiant energy from the sun in winter through passive solar design.

Conventional approaches often don't take full advantage of the natural laws. Homes are oriented any which way, regardless of where the sun's heat is; they incorporate only as much insulation as they are required to by building codes, and too often they are drafty or hotter in some areas than others because natural convection has been ignored. We overpower the natural laws of thermodynamics by using more energy in heating and cooling equipment than we need to, resulting in unnecessary energy costs.

Air Movement

Ventilation

Ventilation is the way we manage the air inside the house. In bathrooms or kitchens we have exhaust fans to eliminate the humid air at the point of highest concentration. In other parts of the house, we typically move air around with the furnace fan. Ultimately, what we want is to control how much air enters the house, where it enters the house, and what we do with it once it is there.

Grandma's house never had ventilation problems because it probably wasn't insulated and it exchanged air with the outdoors all the time. Air sealing houses was never even a consideration. If water got into the wall cavities, it dried right out because there was nothing to keep it from evaporating. In today's homes, the intention is to make them as airtight as possible with as much insulation as possible and to control air movement either by designing for natural convective ventilation or by using mechanical means.

If the house is very tight, it is important to bring in fresh air, especially if there are gas appliances such as furnaces, water heaters, gas ovens, or clothes dryers. A fireplace makes it even more critical to bring in "make-up" air to replace the air that is used in combustion. There are many ways to address this issue, using varying degrees of technology. The first is to keep a window open in the area

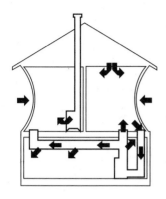

Fig. 4.3a: Depressurized house: supply ducts leak air outside living space; return take more air from inside than leaky supply ducts can replace; and air leaks in through holes in house air barrier.

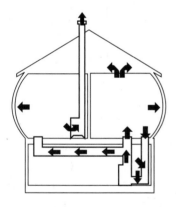

Fig. 4.3b: Pressurized house: return duct leaks take air from basement instead of from house; supply ducts add more air than leaky return ducts remove; and air leaks out through holes in house air barrier.

Credit: Jill Haras, redrawn by Jeremy Drought.

where fuel is burning, for instance, in a room with a fireplace (although it is not so romantic to have a cold draft wafting over the bearskin rug while you are lying there with a glass of wine). The second is to have ventilation built into the combustion appliance, such as a vent that supplies combustion air right to the fireplace. Sealed combustion furnaces and water heaters are designed that way.

Other ways to bring in fresh air are to have a fresh air vent into the return air duct in a forced air system. Many commercial buildings use this approach. On the high end, heat recovery ventilation systems are tied into exhaust fans in the bathroom and kitchen. When these fans are used or when a timer activates the equipment, fresh air is drawn in as air is exhausted. The air streams pass through an air-to-air heat exchanger that transfers the heat from the out going air stream to the incoming fresh air. This allows you to have fresh air without paying the energy penalty for exhausting conditioned air and reheating incoming cold air.

Air Pressurization

Good heating, ventilation, and air conditioning design creates balanced supply and return ducts so that there is no positive or negative pressurization in the home. Just as a balloon expands because you pressurize the air inside it, a home can be pressurized one way or the other. When you turn on an exhaust fan, you pull air out of the house creating a slight negative pressure.

Air always wants to be the same pressure everywhere, so when you turn on the fan, air comes in through leaks around doors, windows, or other penetrations in the envelope. This is called *infiltration*, but we experience it as a draft.

When the house is tight and we turn on a fan or the dryer, there are few leaks around the penetrations so air must come from somewhere else. Air will always follow the path of least resistance so the furnace, water heater, or fireplace flue becomes a likely candidate for make-up air. The problem is that it can backdraft carbon monoxide (CO) from the combustion gas into the house. This then becomes dangerous because you can't smell or see carbon monoxide — and it can be deadly.

The ideal situation is to have the house slightly pressurized. This helps keep out drafts, it creates resistance to external gas such as radon from entering the house and it reduces the risk of backdrafting. Heat recovery ventilators often create a positive pressure indoors.

Water Movement

Hydrodynamics covers the laws of water movement. Water and air both act according to fluid dynamics, but with different densities or viscosities. Water always wants to move or change states. Changing states means that it is converted from ice to water to steam depending on temperature. What we are most concerned about are the movements of water and humid air.

Moisture, like energy, moves from higher concentrations to lower concentration. Any porous material will act like a sponge. The drier sponge will absorb moisture until it is as moist as it can get. The same is true with building materials; wood is porous and will absorb moisture, which is why we protect our buildings with materials that don't absorb water such as shingles or siding.

Water is very insistent, and will migrate through any material or flow through any cracks it can whenever it has the opportunity. So if your foundation is not waterproofed sufficiently, water will always find a way to migrate through the porous concrete and into your basement. Once it is inside it is a pain to deal with, so keeping it out in the first place is the most important thing.

A roof either works or it leaks. The same is true for your siding material, although we are much more aware of leaky roofs than leaky siding. Roofs have two membranes, roofing felt or tar paper and a protective covering such as shingles or roof tiles; the main waterproofing layer is the felt paper, while the shingles are secondary, to protect the felt paper from degrading in the weather or from ultraviolet sunlight.

Siding should have the same consideration as roofing for keeping water out. Siding has more penetrations in it such as windows and doors and every one has the potential for leakage. The problem with leakage around windows is that the water gets into the wall framing and often never gets inside the drywall, so we don't see it. If the water has no way to evaporate and dry out, it causes rot or mold. Once it does become visible, it is often far too late, and the whole wall may need to be replaced. So once again, it is very important to keep the water out in the first place with good flashing around doors and windows and some form of house wrap to direct the water flow down and out away from the walls. This is called a drainage plane.

Moisture can also come from the soil around the house. If you live with a high water table, water can push its way up through the basement floor. If you live with a crawlspace, the moist soil can release moisture that is then trapped in the floor framing or distributed through the house by the ducts running through the crawlspace. The dirt floor of the crawl space should be sealed with sheet plastic, taped at the seams and around the perimeter, and the ducts should be sealed with mastic.

Although it is easier to visualize how water flows than how moist air moves, humid air is another way moisture gets into walls. In most climates, air inside the house contains more moisture than outside air. This is because people have the bad habit of breathing! We also water plants, boil water for dinner, and take showers. All these activities put moisture into the air. If the walls and ceiling are not protected with a vapor barrier, then the moisture can migrate through electrical outlets, recessed ceiling fixtures, or through the paint into the drywall, and ultimately get trapped inside the walls. Once there it can cause the same rot or mold problems.

Alternatively, in a hot, humid climate, there is more moisture outdoors than indoors. In these situations, a vapor barrier does not make sense; you need to ensure the walls can "breathe," or allow moisture to evaporate.

Still, we want some moisture in the air. Believe me, living in the high desert of Colorado, indoor moisture is a great thing — just not too much of it. Relative humidity is the measure of how much water is in the air at a given temperature. Warm air can hold more water than cold air. Most people are comfortable with a relative humidity of 30–50 percent. Above that, moisture problems can develop. Below 30 percent, hard wood floors or wood trim may shrink and crack (to say nothing of your nasal passages!).

This is just an overview of the wonderful world of building science. By having a basic understanding, you can make your home much more comfortable, have a better idea of what is going on when it isn't, and be able to have constructive conversations with your design team and contractors to assure that you are getting the most for your money. Living with nature starts by obeying the laws of nature. Green building is derived from living more closely with the natural processes that surround us, and making our homes a natural part of the larger ecosystem.

Part II

Changing the World
One Room at a Time

Millions of Americans are renovating their homes every year. Whether because of changing lifestyles or simply because houses are getting old and outdated, more money is spent each year on renovation than on new home construction. Buildings are responsible for 40 percent of energy flow and 40 percent of material use worldwide. The US Environmental Protection Agency (EPA) has stated that "indoor air pollution in residences, offices, schools, and other buildings is widely recognized as one of the most serious potential environmental risks to human health."

From another perspective, remodeling is an opportunity to make a difference in the world. As increasing numbers of people upgrade furnaces, cabinets, and toilets, they can have a positive impact on the world, including less fossil fuel pollution, less resources depletion, and lower health risks. Reusing existing buildings may be one of the "greenest" things we can do.

We are not suggesting that installing solar hot water heaters will stop greenhouse gas emissions from reaching what some experts predict will be levels higher than at any time in the last 50 million years. Remodeling your kitchen with certified sustainable wood will not stop the annual destruction of an area of biodiverse, old growth forest equal to the size of Nepal (*Time* magazine, November 1997, page 13). And installing a low-flow toilet will not replenish our dwindling water supply (*Time* magazine, November 1997, page 19). But green remodeling is energy-efficient, resource-efficient, healthy for occupants, and affordable to create, operate and maintain. Room by room, it is a way you can make a positive difference in your home.

If more and more people renovating use green remodeling strategies, the remodeling industry would have a significant role in saving the air, land, and waters we are degrading at unprecedented rates.

CHAPTER 5

Changing the World ...

Energy

NO MATTER WHAT KIND OF REMODELING YOU ARE DOING, whether it is a kitchen, bathroom, or guest room, you should first think about reducing energy use. Over the past 60 years, our access to inexpensive energy has allowed building design to ignore location and orientation. Unfortunately, this tradeoff of historical design wisdom for standardized building comes at great cost to the environment. We are using our natural resources at an unsustainable rate. According to Paul Hawken, author of *The Ecology of Commerce*, "Today's population uses in one day what it took nature 10,000 days to create." To say it in another way, we are living on our energy capital (stored petroleum and coal) and squandering our income (solar energy). If you ran your company on your capital savings and ignored current income, you wouldn't be in business very long. Green building is a step toward reversing that trend.

Based on 1998 figures, the heating, cooling, and lighting of buildings consumes 36 percent of the energy consumption in the US. A significant portion of this energy is in the form of electricity: residential buildings alone consume 35 percent of all electricity in the US.[1] However, the energy that buildings require starts accumulating long before the buildings and homes are even in existence. The energy required to extract, transport, manufacture, and then re-transport materials to the point of use requires a substantial amount of energy at a significant cost to the environment. The sum of all the energy required by all the materials and services (including the costs of upkeep and maintenance) that go into constructing a building is called the *embodied energy*. The unit of measure for embodied energy is British thermal unit per pound (Btu/lb.). It is highly dependent on factors such as geographical location and the technology used during the manufacturing process. For example, stones excavated from a nearby hillside for a new patio have lower embodied energy than stones that must be transported from another state. Embodied energy figures give us a realistic base

Table 5.1: Energy Required to Produce from Virgin vs. Recycled Materials

	Energy required to produce from virgin material (million Btu/ton)	Energy saved by using recycled materials (percentage)
Aluminum	250	95
Plastics	98	88
Newsprint	29.8	34
Corrugated Cardboard	26.5	24
Glass	15.6	5

Source: Roberta Forsell Stauffer. "Energy Savings from Recycling." Resource Recycling Magazine. January/February, 1989.

for comparison as we assess different products or technologies for use in our homes.

To better understand embodied energy, let's consider a brick in your exterior wall. Where did it come from? First, clay had to be extracted from the earth. Then it was transported to the brickworks where the clay was molded into a brick form and fired in a kiln. Eventually the brick was again transported twice more — to a retailer and then again to your building site — before the brick was put into place. But this is only the direct embodied energy of the brick. Embodied energy also includes indirect energy, including mining equipment to extract the clay, trucks to transport the clay, and kilns — anything that had a proportion of its energy invested in that brick.

The embodied energy in recycled building materials is generally much less than the embodied energy in materials produced from raw, or new, materials. Although using recycled materials can involve transporting, cleaning, and sorting, the total energy requirements are still far less than the energy used in extracting and refining a virgin resource.

This section will help you assess the embodied energy that goes into your home, the products you use, and the way you live. In this way you can understand and appreciate the complexity of construction, and its profound affect on everything around us. Our goal is to help you make appropriate choices when planning your remodeling project.

Effects of Fossil Fuel Use

Green building will help wean us off our dependence on fossil fuels. Currently, the US relies on fossil fuels — oil, coal, and natural gas — for 86 percent of its energy needs,[2] despite their polluting effects. Burning these fuels spews tons of fine particles, sulfur dioxide, toxic metals, and other pollutants into the air. The Union of Concerned Scientists (UCS) estimates that fine particles alone may cause 64,000 deaths a year, or more deaths than result from auto accidents. In addition, drilling for oil and natural gas and mining for coal harms the environment by polluting natural surroundings and disrupting local wildlife populations.[3] Given that the building sector is the second largest user of energy, changes in current building behavior are critical to reducing fossil fuel emissions and environmental damage, including the larger international issue of global climate change.

> ## *Is Global Climate Change Really Happening?*
>
> *Have you stepped outside of your air-conditioned homes lately? The fact of the matter is that we have had the hottest temperatures in decades over the last ten years. We have also seen erratic weather patterns, where we have had extreme cold, flooding in eastern Europe, severe drought in the western United States for at least five years, and major tropical storms. In Nicaragua, 40,000 people died as a result of mudslides and extreme weather; 10,000 people died during heat waves in France. These extreme weather patterns are characteristic of global climate change.*
>
> **Claudine Schneider, former Congresswoman**

Global Climate Change

Green building directly addresses the single most significant challenge of our generation — global climate change. The greenhouse effect is primarily a function of the concentration of water vapor, carbon dioxide, and other trace gases in the atmosphere that absorb the terrestrial radiation leaving the surface of the earth. Atmospheric concentration changes in these gases can alter the balance of energy transfers between the atmosphere, space, land, and the oceans. With everything else constant, when greenhouse gas concentrations in the atmosphere increase, there is a net increase in the absorption of energy on the earth.[4] This causes a rise in surface, ocean and air temperature as more heat is transferred to the ground.

Water vapor, carbon dioxide (CO_2), methane (CH_4), nitrous oxide (N_2O), and ozone (O_3) are all naturally occurring greenhouse gases, but can increase to destructive levels as a result of human activities, such as construction. Over the last 250 years, carbon dioxide concentrations have increased by 31 percent, methane by

131 percent, and nitrous oxide by 17 percent. Moreover, each greenhouse gas differs in the way it absorbs heat in the atmosphere: methane traps over 21 times more heat per molecule than carbon dioxide, and nitrous oxide absorbs 270 times more heat per molecule than carbon dioxide.[5] That said, the global mean surface temperatures have increased between 0.5 and 1 degree Fahrenheit since the late 19th century, which further correlates to a four- to eight-inch rise in global sea level. This current trend will likely accelerate the rate of climate change over the next centuries. Some climatologists expect the average global surface temperature to increase between 2.2 and 10°F (1.4 and 5.8°C) in the next 100 years.[6]

Scientists explain that some of the heat due to the greenhouse effect raises the air temperature a bit, but more of it causes increased evaporation of water. As a result, the extra moisture may disrupt weather patterns, producing stronger, longer-lasting, more frequent storms in some areas and droughts in others. Even in the short term we can expect to see these more extreme weather patterns directly caused by global climate change.

"Great!" you may be thinking. "If we keep burning fossil fuels to build homes and provide for our energy, maybe I can wear my swim suit in December!" Think again. Even in the short term, global warming is predicted to destroy coastal wetlands. It will cause unusually frequent but long-lasting bouts of severe weather: flooding in central Europe, vicious cyclones in South Asia, and freak spring snowstorms on the American plains. Global warming may also stress crop production, increase the frequency of diseases like malaria carried by airborne insects, threaten wildlife species, and disrupt entire ecosystems around the world. Low-income communities have the fewest resources to cope with climatic changes, and unfortunately, those communities are often situated near toxic, greenhouse-gas emitting facilities.[7]

In the eyes of most of the world, US leadership and integrity on issues of global climate change have disappeared. Concern with only those issues that are of specifically American interest has replaced historical global concerns, and that is costing the United States an enormous amount of good will. Former US Congresswoman Claudine Schneider thinks that this will result in higher tariffs on some American goods. In other words, the average American will have to pay more for various goods and services because the Europeans, Japanese, and others may eventually say, "All right, we are going to levy a tax on these goods coming from the U.S. because they are using more resources and more energy, and that is creating more of a problem for all of us."

In addition to the cost of political goodwill, there are bottom-line economic costs related to climate change. The US is spending billions of dollars on insurance to cover increasingly devastating weather damage, and also spending more on health care costs related to pollution-induced ailments such as asthma and allergies. As climate change increases, so too will virus-based diseases (such as West Nile disease), for which we have no remedy and which migrate north as a result of weather and habitat changes. Given all the negative ramifications of global warming, even the chief executive of British Petroleum has admitted it would be "unwise and potentially dangerous to ignore" the threat of global climate change.[8]

Green building helps us take personal responsibility for global climate change by clarifying the link between our actions and rises in temperature. Today, confusion masks the problem. For example, the term "global warming" can be misleading because we may still have record low temperature days while the earth is steadily warming. Adding to the confusion, high atmospheric winds carry air pollution long distances: how is someone who lives in the northeastern US supposed to identify their air pollution with Midwestern factories exhausting gases into their air from thousands of miles away? Perhaps even folks next door to coal-fired power plants don't make the connection: the Environmental Protection Agency estimates that we spend 90 percent of our lives in temperature-controlled buildings that shelter us from the elements. Furthermore, global climate change is the combined effect of human activities around the world: what difference does remodeling a house with inefficient incandescent lights make if factories are spewing out tons greenhouse gases in Europe?

Keep in mind the warming is global — and therefore everyone needs to take responsibility. This book helps you to understand the connection between your actions (how you remodel) and their consequences (global climate change). Each person who is remodeling *can* change the world for the better! In the following chapters, you will learn how to remodel your home to save you money and minimize its detrimental energy impact on the environment in two fundamentally "green" ways: use less energy and use renewable resources for energy.

Using Less Energy

In addition to investing in renew-able energy sources, as homeowners we can make simple changes to our homes that save energy — always the cheapest and most environmental solution! It is estimated that a whopping 43 percent of

The US Department of Energy estimates that buildings annually consume $20 billion more energy than would be necessary if the buildings were improved.

Source: Stephen Andros, "Green Buildings Help Cut Huge Energy Consumption," The Phoenix Business Journal, November 4, 2002.

American energy use is wasted.[9] The US Department of Energy estimates that we could save anywhere from 50 to 94 percent of our home energy consumption by making energy-saving changes.[10]

Green building reduces energy consumption in many ways. First, we can decrease the embodied energy of the building through efficient design, by using recycled and local materials, and by recycling construction waste. Second, green building design reduces a building's energy consumption over its lifetime. Installing ceiling insulation and double-glazed windows in every US home can save more oil than the Arctic National Wildlife Refuge can produce at its most optimistic projections, at about $\frac{1}{20}$ the cost.[12] Strategically placing windows and skylights can eliminate the need for electrical lighting during the day, which is often when electricity is in highest demand from utilities. A whole house fan can cool the house over night, rather than relying on air conditioning. Additionally, houses can maximize passive heating and cooling. South facing windows with overhangs can reduce heating costs by 20 to 30 percent, and prevailing breezes, shading, and natural plantings can keep houses cool in the summer using the same physics that cause global warming. This list only scratches the surface of the possibilities for reducing a building's energy requirements. The financial benefits are obvious: less energy leads to a lower energy bill. Additionally, decreasing energy consumption, and thus reducing alterations in the global climate, could help prevent further environmental degradation. Keep in mind that it is the impacts of energy use that we are trying to avoid — not the energy itself. In other words, reducing the use of specific non-renewable, polluting energy sources (for example, coal or oil), should be a higher priority than increasing the use of renewable energies such as solar-generated electricity.

Use Renewable Sources of Energy

Conventional Sources of Energy

When we discuss "renewables," we are referring to solar, wind, geothermal and biomass energy. Like renewables, nuclear power plants do not burn fossil fuels

Energy Efficient Design Savings

The study, *Greening the Building and the Bottom Line* by Joseph Romm of the US Department of Energy and William Browning of the Rocky Mountain Institute (RMI), highlights case studies of several companies that invested in energy-efficient designs and thereby experienced significant savings.[11] Further justifying the investment in retrofitting, they found compelling evidence that daylighting (a design feature which allows the use of natural light, rather than artificial light during daytime hours), improved the effectiveness of heating, ventilation, and air conditioning (HVAC), as well as the quality of indoor air. This resulted in increased productivity, fewer worker errors, and less absenteeism in many cases studies. For example:

• Boeing's "Green Lights" effort reduced its lighting electricity use by up to 90 percent, with a two-year payback and reduced defects.

• Lockheed's engineering development and design facility saved nearly $500,000 per year on energy bills and gained 15 percent in productivity, with a 15 percent drop in absenteeism.

• West Bend Mutual Insurance's new building yielded a 40 percent reduction in energy consumption per square foot and a 16 percent increase in claim-processing productivity.

Because labor costs are such a large share of total costs (workforce accounts for approximately $130 per square foot, 72 times more than energy), a one-percent increase in worker productivity can result in savings to a company that exceed their total energy costs. There are more and more cases similar to those documented by RMI, and as a result, companies are starting to invest in energy efficiency for the reasons suggested above: reduced energy expenditures and increased worker productivity.

and therefore do not emit substances that harm air quality or cause climate change. In fact, substituting nuclear energy for fossil fuel energy has significantly reduced US and global emissions of carbon dioxide, the chief greenhouse gas, and other pollutants. Moreover, radiation from nuclear plants is not an issue — nuclear plants produce only a small fraction of the radiation experienced by the US population. One report estimates that New York's six nuclear power plants cause approximately 0.5 to 1.5 statistical cancer deaths per year. Extrapolated to the US as a whole, this data implies 8 to 30 annual statistical deaths related to nuclear radiation that are concentrated among individuals who work in the power plants.[13]

The primary concern with nuclear energy is disposal. State and federal documents indicate that every dump ever used to store low-level nuclear waste — a total of six — has leaked.[14] In addition, the Congressional Research Service (CRS) reports that transporting nuclear waste to the proposed Yucca Mountain storage facility in Nevada could result in 154 truck and 18 rail accidents per year, a small number of which might release radioactivity.[15] Given that plutonium is radioactive for 250,000 years, one spill could stay on this earth and cause harm to people for longer than our species has inhabited the planet! Although nuclear power itself does not harm air quality, cause climate change, or emit harmful radiation from power plants, the waste from nuclear power is a serious hazard to human and environmental health — a risk not worth taking, given cleaner alternatives.

Hydropower has both positive and negative aspects associated with energy production, but in general is not a viable alternative. Dams needed to generate the power severely alter physical and chemical characteristics of the water and disrupt ecosystems both upstream and downstream. Scientists at Oak Ridge National Lab hold federal hydroelectric dams primarily responsible for reducing Northwest salmon from 16 million to 300,000 wild fish per year. Furthermore, dams are disruptive to human communities: one million people had to be relocated from an area inundated by the Three Gorges Dam in China.[16] Even though the US Department of Energy (DOE) defines hydropower as representing 42 percent of renewable energy production,[17] we do not recommend hydropower as an alternative to fossil fuels.

For building green, we do not need fossil fuels or hydroelectric or nuclear power — we need the services they provide. Most often we want heating, lighting, energy, and fuel, and this we can obtain from other renewable sources — such as wind, sun, and biomass. As Amory Lovins, president of the Rocky Mountain Institute, has said for many years, "People want hot showers and cold beer; they don't care where the energy came from." Renewable energy just needs to prove better or cheaper...

Wind Energy

Wind power is a realistic economic alternative today. Since 1983, prices for wind energy have dropped by an extraordinary 85 percent,[18] exceeding the most optimistic expectations from renewable proponents. Current state-of-the-art wind power plants are generating electricity at less than five cents per kilowatt hour and costs are continuing to decline as more and larger plants are built and advanced technology is introduced. According to Stanford University researchers in a 2001 *Science* article, the direct cost per kilowatt hour of power generated by winds of at least 14 miles per hour is 2.9 to 3.9 cents per kilowatt hour; one quarter of wind monitoring sites are capable of these wind gusts. This price is cost-competitive with new coal plants

Wind Energy Facts

- Denmark, Germany, and some regions of Spain now have 10 to 25 percent of electricity generated from wind power.
- A single one-megawatt wind turbine displaces 2,000 tons of CO_2 each year — equivalent to planting a square mile of forest — based on the current average US utility fuel mix.
- To generate the same amount of electricity as a single one-megawatt wind turbine using the average US utility fuel mix would mean emissions of ten tons of sulfur dioxide and six tons of nitrogen each year.
- To generate the same amount of electricity as a single one-megawatt wind turbine for 20 years would require burning 26,000 tons of coal (a line of ten-ton trucks ten miles long) or 87,000 barrels of oil.
- To generate the same amount of electricity as today's US wind turbine fleet (4,685 MW) would require burning 6.1 million tons of coal (a line of 10-ton trucks 2,300 miles long) or 20 million barrels of oil each year.
- 100,000 megawatts of wind energy will reduce CO_2 production by nearly 200 million tons annually — the amount of wind energy the European Wind Energy Association (EWEA) claims can be installed in Europe by 2010.

producing power at 3.5 to 4 cents per kilowatt hour and new natural gas plants producing power at 3.3 to 3.6 cents per kilowatt hour.

Wind is the world's fastest growing energy source on a percentage basis, growing 32 percent annually for the last five years[19] and on track to grow more than 25 percent in 2003.[20] A modern windmill can produce the energy used for its own production within just three months.[21] Three wind-rich states — North Dakota, Kansas, and Texas — have enough harnessable wind to meet our national electricity needs.[22] Globally, windmills can cover more than half of all energy consumption[23] without adding to air pollution, greenhouse gases, or other types of pollution or environmental damage such as that produced by strip mining or oil spills.

Fig. 5.1: America's wind production is doubling every two years. Credit: David Johnston.

Wind has myriad benefits beyond its ability to supply large amounts clean electricity cheaply, including national security, new wildlife-friendly models,

"

The Future of Wind Turbines

Once wind turbines are in wide use, there will be a large, unused capacity during the night when electricity use drops. Turbine owners can turn on the hydrogen generators, converting wind power into hydrogen, ideal for fuel cell engines. John Deere & Company is working on wind turbines that generate hydrogen to use in hydrogen-consuming farm equipment.

Sandy Butterfield, Chief Engineer, Renewable Energy Lab, Golden, CO **"**

pleasing aesthetics, fast installation, job creation, reliability, and predictability. In terms of national security, wind turbines are widely distributed, unlike a nuclear or coal plant where a single location can be targeted. Many people argue that a more distributed power structure such as one relying on wind power might have avoided blackouts in New York and New England caused by an outdated, centralized system. If 30 percent of America's electric power needs were met by wind, the US would be able to get rid of 60 percent of its coal dependence.[24] Today's larger, slower-turning blades are also less of a hazard for birds, that can be killed while attempting to circumnavigate the smaller, faster turning blades of older models. Turbines are being placed offshore, thereby reducing aesthetic objections (although inland they look and smell better than any power plant).[25] Wind power is the fastest of all technologies to install. The turbines can be built quickly to respond to electricity shortages and are a feasible option for developing countries or rural areas where fossil fuel power plants prove too expensive.

The European Wind Energy Association (EWEA) estimates that every megawatt (MW) of installed wind capacity creates about 60 person-years of employment and 15 to 19 jobs, directly and indirectly. A typical 50-megawatt wind farm, therefore, creates some 3,000 person-years of employment. The wind industry is likely to be one of the largest sources for manufacturing jobs in the 21st century. This is especially relevant for rural areas where it can provide a source of skilled-work income for sometimes hard-pressed farmers.[26] In addition, wind offers a "double cropping" benefit for rural communities. In other words, a farmer can grow crops while leasing his wind rights and earn $2,000 to $4,000 per year or more for housing a single utility-scale turbine on his property. The income-generating wind turbine is essentially a second, or double, crop.[27] Reliability has also increased — the availability of utility-scale machines is typically greater than 98 percent.[28] Wind is also inflation-proof, meaning that once a wind plant is built, the cost of energy is known and is not affected by shifting fuel market prices.[29]

Forecasting is expected to dramatically reduce the impact of wind's variability on utility operations, making integration into the grid easier.[30] It seems the answer to our energy problems may indeed be blowing in the wind.

Solar Energy

By far the largest part of the energy on Earth comes from the sun. The sun gives off so much energy that it is equivalent to a 180-watt bulb perpetually burning for every square meter (nine square feet) on Earth. This solar energy influx is equivalent to about 7,000 times our present global energy consumption. In other words, there is tremendous potential in solar energy to provide a significant portion of our heating, lighting, electrical, and mechanical power needs — 7,000 times our energy needs. Just by covering an area 291 by 291 miles square with solar cells, this 0.15 percent of the Earth's land mass could supply all our current energy requirements.[31]

But you don't need solar cells to take advantage of the sun's energy for your home. *Passive solar heating and cooling* represent an important strategy for displacing traditional energy sources in buildings. Anyone who has sat by a sunny, south-facing window on a winter day has felt the effects of passive solar energy. Passive solar techniques make use of the steady supply of solar energy by means of building designs that carefully balance their energy requirements with the building's site and window orientation. The term "passive" indicates that no additional mechanical equipment is used, other than the normal building elements. All solar gains are brought in through windows, with some use of fans to distribute heat or effect cooling.

All passive techniques use building elements such as windows, walls, floors, and roofs, in addition to exterior building elements and landscaping, to control heat generated by solar radiation. Solar heating designs collect and store thermal energy from direct sunlight. Passive cooling minimizes the effects of solar radiation through shading or generating air flows with convection ventilation.

Fig. 5.2: Designing to use natural light can eliminate the need for mechanical heat.
Credit: David Johnston.

Passive Solar Case Study

Susan Jones is a homeowner in California who has noticed a substantial improvement in comfort and a decrease in energy bills since she incorporated passive solar design into her home renovations. Before, "It would get to be 110 degrees in the house; it was amazing. Our energy bill was about $100 a month." Susan superinsulated her home, replaced leaky, single-pane windows with double-pane windows, and added a reflective roof. Now, "It can be 30 degrees outside and it stays 60 degrees inside; it is really nice. Our energy bill is only about $10 a month."

Beyond her comfort and financial gains, Susan feels emotionally fulfilled knowing that she thoroughly researched all the materials in her home to insure that every detail minimally impacted the environment and the health of the products' manufacturers.

"I feel a new sense of connectivity with nature....When I started to research conventional building materials and looked at where they came from, who made them, how they were made, I started to look at everything differently."

Another solar concept is daylighting design, which uses natural light to illuminate rooms during the day and contributes greatly to energy efficiency by eliminating the need to turn on lights. The benefits of using passive solar techniques include simplicity, low price, and the design elegance of fulfilling one's needs with materials at hand.

Photovoltaic (PV) cells convert sunlight into electricity for your home. They are usually made of silicon; they contain no liquids, corrosive chemicals or moving parts. Moreover, PV cells require little maintenance, do not pollute, and operate silently. Photovoltaic cells come in many sizes, but most are ten centimeters by ten centimeters (3.94 square inches), and generate about half a volt of electricity. A bundle of PV cells that produce higher voltages and increased power is referred to as a PV module, solar collector, or array. A module producing 50 watts of power measures approximately 40 centimeters by 100 centimeters (15.75 inches by 39.37 inches). PV modules can be retrofitted on to a pitched roof above the existing roofing, or the tiles replaced by specially designed PV roof-tiles or roof-tiling systems.

PV modules, like flashlights or cars, generate direct current (DC), but most home electric devices require 120-volt alternating current (AC). A device known as an inverter converts DC to AC current. Inverters vary in size and in the quality of electricity they supply. Less expensive inverters are suitable for simple loads, such as lights and water pumps, but models with good quality waveform output are

Photovoltaic Facts

• Worldwide photovoltaic installations increased to 340 megawatts in 2001, up from 254 megawatts in 2000. In 1985, annual solar installation demand was only 21 megawatts. The total on-grid market segment grew to almost twice the size of the off-grid market in 2001. That is the equivalent of one small conventional energy power plant.

• Of the global demand for solar photovoltaics, over 30 percent is accounted for by Japan, 20 percent by European countries, and less than 10 percent by the US.

• Nearly 45 percent of the world's solar cell production is manufactured in Japan. The US is second with 24 percent (of which 70 percent is exported), with Europe third, at 22 percent.

• Two billion people in the world have no access to electricity. For most of them, solar photovoltaics would be their cheapest electricity source — but they cannot afford it.

Source: Solarbuzz, "Fast Solar Energy Facts," cited September 26, 2003, <www.solarbuzz.com>.

needed to power electronic devices such as TVs, stereos, microwave ovens, and computers. In grid-connected systems, PV supplies electricity to the building and any daytime excess may be exported to the grid. Batteries are not required because the grid supplies any extra demand. However, if you want to be independent of the grid supply you will need battery storage to provide power outside daylight hours.

Between 1987 and 1998, the annual number of US PV shipments in the US grew 640 percent, with a 20.5 percent average annual increase.[32] The cheapest photovoltaic cells have become three times as effective since 1978.[33] Back in the 1970s the cost of PV cells was $70 per watt of production;[34] today, residential solar energy system typically costs about $8 to $10 dollars per watt. In some areas, government incentive programs, together with lower prices secured through volume purchases, can bring installed costs as low as $3 to $4 dollars per watt (10–12 cents per kilowatt hour). Without incentive programs, solar energy costs (in an average sunny climate) range from 22 to 40 cents per kilowatt hour.[35] Scarcely a month goes by without another advance in either PV cell design or manufacturing technology — by 2030, the price is expected to drop to 5.1 cents per kilowatt hour.[36] PV demand has been stimulated in part by government subsidy programs (especially in Japan and Germany), and by equipment rebate policies and tax credits for utilities or electricity service providers (e.g., in Switzerland and California). The central driving force, however, comes from the desire of individuals or companies to obtain their electricity from a clean, non-polluting renewable source, for which they are prepared to pay a small premium. The greater the demand for PV, the faster the price will come down.

PV systems are appropriate for electric devices, but water heating or other heating is most efficiently produced by *solar water heaters*. They convert up to 60 per cent of the sun's energy into heat used for domestic hot water, pool heating and space heating needs.[37] There are two types of systems: passive and forced circulation. A passive water heater consists of a water tank located above a solar collector. As water in the collector warms, water flows by natural convection through the collector to the storage tank. A forced circulation system requires a pump to move water from the storage tank to the collector. Most solar water heaters in the United States are the forced circulation type.

There are several types of solar collectors. Most consist of a flat copper plate with water tubes attached to the absorber plate. As solar energy falls on the copper plate and is absorbed, the energy is transferred to the water flowing in the tubes. Integral collector and storage systems combine the function of hot water storage and solar energy collection into one unit. Solar collectors are typically roof-mounted, with hot water storage tanks inside the house. They are often connected to a conventional water heater for back-up.

Solar water heating systems are efficient, clean, easy to install, and virtually maintenance-free. And since hot water counts for as much as 40 percent of the energy requirements of an average house, solar water heating systems can cut the costs for heating hot water by 40 to 60 percent.[38] An active, flat-plate solar collector system will cost approximately $2,500 to $3,500 installed, and will produce about 80 to 100 gallons of hot water per day. A passive system, typically used in climates that don't freeze, will cost about $1,000 to $2,000 installed, but will have a lower capacity. If the monthly cost of financing the system is less than the net savings, a solar water heating system may result in immediate positive cash flow.[39]

Overall, solar energy has a bright future. Passive heating obviously makes economic and environmental sense today, and solar systems are cheaper, simpler, and more reliable than ever for homes. All types of solar applications are expected to become commonplace when and if the true costs of fossil fuel use — including external costs like pollution, health risks, and military protection for foreign oil sources — become reflected in its price.

Biomass Energy

After hydroelectricity, biomass is the most widely used renewable source of energy, representing three percent of energy consumption in the US[40] In contrast to other sources of renewable energy that rely directly or indirectly on sunlight and its effects on weather patterns — such as wind power, solar cells and hydropower — biomass energy comes from stored solar energy in plants. Electricity from burning biomass (crops and crop waste) is also predicted to have substantial growth. Unfortunately, biomass can cause respiratory infections and various pollution problems, including sulfur, nickel, cadmium, and lead pollution. For some places in the world, however, growing biomass may turn out to be sensible since production can take place on poor soils, help prevent erosion, and even help restore more productive soil. Others argue it is not likely biomass will provide a major part of global consumption because the total agricultural biomass production from stalks and straw, constituting half the world's harvest in mass, only makes up about 16 percent of the current agricultural production. Still, Shell Oil, the most successful company in the oil industry, expects that biomass will provide between 5 and 10 percent of the world's energy within 25 years, possibly rising to 50 percent by 2050.[41]

Geothermal Energy

Like biomass energy, geothermal energy can be used at all times. Produced by the earth's natural subterranean heat, it is a vast resource, most of which is deep within the earth. Geothermal energy can be economically tapped when it is relatively close to the surface, as evidenced by hot springs, geysers, and volcanic activity. (The "Old Faithful" geyser in Yellowstone National Park is an example of geothermal energy.) In contrast to oil fields, which are eventually depleted, properly managed geothermal fields keep producing indefinitely.[42]

In the home, geothermal energy is used for heat pumps. Pipes are drilled into ground water that stays a relatively constant temperature. That is then used as the heat source for a pump that extracts the heat and blows it into the building or dumps excess heat in summer and cools the home. The cost of geothermal energy is currently priced

> **❝** Just as the Stone Age did not end for lack of stone, the oil age will not end for lack of oil. Rather, it will be the end because of the eventual availability of superior alternatives.
>
> **Bjorn Lomborg, *The Skeptical Environmentalist: Measuring the Real State of the World.* ❞**

Table 5.2: Major Life-Cycle Environmental Impacts of Energy Sources for Electricity Generation

	Air Pollution	Climate	Land Use/ Degradation	Water Use & Quality	Wildlife	Radiation
Coal	Very High	Very High	High	Very High	High	Low
Oil	High	High	Moderate	Moderate – High	Moderate – High	Near 0
Natural Gas	Very Low – High	Moderate – High	Low – Moderate	Near 0 – Low	Low	Near 0
Biomass	Near 0 – Moderate	Very Low – High	Near 0 – High	Very low – High	Low – Moderate	Near 0
Wind	Near 0	Very Low	High	Near 0	Near 0 – High	Near 0
PV	Near 0	Low	Very High	Near 0 – High	Near 0	Near 0
Geothermal	Near 0 – Very Low	Very Low – Low	Very Low	Near 0	Near 0	Near 0
Hydroelectric	Near 0	Low	Very Low – High	High	Very High – High	Near 0
Nuclear	Near 0	Very Low	Very Low	High – Low	High	High

This table is qualitative, labeling impacts in order from worst to best as high, moderate, low, or near zero. It should be read vertically; it does not attempt to compare the severity of different categories of environmental effects (e.g., air and water pollution). Finally, the life cycle impacts are not only based on power plant operation, but but also fuel production and transport, waste disposal, and other operations. Therefore, no cell in the table is empty, because even very clean energy sources like solar and wind require energy at some point in their cycle — for instance, for manufacturing — and this energy itself has environmental impacts.

Source: Adam Serchuk. The Environmental Imperative for Renewable Energy: An Update.
Renewable Energy Policy Project. REPP-CREST. April, 2000.

at five to eight cents per kilowatt hour, and is expected to drop to four to six cents per kilowatt hour with more industry experience and improved drilling technology.[43]

Renewable Energy Outlook

Compare the $50 billion the US spends on safeguarding oil supplies in the Persian Gulf with the $150 billion Americans could save annually by switching just 20 percent of our energy production to renewables.[44] A study by the Economic

Research Associates found that by switching to renewables, Colorado residents alone would gain 8,400 jobs and enjoy a $1.2 billion energy savings. Co-op America's Solar Catalyst Group found that California could create up to 15,000 new full-time jobs by producing up to 500 megawatts of new solar photovoltaics (PVs) per year by 2008.

Given this potential savings, why do fossil fuels still supply 85 percent of US energy needs, and renewable energy sources less than 3 percent?[45] Importantly, there exist powerful economic incentives to "disinvest" in renewables. We don't have to pay for resource depletion or air pollution, to say nothing about the tax incentives and corporate welfare the fossil fuel industry enjoys; therefore, the true price of fossil fuels is not reflected on our energy bills.

Additionally, the price of fossil fuels does not reflect the heath treatment costs, higher insurance rates, missed work, and lost life resulting from air pollution. Studies by the American Lung Association indicate that annual US health costs from all air pollutants may amount to hundreds of billions of dollars.[46] Until we pay for pollution, waste, carbon fuels, and resource exploitation (all of which are presently subsidized), there is little to encourage us to install unfairly expensive solar panels or other renewables.

Despite economic obstacles, the global wind industry is growing by 25 percent annually while the markets for oil and coal are expanding only 1 to 2 percent per year. Although the oil and coal markets are a significantly larger percentage of our energy use, meaning that, statistically, a one percent growth reflects a significant increase, a 25 percent growth in wind is still substantial growth. People are realizing that if we are going to approach our future with environmental and economic foresight, it would be wise to reduce our consumption of fossil fuels and invest in renewable energy production.

There are great advantages to using renewable energy: It pollutes less, makes a country less dependent on imported fuel, requires less foreign currency, and has almost no carbon dioxide emissions. Many of the renewable technologies are cheap and easy to repair. Renewable energies can be the most cost-effective method of bringing electricity to developing countries or remote villages that do not have reliable or wide-reaching energy infrastructure. Surpluses of

> *Oil scarcity may be the weakest reason for making the transition away from oil. Profit, climate protection, security, and quality of life are all more relevant and defensible.*
>
> **Amory Lovins, co-founder of the Rocky Mountain Institute**

Ozone Depletion from Construction

In 2000:

- 60 percent of ozone-depleting substances, including chlorofluoro-carbon (CFC) and less-destructive hydrochloro-fluorocarbon (HCFC), were used for building and construction systems in the US.

- 7,000 tons of CFCs were used to replace leaking or otherwise emitted refrigerants from older equipment in buildings.

- 120,000 tons of HCFC were used for new and existing building equipment.

- 75,000 tons of HCFC were used for foam building insulation.

Source: US Green Building Council, "Industry Statistics,"
cited September 25, 2003, <www.usgvc.org>.

wind, solar, and geothermal electricity on long-term contracts can guarantee the price, something those relying on oil or natural gas cannot do. And, once we get cheap electricity from these renewables, we can use it to electrolyze water, splitting the water molecule into its component elements of hydrogen and oxygen. Hydrogen, an exceedingly environmentally friendly energy-carrier that leaves behind only water, can later be used in electricity production.[47] Although renewables were once referred to in *The Economist* as "alternative" pet projects for "bearded vegetarians in sandals,"[48] these energy sources are quickly becoming recognized as economically competitive, more socially just, and environmentally sustainable energy solutions.

Natural Resources

Green building not only saves energy, but helps reduce the three billion tons of raw materials that are turned into foundations, walls, pipes, and panels every year. In fact, the building construction industry is the biggest user of materials, including steel and cement.[49]

Reducing the need for raw coal minimizes one of the most environmentally-destructive processes of the entire energy sector — mining. Mining often involves mountaintop removal and produces acid mine drainage caused by exposing iron- and coal-bearing rocks to water. Waste from uranium mines and milling operations constitutes the largest source of low-level radiation in the US; moreover, a disproportionately large fraction of this waste resides on Native American lands.[50] According to the Worldwatch Institute, mining for building materials such as copper and steel is responsible for 50 percent of CFC (chlorofluorocarbon) production in the U.S., thereby destroying the ozone layer that protects us from the sun's harmful radiation. Mining also generates 33 percent of the carbon dioxide (CO_2) emissions that contribute to global warming.[51]

Table 5.3: Remodeling Project Comparisons

Type of Remodeling Project	Geographic Location	Size of Project (Sq. ft)	Total Waste (lbs)	Generation Rate (lb/sq. ft.)	Average Generation (lb/sq. ft)
Kitchen and Room	Maryland	560	11,020	19.68	
Bathroom	North Carolina	40	2,883	72.10	
Totals		**600**	**13,903**	**23.17**	
Kitchen	Oregon	150	9,600	64.00	
House	Oregon	1,330	26,000	19.55	
Totals		**1,480**	**35,600**	**24.05**	
New Roof	Maryland	1,400	4,464	3.31	3.31

Source: Environmental Protection Agency. Characterization of Building-Related Construction and Demolition Debris in the United States. Report No. EPA530-R-98-010. Municipal and Industrial Solid Waste Division. Franklin Associates, June 1998. Chapter 2.

These disheartening facts do not even consider building waste. Alex Wilson, editor of *Environmental Building News*, describes construction and demolition (C&D) waste as "one of the most daunting challenges we face in the construction industry." Disposal costs are high, resources are needlessly wasted, and we are running out of landfill space. Even though there has been considerable media attention given to the solid waste crisis, Wilson comments that "it is remarkable how little we really know about C&D waste."[52] There are few reliable statistics on quantities of C&D waste generated nationally, and just a few studies of the composition of this waste.

It is especially difficult to assess total waste generation for renovations because of the wide variation in the types of remodeling jobs. Table 5.3 above, shows the results of five waste assessments that have been made at residential sites in the US, showing a wide variety of generation rates on a square-foot basis.

The National Association of Home Builders (NAHB) avoided this discrepancy by estimating the amount of material produced by the type of remodeling project. In the US, the major waste generated during remodeling activities stems from kitchens, bathrooms, and room additions. Annually, there are approximately 1.25 million major kitchen remodeling jobs (complete tear-out), with average waste generation of 4.5 tons per job. Americans perform 1.5 million minor kitchen remodeling jobs (facelift, cabinet replacement, etc.), that generate 0.75 tons of waste per job. Major bath remodeling (1.2 million per year) produces on average 1 ton of waste material each, and 1.8 million minor bath remodeling jobs produce on average 0.25 tons of waste each. Room additions, estimated at 1.25 million per year, produce about 0.75 waste tons apiece. From these calculations, NAHB estimated total residential renovation waste generation, from improvements or replacement projects, to be 31.9 million tons per year.[53]

In the US, construction and demolition debris account for 20 percent of all landfill waste; 43 percent (58 million tons) of this total is from residential construction, demolition, and renovation projects.[54] The number of landfills in the United States is steadily decreasing — from 8,000 in 1988 to 2,300 in 1999.[55] Creating new landfills is limited due to the protests of area residents near proposed sites. As a result of landfill limitations, the disposal costs are soaring to an average of 2 to 5 percent of the overall budget costs,[56] or $511 per house for construction disposal.[57] These costs drive more and more people to illegally dump construction and demolition waste.

Reduce, Reuse, Recycle

Green building provides myriad ways to dramatically reduce your waste and the costs associated with disposing it. Did you know that 85 to 90 percent of construction disposal is recyclable? If you plan with the 3 R's of waste reduction (reduce, reuse, and recycle) in mind, you will definitely see economic benefits. To this end, green remodeling encourages three key steps: planning ahead of time; reusing materials wisely; and recycling building waste.

The first step is planning. For minimizing the amount of waste generated, the following main areas should be focused upon: dimensional planning and design, material use and recycling, and use of modular/pre-constructed elements along with other resourceful building techniques. Since you may need to outsource this work, it is essential that the design team establish the waste reduction goals in contractual form with the subcontractors. Since contracts are often sidestepped (either purposely or inadvertently), it is the job of the construction manager to oversee all work and verify its successful completion.

Poor planning and design results in insulation leaks, moisture, rot, insect infestation and added waste — leading to added costs ranging from higher energy and waste removal bills to the worst-case expense of evacuating and demolishing

Case Study: One Moldy, Rotten House

In 1999, Melinda Ballard paid a plumber to repair a leak. A few months later, the hardwood floors began to warp and buckle. Soon mold grew and started to destroy their furniture and walls. Melinda and her family began coughing up blood and suffering memory loss. Finally, they had to evict themselves from their "dream home." Melinda's insurance company had to pay $32 million for material damage to the house. Still, the lawsuit verdict did not pay for the health damage to her husband, who lost his job and now goes to cognitive therapy four times a week as a result of their moldy home. Although this is a particularly extreme case, it clearly shows how cheap repairs and renovations can end up costing consumers and the environment more than do well-made homes.

Source: Lisa Belkin, "Haunted by Mold," *The New York Times Magazine*, August 12, 2001.

a mold-infested house. The extra energy and resources needed for repairing poorly planned buildings also contribute to the deterioration of the quality of our water, air and land.

Second, as you will read, green remodeling helps you use resources more wisely. We can build structures durably out of energy-efficient materials, including reused and recycled products. Some examples are engineered lumber, which reduces the amount of material needed by as much as 50 percent without sacrificing strength; walls built from insulation sandwiched between panels of oriented strand board; and recycled-content building products such as carpet, decking, cellulose, and fiberglass insulation.

Third, recycling waste helps to keep resources out of landfills. As an incentive to builders, many local municipalities are beginning to collect used construction materials at solid waste transfer stations with little or no tipping fee. Some non-profit recycling or reuse organizations will come to the site and load waste materials, often offering the builder a tax receipt for the donation.

Although there would seem to be no reason not to reduce, reuse, and recycle, the construction industry is a conservative one. New green technologies, products, and procedures typically take time to establish their stronghold. However, in our capitalist society the construction industry will adapt more quickly to a change in demand for green products. Homeowners who choose to make a difference and embrace green remodeling, therefore, have a huge potential to green the building industry and save precious natural resources.

Fig. 5.3: Weeping tree.
Credit: David Johnston.

Wood Resources

Residential construction accounts for more than 50 percent of the wood consumed in the United States;[58] 30 percent of softwood lumber (pine) consumption is accounted for by remodeling and repair alone.[59] Wood is durable, beautiful, and renewable. Compared to steel and concrete, wood also uses less energy; creates less air, water, solid waste, and greenhouse gas pollution; and uses fewer ecological resources.[60] However, we still waste wood unnecessarily and we need to be more reflective about the types of wood that we use. In particular, wood harvested from virgin, endangered, or old-growth forests should be avoided.

Old-growth forests take hundreds and even thousands of years to reach maturity and are home to innumerable plants and animals found nowhere else. To date, we have harvested over 97 percent of North American old growth forests.[61] Of that, one third goes into lumber, plywood, particleboard, and other structural building material.

Although the extraction process for wood is less polluting than that for many other building materials, we must consider the quantity of wood extracted compared to steel and plastics. Global consumption of industrial timber (approximately 1.66 billion tons per year) exceeds the use of steel and plastics combined. The average US home requires about 15,000 board feet of lumber,[62] the equivalent of harvesting an acre of trees. Humans are deforesting the world at a rate of 37 million acres a year (approximately the land area of Finland).[63] As a result of this high rate of timber consumption globally, forests are not just declining in area, but also in quality. As recently as 20 years ago, the average old-growth tree harvested from US national forests was 24 inches in diameter. Today the average is 13 inches! Globally, we also lose an estimated 27,000 species annually because of habitat loss, and the building industry is largely responsible.[64]

Although the US is home to only 4.5 percent of the global population, it is responsible for over 15 percent of the world's consumption of wood.

Source: "World Primary Energy Consumption and Populations by Country/Region," US Department of Energy, August 7, 2000.

Use Less Wood

In recent years, there has been a strong movement towards using wood more efficiently in construction in order to minimize wood consumption and the cost of wasted materials, as well as to optimize energy savings. By understanding the common dimensions available for building materials, we can quickly see what dimensions can be incorporated into design to reduce waste. Plywood, oriented strand board (OSB), and rigid insulating sheathing all come in four-foot by eight-foot dimensions. If you build in two-foot increments (outside dimension to outside dimension), there will be less wood sheet waste because the two-foot dimension is divisible by both the four-foot and eight-foot factors. This is applicable for wall construction, roofs, and overhangs.

For maximum framing efficiency, 24 inches on center (o.c.) spacing between framing members should be used instead of the standard 16 inches o.c. In other words, if you have studs (vertical pieces of wood) spaced 24 inches apart as opposed to spacing them 16 inches apart, you use less wood along the length of any given wall. If your wall is 12 feet long, you can build 24 inches o.c. with seven studs (144 divided by 24 +1). If you build 16 inches o.c., then you have to use ten studs. When you're adding on multiple walls, you can see how the wood savings from building 24 inches o.c. quickly add up.

When you construct with 24 inches o.c., you typically use two-by-six pieces of lumber for framing. These are usually more expensive than two-by-fours used with 16 inches o.c., and the volume of material is roughly the same. However, the 24 inches o.c. two-by-six method uses about 30 percent fewer pieces of wood, translating into labor savings and greater wall cavity space for more insulation. And more insulation equates to more energy savings. The typical R value (rate of heat loss through a material) for two-by-fours is 13, whereas for two-by-sixes, the R value is 19. This difference can save a significant amount of energy!

Additionally, you can also couple the 24-inch frame spacing with other efficient measures, including two-stud corners. Often builders use up to four studs at a corner, but this method merely wastes material, since you only need two. Also, the entire corner is wood with four studs — a "thermal bridge," or area with minimal insulation resulting in greater heat loss. Using two studs enables you to install more consistent insulation. In turn, the two-stud corner lends a higher R value and can save you money on lower energy bills.

Certified Wood

Not all timber is created equal: some is harvested with care and knowledge of valuable ecosystems, while some is the product of clearcuts and deforestation. A forest is very different from a tree farm or a plantation forest. Once a forest has been clearcut, the fundamental ecology is changed, from the micro-organisms in the soil to the diversity of fauna and flora. In other words, a tree farm is much like

Small is Beautiful: 14 Ways to Optimize Space

1. Provide an open plan for the kitchen/dining and living areas. Family members often prefer to spend time in the kitchen, so provide for that in the design. In many cases it also makes sense to extend this open layout to the living area, so that one space serves all three.

2. Avoid single-use hallways. Design houses so that circulation areas serve additional functions — circulation through the living/dining area, or hallways that also serve other functions — library space, for example, or (with adequate separation) laundry.

3. Combine functions in other spaces. By combining functions in certain rooms, space can be optimized. For example, combine a guest bedroom with a home office.

4. Provide built-in furnishings and storage to areas to better utilize space. For example: storage cabinets and drawers built into the triangular space beneath stairways; bench seats built into deep windowsills; library shelves along stairway or hallway walls; and display cases built into wall cavities. Small windows in walk-in closets can make those spaces more inviting and better used.

5. Make use of attic space. A tremendous volume in most houses is lost to unheated/uncooled attic space. Instead, insulate the roof and turn attic spaces into living area — making use of skylights and dormers to bring in light and extend the space. Having some rooms extend right up to the ceiling often makes sense, because variations in ceiling height make the room feel larger. If a standard uninsulated attic can't be avoided, at least design easy access and provide convenient storage areas so that the space can be used.

6. Don't turn bedrooms into living rooms. These are actually primarily used for sleeping and dressing. Keep them relatively small to avoid wasted space.

7. Provide acoustic separation between rooms. A small house will be more acceptable if there are no common walls between bedrooms. Closets can help provide this separation. Also consider insulating interior walls and providing staggered wall studs for acoustic isolation.

8. Provide connections to the outdoors, especially from the master bedroom. This will create a more pleasant house and make a compact house feel significantly larger. Careful placement of windows and glazed patio doors, as well as tall windows that extend down close to the floor help extend spaces to the outdoors.

9. Provide daylighting and carefully placed artificial lighting. Try to provide natural light on at least two sides of every room to provide a feeling of spaciousness and an opportunity for natural cross-ventilation. Incorporate some natural and artificial lighting where the light source is not readily visible to make compact spaces feel larger. Uplighting onto ceilings also makes a space feel larger.

Fig. 5.4: Credit: Kim Master.

Small is Beautiful: 14 Ways to Optimize Space

10. Provide visual, spatial, and textural contrasts. Contrasting colors, orientations, degrees of privacy, ceiling heights, light intensities, detailing, and surface textures can be an important design strategy for creating spaces that feel larger than they really are.

11. Use light colors for large areas. Most walls and ceilings should be light in color to make spaces feel larger. Use dark colors only for contrast and accent.

12. Keep some structural elements exposed. Structural beams, posts, and timber joists can be left exposed, creating visual focal points and texture. Be careful not to let these elements overwhelm the space; too many exposed timbers can make a space feel smaller.

13. Design spaces for visual flow. Careful building design can make small spaces feel larger by causing the eye to wander through a space. A continuous molding line that extends throughout a house somewhat below the ceiling can assist with this visual flow. Continuity of flooring and wall coverings can also tie spaces together visually. With very small spaces, provide diagonal sight lines that maximize the distance and feeling of scale.

14. Provide quality detailing and finishes. By limiting the overall square footage of a house, more budget can be allocated to green building materials and products that cost more (natural granite countertops, linoleum, certified wood flooring, top-efficiency appliances, etc.).

Source: Alex Wilson, "Small is Beautiful: House Size, Resource Use, and the Environment," Environmental Building News, January, 1999, p.11.

a cornfield: it is not a native habitat for the original species that once inhabited the area. Therefore, when we do use virgin wood, we should buy wood that is certified as being sustainably harvested, thus supporting environmentally appropriate, socially beneficial, and economically viable management of the world's forests and encouraging lumber companies to adhere to sustainable forestry guidelines. Certified wood is becoming increasingly commonplace; you can even find it in places such as Home Depot.

Keep in mind there are numerous certification programs, and some are more reputable and reliable than others. The Forest Stewardship Council (FSC) was launched in 1993 by indigenous groups, timber companies and environmental organizations in an effort to standardize the emerging programs. FSC is an international nonprofit organization established for the purpose of creating a verifiable international standard for well-managed forests and a process for tracking and certifying products derived from those forests. Significantly, certification is a third-party process — in other words, the people certifying the forest operations are not forest owners or managers, whose biases may cause them to overlook forest

Fig. 5.5: The FSC logo identifies products which contain wood from well-managed forests certified in accordance with the rules of the Forest Stewardship Council. © 1996 Forest Stewardship Council A.C.

Over the past ten years, 104 million acres in more than 40 countries have been certified according to FSC standards.

Source: Forest Stewardship Council, "About FSC" [online], cited July 20, 2004, <www.fsc.org>.

management inadequacies. Most environmental groups, the U.S. Green Building Council, and progressive businesses recognize FSC as the only environmentally and socially credible certification program in existence at this time.[65]

Rapidly Renewable Wood

All wood is renewable to a degree — which is one reason why wood is such a sought-after material for green construction. However, just as there are more efficient appliances for the home than others, there are also more renewable species of trees than others. Fast-growing trees offer the general benefits of plants (helping replenish oxygen in the air, and removing harmful CO_2), but also offer a consistent supply of material for construction.

There are many different species of wood used in construction that are highly renewable, with varying uses and applications based on geography, building codes, and availability. For instance, bamboo reaches up to a height of 60 feet in the first several months, making it extremely fast growing and renewable. Its strength is unmatched relative to its weight as compared to other construction materials. Although it is rarely used as frame construction in North America, it is commonly used in other areas of the world. Bamboo has successfully been adopted as a green flooring material due to its quick regrowth, and resulting consistent supply. Aspen is another rapidly renewable tree, but its uses are limited to engineered lumber products such as oriented strand board (OSB), because of its limited strength.

Alternate, Under-Utilized Tree Species

In Table 5.4 opposite, the left-hand column lists examples of endangered, vulnerable and rare tree species that you should avoid purchasing. For example, ipê, like many other trees in the tropical rainforests, only occurs in densities of one or two individuals per acre throughout most of its natural range. To meet the orders for hundreds of thousands of board feet of FAS ("fine and select" or "four-side-clear," meaning no knots or defects on all four sides of the board for its entire length), loggers have to log thousands of trees. That means punching roads and skid trails into thousands of mostly pristine acres of old-growth rainforests, as well

Table 5.4: What Trees are OK to Use?

Trees to Avoid[1]	Alternate Species[2]
Temperate:	
Alaskan Cedar	Angico—outdoor applications
Douglas Fir	Arariba—furniture, cabinetry, flooring, interior
Giant Sequoia	Cancharana—interior and exterior, joinery, furniture
Sitka Spruce	Chakte Kok—variety of applications
Western Hemlock	Chechen—furniture and variety of applications
Western Red Cedar	Curupau—heavy construction, outdoor, flooring, turnings
	Granadillo—substitute for Cocobolo or Rosewood; furniture
	Katalox—substitute for Ebony
Tropical:	Peroba—furniture, cabinet, flooring, trim, sashes, doors, turnery
Mahogany	T'zalam—furniture, interior finish work
Rosewood	
Okoume	
Ramin	
Ipê	
Cocobolo	
Ebony	

[1] Prior to specifying any tropical wood, reference the CITES listing of endangered species.

[2] Harvested from forests that have been certified as "well managed" according to standards endorsed by the Forest Stewardship Council.

Source: Rainforest Relief. Woods to Avoid and Alternatives [online]. [Cited October 28, 2003]. Rainforest Relief, 2003. <www.rainforestrelief.org/>.

as damaging or destroying up to 28 trees for every one they target. The canopy is reduced by about 50 percent after loggers take the mahogany, ipê, jatoba, and a few other high-value species for export.[66] The right-hand column in Table 5.4 will help you identify and specify alternative, under-utilized species to expand your design options and extend the forest resource. By using lesser-known species in lieu of the handful of "standards" we now depend so heavily upon, we can alleviate pressure on species that are threatened with extinction from over harvesting. We can also dramatically improve the economics of sustainable forest management by demonstrating the value of the full panoply of forest resources. This, in turn, will provide an incentive to maintain the wide diversity of species in natural forests.

Engineered Wood

In addition to rapidly renewable certified wood, homeowners should consider engineered wood as they remodel their home. Today's building industry is limited to younger, smaller trees that yield little sizable lumber. Much of this new wood tends to be weaker and wetter, with more natural defects and less tensile strength. However, unlike these smaller trees, engineered wood can utilize the strongest fibers. New technologies can take a tree apart and put its fibers back together to take advantage of its natural strengths wherever they are found on the tree. Using trees too small for sawn lumber, they can produce engineered lumber that's bigger and stronger than anything cut from a tree today. The result is a structural system of high quality lumber that's superior to the original log in size, strength, and dimensional stability.

Not only does this process avoid the use of old-growth trees, but the manufacturing process converts as much as 75 percent of a log into structural lumber compared to less than 50 percent by conventional methods — using fewer trees to do the same job. Making the most of under-used fiber, they produce cost-effective, readily available lumber that maximizes underused resources and minimizes environmental impact. In addition, engineered woods are able to use wood from readily available and quick-growing trees such as yellow poplar and aspen.

Significantly, engineered wood should still be certified. Most wood utilized for engineered wood is extracted using clearcut logging practices. Approximately 1.2 million acres are cleared annually to operate 140 chip mills in the Pacific Northwest. According to the US Forest Service, the removal of softwoods is currently exceeding the growth rate in southeastern US states.[67] FSC-certified wood — whether virgin or engineered — ensures proper management and longevity of our precious forests.

Reclaimed/Salvaged Wood

Non-forest sources of wood are another alternative to virgin or old-growth lumber. Reclaimed wood is salvaged from buildings and structures that are being remodeled or torn down. Sometimes logs that sank decades ago during river log drives can even be salvaged. Reclaimed wood is desirable from an environmental perspective because it is not associated with recent timber harvesting, it reuses materials, and it can reduce the construction and demolition load on landfills. Additionally, reclaimed wood is often available in species, coloration, and wood quality not found in today's forests. But, as with other resources, the supply of reclaimed wood is limited, therefore efficient and appropriate use of reclaimed wood is important for its long-term availability.[68]

Wood Treatments

Overall, whether you choose certified rapidly renewable, engineered, or reclaimed/salvaged wood, avoid selecting wood that has been treated with chromated copper arsenic (CCA) or ammoniacal copper arsenate (ACA). Arsenic is a rat poison. Treating wood increases its durability in locations where degradation by rot or insects might occur. However, leaching of chemicals out of the wood into the surrounding environment may occur to a limited extent. Handling the wood can also pose a risk to human health, especially if the chemical treatment is not fully dry. This is particularly a problem for young children who might be playing on a treated wood deck or playground equipment. Roughly 17 percent of all softwood lumber is pressure-treated today, including about 40 percent of all softwood from the southeastern US.[69] From 1985–97, approximately 48 billion board feet of wood products have been treated with CCA. The Environmental Protection Agency classifies wood treated with these substances as hazardous waste.

The most significant environmental concern associated with preservative-treated wood is disposal by incineration. Currently, an estimated 2.5 billion board feet of preservative-treated wood is disposed of annually; as much as 16 percent may be incinerated. Toxins, such as arsenic, may become airborne to a limited extent, but most toxins end up as ash where they are highly leachable.[70] One tablespoon of ash from CCA lumber is enough to kill a cow. Unfortunately, the only environmentally acceptable disposal option for CCA-treated wood is to send it the landfill, and these are filling up.

Although CCA was phased out of production for residential use in December 2003, it will remain an issue for existing homes. To avoid problems with CCA-treated wood, try to ensure that treated wood is disposed of in lined landfills only, so that it does not pollute the soil and water. Better yet, try to reuse treated wood so it does not end up in a landfill in the first place. Use construction details that minimize the use of wood in locations where rot or insect infestation is likely. If wood must be used, a healthier alternative is Alkaline Copper Quarternary (ACQ), which does not contain arsenic or chromium.[71] Borate preservatives are also much less toxic, but can somewhat leach out of wood in wet conditions. You might also consider naturally rot-resistant species like cedar or redwood, but only if they are from a certified forest.

Recycled Plastic "Wood"

Plastic lumber, a newer and increasingly popular replacement for CCA pressure-treated lumber, is most effective because it protects timber resources and prevents the use of highly toxic lumber treatments.[72] It also provides a use for the millions of tons of annual plastic waste.[73] Plastic lumber will not rot, absorb water, splinter, or crack; it is also resilient to shock, making it an extremely durable component in exterior applications. Plastic lumber is most cost-effective in large-dimensions where wood is most expensive,[74] but should not be considered a suitable replacement for load-bearing structural components.[75]

Composite lumber incorporates some of the characteristics of wood with those of plastic lumber. Recycled wood waste fiber is combined with recycled plastic resins to create a product that has some improved strength and aesthetic characteristics. Like plastic lumber, it will not rot, crack, or splinter. Furthermore, wood composite materials generally have a more natural coloring and appearance, although they may still be stained or painted. In general, plastic and composite lumber are great options to consider for decks or other exterior applications.[76]

Water Resources

When remodeling we must also consider another increasingly valuable resource — fresh water. According to the Worldwatch Institute, buildings in the United States use 17 percent of the total freshwater flows.[77] Large quantities of water are required to produce many construction materials: during manufacturing, steel uses 25 times more water than wood.[78] More generally, our water consumption has almost quadrupled since 1940.[79] US indoor residential water use is estimated to average 80 gallons per day in homes without efficient fixtures. Outdoor use varies tremendously, but obviously adds significantly to this number.[80]

Washed Down the Drain

- Older toilets use between 3.5–7 gallons of drinking-quality water per flush.[85]
- Most dishwashers use between 8 and 14 gallons of water for a complete wash cycle.[86]
- Clothes washing in a typical top-loading machine require about 45 gallons of water.[87] Front loading machines can reduce this figure by a third to a half.
- Nationally, lawn care accounts for 50 to 75 percent of outdoor residential water use.[88]
- A typical family of four on public water supply uses about 350 gallons per day at home.[89]

- Each day, US water users withdraw over 300 billion gallons of water from the earth, enough to fill a line of Olympic-size swimming pools reaching around the world.[90]
- Hand washing dishes with the tap running half open uses 25 gallons. Washing and rinsing in a sink or dishpan uses only 6 gallons.[91]
- A dripping faucet wastes 15 to 21 gallons per day.[93]
- Taking a bath uses 36 gallons for a full tub.[94]
- Showering uses 12 gallons per minute, or three gallons per minute with a flow restrictor. (Try a shower with a friend).[95]

In the Middle East, China, India, and the United States, groundwater is being pumped faster than it is being replenished, and rivers such as the Colorado and Yellow River no longer reach the sea year round. Fresh water in groundwater often takes centuries or millennia to build up — it has been estimated that it would require 150 years to recharge all of the groundwater in the United States to a depth of almost 2,500 feet if it were all removed.[81] Issues over water rights create conflict between neighbors, counties, states and countries. Water issues also force us to dam rivers and irrevocably damage countless ecosystems: "In the northern hemisphere, three-quarters of the flow from the world's major rivers has been tamed to quench our thirst."[82] Bjorn Lomborg, author of *The Skeptical Environmentalist: Measuring the Real State of the World*, reasons that our water wells are not going to run dry; in fact, even projected "overestimates" of total water use for 2025 require just 22 percent of the readily accessible, annually renewed water.[83] But when we consider its uneven distribution and the environmental degradation caused by pollution and damming, it becomes clear that better water management practices are essential to providing clean water for everyone while preserving fresh water ecosystems.

Fig. 5.6: Xeriscaping minimizes
water use and maintenance,
saving time and money.
Credit: B.F. Norwood.

Water Quantity

Better water management can begin in the home. While domestic use only accounts for a small fraction (12 percent) of total water use,[84] it is still problematic. For example, rapid urban growth has resulted in depleted groundwater sources in Los Angeles, San Diego, and Tucson, forcing these cities to siphon water from far-reaching places. However, with minimal impact on our current lifestyles, we can reduce our domestic water use by as much as half. Today's water-saving showerheads and faucet aerators save water by creating a more forceful spray, sometimes mixed with air, using less water. Low-flush toilets, required by code in the US for new construction use, use 1.6 gallons of water or less per flush. Composting toilets are also an option; they can save a typical family of four 47,028 gallons of water per year.[96] Additionally, modern dishwashers can fit on a countertop and use only 4.75 gallons of water per wash cycle. Newer front loading washing machines are initially more expensive, but use only 20 to 28 gallons per load as opposed to 45 gallons and save $60 to $100 per year in water and energy costs.[97]

Outside, we can take steps to lower the water needs of our lawns. One solution is to use wildflowers, low-water native plants, and attractive xeriscape designs. Not only do these plants lower water use, they also save time and money because of fewer maintenance requirements. If giving up a traditional grassy lawn is unappealing, simply plant buffalo grass, which requires less water than traditional lawn grasses. Finally, instead of installing a conventional sprinkler system, use drip irrigation. Drip irrigation delivers water directly to the plants' roots rather than wastefully spraying water over a large area, where much of it evaporates.

Water Quality

Although our planet is 71 percent water, humans depend on a mere 0.65 percent of this water for survival — much of which is polluted. In a recent article in *The Washington Post*, it was reported that "about a quarter of the nation's largest industrial plants and water treatment facilities are in serious violation of pollution standards at

What's in Your Water?

- **Arsenic**: An element that gets into the water supply through natural soil deposits or industrial and agricultural pollution. Arsenic causes bladder, lung, and skin cancer; heart and nervous system damage; and skin problems. Very low levels of arsenic have also been found to disrupt hormone functions.

- **Trihalomethanes (THMs)**: These chemicals are formed when chlorine used to disinfect water reacts with organic matter, such as animal waste, treated sewage, or leaves and soil. They can increase the risk of cancer and may damage the liver, kidneys, and nervous system, and increase rates of miscarriage and birth defects.

- **Trazine**: A weed killer used on most corn crops, atrazine can cause organ and cardiovascular damage and is a suspected hormone disruptor.

- **Coliform bacteria**: These indicate the presence of dangerous microbes such as cryptosporidium, which can be life-threatening to people with weak immune systems.

- **Lead**: A toxic heavy metal that can damage developing brains and nervous systems. While lead may not be present at the source, it can leach into your water from lead-containing pipes in your home and in public water mains.

Source: The Green Guide Institute, "Water Filters,"
cited September 29, 2003, <www.thegreenguide.com>.

any one time."[98] Moreover, half the serious offenders exceeded pollution limits for toxic substances by more than 100 percent.

According to the Natural Resources Defense Council, an estimated 7 million Americans are made sick annually by contaminated tap water; in some rare cases, this results in death.[99]

Many Americans worry about drinking tap water: we spend over $4 billion per year on bottled water and over $2 billion per year on in-home water filtration systems.[100] We do not recommend bottled water for several reasons:

- **It's wasteful**. According to a Los Angeles' *Times* report, "If the [plastic water bottle waste] problem continues, enough water bottles will be thrown in the state's trash dumps over the next five years to create a two-lane, six-inch-deep highway of plastic along the entire California coast."[101] In an April 2001 report by the World Wildlife Fund, an estimated 1.5 million tons of plastic is manufactured from petrochemicals each year to package water. The EPA estimates that in 1999 alone, about one million tons of plastic bottles ended up in US trash bins.

- **It's just tap water**. National Resources Defense Council and *Consumer Reports* have found that some bottled water is simply tap water. In can be worse for you, in fact: tests by *Consumer Reports* in August 2000 showed that plastic water containers can leach phthalates and other chemicals into the water, and bottles that are left open for long periods are subject to bacterial growth and contamination. Unlike tap water, regulations allow bottled water to contain E. coli or fecal coliform and don't require disinfection for cryptosporidium or giardia.
- **It's expensive**. Bottled water can cost from 240 to 10,000 times more per gallon than water from the faucet, according to calculations by the Natural Resources Defense Council. In the US, we pay more for bottled water than we pay for gasoline.

Water filters are the best option for contaminated water. First, you need to figure out what's in your water. In a May 2002 study of the nation's stream water by the US Geological Survey, scientists discovered chemicals found in drugs, detergents, disinfectants, insect repellants, plastics, and personal care products, including 33 suspected hormone disruptors. If your water source is "good," the easiest and least expensive option is to select a charcoal filter that removes the chlorine taste and smell, as well as excess minerals and possibly some heavy metals. For water that is contaminated with pollutants such as pesticides and harmful chemicals, consider systems for reverse osmosis and distillation. The Environmental Protection Agency (EPA) estimates that the cost of providing state-of-the-art drinking water via water filtration systems would cost the average household only $30 per year. (Learn more about water filtration in the Plumbing section of Chapter 7.)

Aside from using water filters, there are several other options. You can boil water for a minute to kill bacteria and parasites. You can drink and cook with only cold water and let the water run for one minute in the morning to decrease lead from pipes in old homes (although this option wastes water). Or you can leave tap water in an open container in the refrigerator for few hours to dissipate chlorine and trihalomethanes (THMs).

Ultimately, the best defense against water pollution is to protect the rivers, streams and wetlands that are the source of your drinking water. Campaign to limit development around reservoir watersheds. Refuse to use yard pesticides and fertilizers that can run off or seep into groundwater. Dispose of leftover paints and other household chemicals through your community's hazardous waste collection program — don't pour them down the drain. Use biodegradable cleansers and detergents, and buy organic food that is grown in accordance with watershed-protecting farming practices. Small actions add up to a big difference!

Health and Indoor Air Quality

It is estimated that we now spend more than 90 percent of our time indoors.[102] Our health and well-being are notably affected by the large amount of time we spend indoors. In part because of dust, mold, lead, and asbestos in the home, more than 38 percent of Americans suffer from allergies,[103] resulting in a variety of symptoms ranging from runny noses and fatigue to learning problems and epilepsy.[104] Moreover, asthma, which is linked to animal dander, paint fumes, dust mites, cockroaches, molds, pollens, and indoor secondhand smoke,[105] is on the rise. From 1980 to 1994, the number of asthma cases grew 75 percent[106] to over 17 million.[107] We are just beginning to understand the connections between many more ailments — some life threatening — and your home.

Indoor air quality represents a considerable challenge for several reasons. First, the current method of determining IAQ is not related to any health or productivity measure. Rather, it is simply a function of the percentage of people who are dissatisfied with a space: if 20 percent are dissatisfied, the building has poor indoor air quality. Not only does this measure ignore health care costs, productivity, and crippling law suits, but it is technically inaccurate.[112]

Second, pollution laws in general focus more on the sources of the greatest amounts of pollution, rather than on the sources of the greatest amount of exposure to pollutants. Smoke stacks emit large amounts of pollution, but most of us do not live next door to one. Our homes, on the other hand, emit relatively small amounts of pollution, and yet they are often the places where we face the greatest exposure to health-threatening pollutants.

Third, in recent years we have become more concerned with energy efficiency and sealed up our homes more tightly without consideration for proper ventilation. In effect, as cabinets, carpets, furniture, stoves, and many other common household items contribute to indoor air, they build up to unhealthy concentrations because they can not escape from the house as they might in a leakier home.

By remodeling green, you can prevent serious health issues related to indoor air. It is important to understand the sources of your exposure to toxic pollutants

Indoor Air Pollution

- The US EPA ranks indoor air pollution among the top five environmental risks to public health. Unhealthy air is found in up to 30 percent of new and renovated buildings.[108]
- According to the World Health Organization (WHO), indoor air pollution causes about 14 times more deaths than outdoor air pollution, or about 2.8 million lives each year.[109]
- Of all the hundreds of chemicals regulated by the Environmental Protection Agency (EPA), only ozone and sulfur dioxide are more prevalent outdoors than indoors.
- According to the US EPA, the medical and lost-productivity costs of workers breathing poor air amounts to tens of billions of dollars each year in the United States alone.

- The National Contractors Study indicated an average indoor air quality (IAQ) productivity loss of 10 percent from IAQ on its review of 500 studies.[110]
- A study at Cornell University showed that poor lighting results in a 10 percent worker productivity loss.[111]
- Green designs resulting in productivity gains of 1 percent can provide savings to a company greater than the savings from reduced energy consumption.

in your home, including showers (chloroform), stoves (carbon dioxide, respirable particles), and carpets (lead, pesticides). Additionally, there exist many often-unrecognized indoor pollutants that you should be aware of as you remodel. We have organized all these pollutants into six overarching categories to clarify the indoor air quality issues in your home and allow you to make simple, healthy remodeling choices:

1. Particulates (lead, asbestos, fiberglass, dust)
2. Combustion gases (carbon monoxide, nitrogen oxide, hydrocarbons)
3. Volatile Organic Compounds (VOCs — formaldehyde, pesticides, vinyl chloride, soil gases)
4. Radioactive contaminants (radon gas)
5. Environmental tobacco smoke
6. Moisture and mold

Particulates

Particles, sometimes called particulates, are small specks of solid matter. Common household dust can include microscopic particles from fabrics, soil, plants,

Sick Building Syndrome

When I was working as an independent contractor, I got a call from a homeowner who was at his wit's end. He recounted that for several months his kids had been constantly sick with headaches, fever, coughs, and other flu-like symptoms. The doctors didn't seem to be able to help the kids and they were missing a lot of school. During the same period, his wife had two or three migraine headaches a week, and was losing a lot of sleep. All of them were exasperated. He asked me if these conditions could possibly be from something in his home.

I asked him many questions about their home and lifestyle and if they had done any remodeling. Nothing seemed relevant to their situation. Finally, he told me he had gotten a bonus at work and had bought built-in shelves and desks for his kids and their master bedroom. He hadn't mentioned it because he couldn't see how that made any difference. I asked him to go to the rooms and pull out a shelf and describe to me what he saw. It turned out to be particleboard with a melamine veneer.

I suggested that he try an experiment: take everything off the shelves and remove them for the weekend since all the shelf edges were not sealed and I suspected they were offgassing formaldehyde. I asked him to call me a week later to report if there was any difference. He called and said that the kids' symptoms had decreased and he was encouraged. The next weekend he removed all of the built-in cabinetry. He called me two weeks later and said all of their symptoms were gone. The kids felt great and his wife hadn't had a headache since he took the shelves out.

insulation, human and animal dander, food, dirt, paint, plastic, soot and cigarette smoke. The particles themselves can carry harmful chemicals, such as lead. In addition, particles can carry dust mites, or tiny insects that live on the dust itself. Many of these are biological allergens that can cause reactions ranging from sneezing and running eyes to heart palpitations, internal pains, and loss of muscle control.[113]

Particulates are especially dangerous to small children, who play on floors, crawl on carpets, and regularly place their hands in their mouths. Infants are particularly susceptible: their rapidly developing organs are more prone to damage, they have a small fraction of the body weight of an adult, and they may ingest 5 times more dust — 100 milligrams a day on average. Carpets act as a reservoir for particulates, especially plush and shag carpets. Vacuuming hard-surface floors will remove almost 100 percent of dust, while vacuuming a carpet typically removes only 30 to 60 percent of the dust. However, wiping one's feet on a commercial grade doormat appears to reduce the amount of lead in a typical carpet by a factor of 6.[114] A central vacuum that vents outdoors and a vacuum with a high-efficiency filter are the best options to keep particles to a minimum.

Lead paint.

Twenty million lead-painted houses have too much lead dust or chippings — about 20 percent of all housing in the US. Homes in the Northeast, Midwest, and Western states have more lead than those in the South.

Source: Debra L. Dadd, The Nontoxic Home and Office, *Jeremy Tarcher, 1992, p. 165.*

Lead

Lead has long been recognized as an especially dangerous particulate. It impairs mental and physical development in fetuses and young children and decreases coordination and mental abilities. Additionally, lead damages the kidneys and red blood cells, and may increase high blood pressure.[115] The margin of safety between measured blood levels and the levels causing clinical symptoms is remarkably small; only a slight increase in lead concentration in children's blood levels is necessary to cause anemia and the onset of mental losses.[116] Low exposures can be even more dangerous than higher exposures because there are no visible symptoms, but the subtle learning and behavior problems can go on untreated for years.[117]

Children, especially between the ages of six months and six years, are at high risk because they are more likely to ingest lead particles, and because harmful effects begin at lower blood levels. Lead enters the body when an individual breathes or swallows lead particles or dust once it has settled. There are many ways that humans can be exposed to lead: through air, food, contaminated soil, deteriorating paint, and dust. Additionally, pipes, solder in pipe joints, and home plumbing can release lead into your drinking water; in some instances, this may account for up to 40 percent of your family's lead exposure.[118]

If your house was built before 1950, you almost certainly have some paint with high lead content — the most significant source of lead exposure in the US today.[119] According to a 1990 study by the US Department of Housing and Urban Development (HUD), 75 percent of all private housing built before 1980 has some lead paint.[120] If the house has been repainted, the old paint has been sealed in. However, if the new paint deteriorates or is improperly stripped, it can create a new hazard.

It is especially important for children and pregnant women to avoid drinking, eating, or breathing lead because growing children and developing fetuses are the most vulnerable to lead poisoning which can cause irreversible brain damage and behavior and learning problems, and delay mental and physical growth. Kids tend to get lead on their hands and then put their fingers and lead-ridden objects in their mouths, so they are likely to have higher exposure. The US Centers for Disease Control (CDC) estimates 890,000 children in the US between the ages of 1 and 5 have elevated lead levels in their blood.

To avoid lead in your home, have old paint lab-tested for high lead content before major renovations. Leave paint undisturbed if it is not accessible to children and if it is in good condition. Replace doors, windows, or trim covered with lead paint, or strip and repaint them away from the house. If you want the paint removed on-site, have it done by a trained contractor. The work will be dusty, so you should move the family out the house during the work, and clean up thoroughly before moving back in. Also, have your water tested for lead content. If the results show high lead levels, replace old plumbing or install a point-of-use water purification device, such as a reverse osmosis system or distiller.

Asbestos

Asbestos is a naturally occurring particulate with long, flexible fibers that can lodge in your lungs. There are no immediate symptoms, but prolonged exposure can cause asbestosis (severe impairment of lung function), or cancer of the lung or lung cavity. Asbestos can also cause cancer in the digestive tract if fibers are inadvertently swallowed. Smokers are at a higher risk of developing asbestos-induced lung cancer.

In 1974, after the elevated risk to asbestos workers was documented, asbestos was banned for interior use in the US. It still exists in many homes today and is difficult to remove, given that if any of the products mentioned above are damaged, improperly removed, or get disrupted in any number of ways (like sanding), it is likely that the asbestos will become airborne. However, most asbestos is embedded in other materials, such as ceiling tiles or pipes, and is not as dangerous in that form.

Combustion Gases

Appliances that burn fuel, such as furnaces and fireplaces, commonly leak at least small amounts of the gases created during combustion. They will emit larger amounts when the appliances are not working properly, or when another appliance or system, such as a bathroom fan or range hood, draws air from the house and causes normal chimney flow to be reversed, known as backdrafting. This negative pressure causes gases to be drawn back down the chimney or exhaust duct, and into the house. For example, leaky return-air ductwork, fireplaces, down-draft kitchen

Asbestos Alert

Asbestos is only detectable under the lens of a microscope and is only problematic when the mineral fibers become airborne. Because asbestos is cheap, fireproof, and a good insulator, it was used in a multitude of products, including:

- **Vinyl floor tiles and vinyl sheet flooring**. French scientists found that heavily trafficked floors released significant amounts of fibers. Fibers can also be released if the tiles are sanded, cut, dry scraped, cleaned abrasively with a broom or vacuum, or otherwise damaged.

- **Patching compounds and textured paints**. Asbestos has not been allowed in textured paints since the Consumer Product Safety Commission banned the use of asbestos in patching compounds in 1977. Still, scraping or sanding these materials in older homes will release asbestos fibers.

- **Ceilings**. A home built between 1945 and 1978 may have had asbestos-containing material sprayed or troweled onto the ceilings or walls.

- **Pipe and Duct Insulation**. If your home was built or repaired between 1920 and 1972, the hot water pipes and furnace ducts may be covered or wrapped in asbestos-containing material, such as asbestos paper tape.

- **Wall and ceiling insulation**. Asbestos insulation is typically "sandwiched" between plaster walls in homes constructed between 1930 and 1950.

- **Fireproofing**.

- **Acoustical Materials**.

- **Wood Stove Door Gaskets**.

Backdrafting

Rob deKieffer and Rich Moore, working with Sun Power, Inc., tested the whole-house pressure of 100 homes in Colorado. They found that negative pressure is common in homes with forced-air systems, especially in basements whenever the furnace is operating.

Source: Steve Andrews, "A House Under Pressure," Home Builder, January 1997, pp. 13–14

exhaust fans, undersized cutouts through floor plates for return-air drops, normal exhaust fans (kitchen range hoods and bath fans), dryers, central vacuum systems, large combustion air ducts located on the sheltered side of a home, and large penetrations into attics, such as whole-house fans,[121] can all cause backdrafting.

Carbon Monoxide

Although you can't see, taste, or smell carbon monoxide (CO), this combustion gas is responsible for a larger number of severe chemical poisonings than any other single agent.[122] When breathed, carbon monoxide preferentially binds to blood hemoglobin, displacing oxygen at the binding site and thereby depriving the body of oxygen. Early symptoms of carbon monoxide poisoning include persistent

Tips to Avoid Carbon Monoxide Poisoning

- Never burn charcoal inside a home, garage, vehicle, or tent.
- Never use unvented fuel-burning camping equipment inside a home, garage, vehicle, or tent.
- Never leave a vehicle running in an attached garage, and minimize the amount of time the vehicle is in the garage when you start it each morning, even with the garage door open. Move the vehicle out as soon as possible after starting.
- Have your fuel-fired appliances serviced every one to two years.

- Never use gas appliances such as ranges, ovens, or clothes dryers for heating your home.
- Never operate unvented fuel-burning appliances in any room without adequate ventilation or in any room where people are sleeping.
- Do not use, or service, gasoline-powered tools, and engines indoors or in attached garages.

Source: American Lung Association of Minnesota, <www.healthhouse.org>, cited October 1, 2003.

headaches, nausea, fatigue, blurred vision, rapid heartbeat, loss of muscle control, and flu-like symptoms that clear up upon leaving the house. More severe exposures cause vomiting, collapse, coma, and death, depending on the degree of oxygen deprivation. Suicidal or accidental death from running a car in an enclosed garage or from using an unvented, poorly tuned combustion appliance indoors results from depression of the central nervous system to the point where breathing is stopped. The Consumer Products Safety Commission reports that more than 200 people in the United States die from CO poisoning every year.[123]

In addition to carbon monoxide, fuel-fired appliances also emit other combustion by-product pollutants, including formaldehyde, nitrogen dioxide, sulfur dioxide, carbon dioxide, hydrogen cyanide, nitric oxide, benzo(a)pyrene, and vapors from various organic chemicals. At low levels produced from average use from combustion appliances, possible symptoms from exposure to these by-products include eye, nose, and throat irritation; headaches; dizziness; fatigue; decreased hearing; slight impairment of vision or brain functioning; personality changes; seizures; psychosis; heart palpitations; loss of appetite; nausea and vomiting; bronchitis; asthma attacks; and breathing problems.

The most effective way to avoid combustion by-products is to use all-electric appliances — range, heaters, water heaters, and clothes dryers. Self-cleaning electric ovens are one exception because they produce carcinogenic polynuclear aromatics that are on the EPA list of priority pollutants. In general, however, indoor air pollution studies show that all-electric homes have significantly lower concentrations of combustion by-products than do homes with gas appliances.

You can still use gas appliances and protect yourself somewhat by ensuring that fuel-burning appliances like furnaces, hot water heaters, stovetops, and fireplaces are properly ventilated. If you are replacing your furnace or water heater, consider using a sealed combustion unit that isolates the appliance from the indoor space. During cooking, for example, a hood fan can remove up to 70 percent of pollutants produced.[124] If possible, put your gas appliances in a space outside of the living area, venting the fumes to the outside and placing a tight seal between the appliance and the living space to prevent gases from spreading throughout the home. Use a new-model gas stove with low-heat-input gas pilot light and non-gas ignition systems, which produce significantly lower pollutants than do older stoves with pilot lights.

For wood stoves and fireplaces, make sure they are installed and fitted properly. Have your chimney inspected for creosote buildup when the weather starts getting cold, or periodically throughout the year if you frequently use your fireplace. Fix cracks or leaks in the stovepipe and keep a regular maintenance schedule to keep the chimney and stovepipe clean and unblocked. Have your fireplace retrofitted with an insert that draws air from the outside and has airtight doors that can be closed when you leave the room.

Finally, beware of negative air pressure and downdrafts indoors that can pull pollutants into your living space instead of up the flue. Eliminate the problem by installing only sealed combustion appliances. If this proves too expensive, reduce the backdrafting by selecting only fan-driven, draft-induced water heaters and eliminating any wood or gas fireplace that isn't sealed combustion. Other alternatives include carefully sealing all ductwork with mastic, installing a return-air duct in all the bedrooms, and selecting a standard kitchen range hood rather than an excessively powerful downdraft exhaust cooktop, which can suck too much air out of the house too fast, causing negative air pressure. Keep in mind that "leakier" homes are *not* a solution; this merely makes the home uncomfortable and more expensive to heat and cool. Moreover, if leaks are located in the wrong places, they can add to unbalanced pressure and introduce combustion gas issues in the home.

Tighter Buildings and Formaldehyde Risk

In past decades, energy-conserving airtight buildings that use urea-formaldehyde foam insulation (UFFI) have become common. The Consumer Product Safety Commission (CPSC) banned use of UFFI in residences and schools in 1982, after receiving numerous complaints that exposure to this insulation caused respiratory problems, dizziness, nausea, and eye and throat irritations, ranging from short-term discomfort to serious adverse health effects and hospitalization. The ban was later overturned by the US Court of Appeals, but the CPSC continues to warn consumers that evidence exists to indicate the substantial risk associated with UFFI.

Volatile Organic Compounds (VOCs)

Many of the hazardous chemicals in modern houses are members of a large family called volatile organic compounds, or VOCs. The distinctive smell of a newly remodeled house is primarily composed of offgasing toxic volatile organic compounds. They are commonly found in plywood, particle board, wood paneling, carpets and carpet padding, insulation, paints, finishes, adhesives, heating fuels, solvents, waxes, polishes, and many other household products.

VOCs found in the home include formaldehyde, benzene, xylenes, toluene, and ethanol. There are several main concerns with VOCs. First, VOCs contribute to pollution outside by reacting with sunlight to form ground-level ozone, a major component of smog. Second, VOCs' effect on indoor pollution is even worse — the Environmental Protection Agency's total exposure assessment methodology (TEAM) studies have found indoor levels of common organic compounds to be two to five times higher than those found outside. Third, producing many of the binders and solvents in wood finishes creates significant amounts of hazardous wastes. If leftovers are discarded into the trash, they contribute to air pollution. When poured down the drain, they contribute to water pollution and clog your sinks and water pipes in the process. And finally, petroleum-based finishes (like most wood finishes) contribute to the depletion of this non-renewable resource and increase our nation's dependence on imported oil.

While some building products now report the parts per million of VOCs on labels, this information can be misleading. Fewer parts per million is certainly better, but chemicals like dioxin are not safe in any detectable amount.

VOCs and Offgasing

VOCs are characterized by the fact that they release vapors at room temperature. The easiest example to understand would be the application of interior house paint. When you open the can of paint and get out the brush to begin, the paint is wet and in a liquid form. You apply the wet paint and several hours later, the paint is dry. During the drying period, the volatile solvent in the paint vaporizes to a gas, known as "offgasing," leaving the non-volatile portion of the paint on the wall, and dry. Most VOCs are released into the air during this offgasing period. You should open windows and use fans to move the VOCs out of your indoor air as quickly as possible. Turning up the heat also makes the paint dry and offgas faster. The rate of offgasing dissipates dramatically within a few days. However, the paint may continue to offgas small amounts for the lifetime of the paint, especially as the paint begins to age and chip small flecks of paint that can become trapped in your rug. It is best to avoid VOCs altogether when you are buying paints or any other products for your home.

Formaldehyde

Another one of the more lethal VOCs in the home is formaldehyde. It comes primarily from the adhesives in pressed wood products, such as particleboard, hardwood paneling, and medium density fiberboard. It is also found in urea formaldehyde foam insulation (UFFI), combustion sources, environmental tobacco smoke, durable press drapes, some other textiles, and glues. Building materials can offgas formaldehyde for five or more years following manufacture. The amount of formaldehyde released from building materials increases as temperature increases — for example, during the winter when indoor heaters are on and there is little ventilation from windows.

Several studies have shown that a significant portion of the public is exposed to formaldehyde at levels high enough to produce symptoms such as eye, throat, and nose irritation; wheezing and coughing; fatigue; skin rash; and severe allergic reactions. Symptoms have occurred at levels as low as 0.05 parts per million — the level that has been proposed as the California indoor air quality standard.[125] Over time, some people develop heightened sensitivities to formaldehyde — in other words, once a person is exposed, even minute exposures to formaldehyde can induce health problems.

The long-term effects of prolonged exposure to low levels of formaldehyde — levels typically found in mobile homes or energy-conserving buildings — are controversial. Still, the EPA has concluded that formaldehyde is a probable human carcinogen on the basis of experimental studies and human epidemiological studies and laboratory tests. In the human epidemiological studies, they found increased incidence of brain tumors, leukemia, and cirrhosis of the liver among workers exposed to formaldehyde. Laboratory studies indicate that formaldehyde causes nasal cancer in rats and that it appears to cause mutations in bacteria, yeasts, fruit flies, and mammalian and human cells.[126]

To reduce your exposure in the home, keep in mind that medium-density fiberboard emits about three times more formaldehyde than particleboard. And plywood, although made with a formaldehyde resin, is preferable to particle-board. Also, exterior grade plywood emits less than interior grade. "Swedish" hardwood floor finish is notorious for high formaldehyde emissions: it is banned in Sweden and many other areas.[127] The best option is to choose pre-finished solid wood flooring or use water-based finishes.

In addition, use air conditioning and dehumidifiers to maintain a moderate temperature and reduce humidity levels. Insist on carpet or carpet pad with little or no formaldehyde content. Also, apply surface barriers (such as low-VOC latex paints and primers) to particleboard or plywood to reduce offgassing. These vapor barriers can reduce formaldehyde emissions by up to 95 percent, but tend to break down after several years and require reapplication. Foil tape can seal the edges and keep fumes from escaping.

Finally, open the windows! This is particularly important after bringing new sources of formaldehyde into the home. Consumer advocate Debra Dadd also recommends spider plants to absorb formaldehyde: "You'll need about 70 Chlorophytum elatum (spider plants) in one-gallon containers to purify the air in an average 1,800-square-foot energy-efficient home (a veritable jungle!), but don't be overwhelmed."

Pesticides

Although some pesticides may technically be considered VOCs, these often odorless and invisible substances have become such a health threat that they warrant a separate discussion. According to the World Health Organization, more than 3 million people get sick, and 220,000 die worldwide from pesticides each year. In the United States alone, pesticides poison 110,000 people each year. More than one-third of calls to animal poison control centers result from pets exposed to pesticides.

Pesticide Facts

- The pesticides and VOCs found indoors are believed to cause 3,000 cases of cancer a year in the US, making these substances just as threatening to nonsmokers as radon or second-hand tobacco smoke.[131]

- A National Cancer Institute study indicated that the likelihood of a child contracting leukemia was more than six times greater in households where herbicides were used for lawn care.[132]

- According to a report in *The American Journal of Epidemiology*, more children with brain tumors and other cancers were found to have had exposure to insecticides than were children without cancer.[133]

- According to the New York State Attorney General's office, 95 percent of the pesticides used on residential lawns are considered probable carcinogens by the EPA.[134]

- 2,4-D — a component of Agent Orange — is used in about 1,500 lawn care products.[135]

- Pesticides have been linked to the alarming rise in the rate of breast cancer.[136]

- Besides causing cancer, pesticides have the potential to cause infertility, birth defects, learning disorders, neurological disorders, allergies, and multiple chemical sensitivities, among other disorders of the immune system.

Pesticides, or biocides, are poisons designed to kill a variety of plants and animals such as insects (insecticides), weeds (herbicides), and mold or fungus (fungicides). Although pesticides were first developed as offshoots of nerve gas during World War II, many people falsely assume pesticide ingredients are now safe for humans. Would pesticides be on the market if they were unsafe? *Yes.* The US EPA approves pesticides based on efficacy, not safety. Out of the hundreds of active ingredients registered with the EPA, less than a dozen have been adequately tested for safety.[128] In two studies of indoor air quality conducted during the late 1980s, investigators found that indoor air contained at least five (but typically ten or more) times higher concentrations of pesticides than outside air.[129]

Pesticides can be absorbed through the skin, inhaled, or swallowed. Many building products and household furnishings such as paints, wood products, and carpets are treated with biocides. Carpets are especially hazardous; if you have kids, you know that infants and small children are likely to touch and crawl around on them. Also, people and pets may track pesticides into the house from the lawn or even the neighbor's lawn, allowing the toxins to become trapped in carpets. Pesticides that break down within days outdoors may last for years in carpets, where they are protected from the degradation caused by sunlight and bacteria. For example, the pesticide DDT (dichlorodiphenyltrichloroethane), was

Understanding Product Labeling

The following are important areas of a product's label:

- **EPA Registration Number**: This number is your assurance that the product has been approved by the US Environmental Protection Agency.
- **Directions**: Follow these carefully!
- **Precautions**: There are three types. "Danger — Poison" is extremely toxic in the form found in the container. "Warning" is less toxic to humans, but you should use extreme care when applying. Finally, "Caution" is the least harmful when you use it as directed.
- **Statement of Practical Treatment**: This includes information about first aid.
- **Classification Statement**: Tells the customer if the product needs a license to be used.
- **Storage Tips**: Always keep products in the original container; do not stockpile. Keep out of reach of children. Store flammable liquids outside your living area; do not store near food or medical supplies, where flooding is possible, or in places where spills could leak into your water supply. It is best to store products in an outside, locked cabinet.

outlawed in the US in 1972 because of its toxicity, yet researchers found that 90 of 362 homes they examined in the Midwest between 1992 and 1993 had DDT in the carpets.[131]

To avoid being exposed to harmful pesticides while remodeling your home, learn to understand the labels on products you buy. You can also reduce pesticide exposure by not treating the soil under the building, and by eliminating standard building products that contain biocides. A well-renovated home will be pest-resistant if it incorporates features like weather tightness; appropriate grading and drainage; ventilation fans and windows that allow cross ventilation to prevent excess moisture buildup from within; dry wood without rot or infestation; exterior wood treated appropriately for prevailing climatic conditions; screens on windows; and ground cover, leaves, chips, wood piles, and other potential insect habitats kept at a distance from the building. If a pest must be eliminated, first see if its current access to nourishment and habitat can be limited. For example, if you have ants, you might clean up crumbs from the floor and counters and caulk the cracks. Second, consider the most benign trapping or killing methods — use the least toxic chemicals you can find, and then only as a last resort!

> *My father had built an office for himself in the basement of our house that was totally enclosed, without any windows or venting. So I put the radon-detecting canister in his office and found that it was off the charts. My father died of small cell carcinoma, the type caused by radon. And I remembered that after working a whole day in the office, he would come home and go to his little office in the basement and do his bookkeeping for many hours, day after day. So the radon exposure was a contributing factor to his ultimate death. Even though I was not able to save my dad, I can hopefully save other lives through promoting awareness of how we live in the midst of all different types of health dangers.*
>
> **Claudine Schneider, former Congresswoman**

Vinyl Chloride

Vinyl chloride is a widely produced chemical VOC used to make polyvinyl chloride resin, which is the raw ingredient of polyvinyl chloride (PVC) plastic. Vinyl chloride is a colorless gas with a characteristic "plastic" odor. It is widely distributed throughout the industrialized world and can be found in municipal drinking waters, #3 plastics, PVC pipes, vinyl flooring, adhesives, swimming pools, upholstery, and wall coverings in your home.

As dramatized in the movie "Blue Vinyl," acute exposure to vinyl chloride can induce the feelings of intoxication, dizziness, weakness, and lightheadedness that many kids are drawn to for the same reasons people are enticed by alcohol and other drugs. These symptoms rapidly disappear when exposure is stopped. However, long-term exposure to vinyl chloride can cause cancer, birth defects, genetic changes, indigestion, chronic bronchitis, ulcers, skin diseases, deafness, vision failure, circulatory changes, and liver dysfunction.[137]

Although many plastics are potentially dangerous, some seem to be relatively safe alternatives to vinyl chloride. For example, melamine formaldehyde plastics, such as Formica™ countertops, are relatively inert. You can also avoid PVC in plastics; instead, use natural materials, like wood and glass.

Radioactive Contaminants

The EPA estimates that 15,000 lung cancer deaths each year in the United States are due to radon exposure, which makes it the second leading cause of lung cancer in the United States after smoking. Smokers have an even higher risk of developing radon-induced cancer; in fact, many smoking statistics may be caused by the combination of

The Environmental Protection Agency estimates that one of every 15 homes in the United States has indoor radon levels at or above the EPA's recommended action guideline level of four picocuries per liter (pCi/L) of air.

Table 5.5: Know Your Plastics

Not all plastics are equal — some are safer than others. Plastics are typically classified by one of seven recycling codes (found on the bottom of containers), indicating the type of resin used:

#	Name	Where It's Found	Good or Bad?
1	Polyethylene terephthalate (PET or PETE)	Used to make soft drink, water, sports drink, ketchup, and salad dressing bottles, and peanut butter, pickle, jelly, and jam jars.	**Good** • Not known to leach any chemicals that are suspected of causing cancer or disrupting hormones. • Widely recycled.
2	High density polyethylene (HDPE)	Milk, water and juice bottles, yogurt and margarine tubs, cereal box liners, and grocery, trash, and retail bags.	**Good** • Not known to leach any chemicals that are suspected of causing cancer or disrupting hormones. • Widely recycled.
3	Polyvinyl chloride (V or PVC)	Most cling-wrapped meats, cheeses, and other foods sold in delicatessens and groceries are wrapped in PVC.	**Bad** • To soften into its flexible form, manufacturers add "plasticizers" during production. Traces of these chemicals can leach out of PVC when in contact with foods. • According to National Institutes of Health, di-2-ehtylhexyl phthalate (DEHP), commonly found in PVC, is a suspected human carcinogen. • Not recyclable.
4	Low density polyethylene (LDPE)	Some bread and frozen food bags and squeezable bottles.	**OK** • Not known to leach any chemicals that are suspected of causing cancer or disrupting hormones. • Not as widely recycled as #1 or #2.
5	Polypropylene (PP)	Some ketchup bottles and yogurt and margarine tubs.	**OK** • Hazardous during production, but not known to leach any chemicals that are suspected of causing cancer or disrupting hormones. • Not as widely recycled as #1 or #2.

continued:

Table 5.5: Know Your Plastics – continued:

Not all plastics are equal—some are safer than others. Plastics are typically classified by one of seven recycling codes (found on the bottom of containers), indicating the type of resin used:

# Name	Where It's Found	Good or Bad?
6 Polystyrene (PS)	Foam insulation and also for hard applications (e.g., cups, some toys).	**Bad** • Benzene (material used in production) is a known human carcinogen. • Butadiene and styrene (the basic building block of the plastic) are suspected carcinogens. • Energy-intensive. • Poor recycling.
7 Other (usually polycarbonate)	Baby bottles, microwave ovenware, eating utensils, plastic coating for metal cans.	**Bad** • Made with bisphenol-A, a chemical invented in the 1930s in search for synthetic estrogens. A hormone distruptor. Simulates the action of estrogen when tested in human breast cancer studies. • Can leach into food as product ages.

Source: The Green Guide. Plastics for Kitchen Use [online]. [Cited October 29, 2003]. The Green Guide Institute, 2003. <www.thegreenguide.com/reports/product.mhtml?id=44&sec=2>.

exposure to radon and smoking. The main risk is not from the radon itself, but from radon's "progency," or "decay products," which directly or by attaching to airborne particles may be inhaled into the lungs.

This invisible radioactive gas seeps up into homes through the earth. Its points of entry include floor drains and sumps, joints where the basement wall and floor come together, cracks in the basement walls and floors, holes in the foundation wall for piping or wiring, exposed earth or rock surfaces in the basement, or well water. It is a breakdown product from uranium-238, which occurs naturally in the subsoil. Although it is harmless when it is outside, inside our homes the gas and its decay products can build up to high levels and when inhaled can cause lung cancer. Energy-efficient, tightly sealed homes are especially vulnerable to this accumulation of radon. And because it takes 1,602 years for only half of the radon atoms to disintegrate, radon concentrations tend to become higher as time goes by.

The EPA and the Surgeon General recommend that all homes test their radon levels below the third floor. In the US, the average indoor radon level is 1.3 picocuries per liter (pCi/L), while average outdoor levels are only 0.4 picocuries per liter. The EPA suggests that action be taken to increase ventilation if tests result in a radon level above 4 picocuries per liter. When looking for a radon test kit or mitigation professional, make sure you choose one that is certified by either the National Environmental Health Association (NEHA) or the National Radon Safety Board (NRSB).

If you are adding on to your home, ask your remodeler about methods of radon mitigation that can be incorporated into the construction. For example, you might place an airtight membrane under carpets, or provide some form of under-slab ventilation. Additionally, consider covering any exposed earth with a polyethylene air barrier, and seal all cracks and joints in the foundation wall and floor slab with caulking or foam. You can also install a self-priming drain or gas trap in the floor drains leading to a sump or to drainage tiles, and remove radon from well water using activated charcoal filters or aeration units. Radon-resistant construction features usually keep radon levels in new homes below 2 picocuries per liter.

Environmental Tobacco Smoke

Environmental tobacco smoke (ETS), also known as secondhand or passive smoke, is the smoke that comes from a burning cigarette, pipe, or cigar, as well as smoke exhaled from the lungs of smokers. ETS is a mixture of over 4,000 chemicals, 200 of which are known poisons, and more than 50 known cancer-causing agents. There is no safe level of exposure to ETS; every year, smoking-related diseases claim approxi-mately 430,700 lives in the U.S.[138]

The best way to deal with ETS is to stop smoking in the house. Otherwise, you can create a separate smoking room, with its own ventilation and air seals to keep the smoke from spreading

In the United States alone, secondhand smoke has been estimated to cause about 35,000 deaths per year from heart disease in nonsmokers, and about 3,000 deaths each year from lung cancer in nonsmokers. Secondhand smoke is responsible for 150,000–300,000 lower respiratory tract infections (i.e., bronchitis and pneumonia) in children under 18 months of age. This results in 7,500–15,000 hospitalizations each year.

Source: American Lung Association, "When You Smoke, Your Family Smokes!" <www.lungusa.org>, October 1, 2003.

Table 5.6: Signs of Too Much or Too Little Humidity

	Too Much Humidity	Too Little Humidity
Typical symptoms	Condensation on windows	Chapped skin and lips
	Wet stains on walls and ceilings	Scratchy nose and throat
	Moldy bathroom	Breathing problems
	Musty smells	Static and sparks
	Allergic reactions	Problems with electronic equipment
Long-Term Effects	Damage to house and contents	Continuing discomfort
	Ongoing allergies	Damage to furniture and other items

through the house. An effective ventilation system should supply outside air and incorporate a particulate filter.

Moisture and Mold

Moisture may not be a pollutant, but high or low humidity in a house can make you feel uncomfortable, and can even cause health problems. High moisture levels create a welcoming environment for bacteria, viruses, molds, and dust mites, increases offgasing, and can cause a room to feel stuffy. On the other hand, low moisture levels can cause high dust levels and respiratory infections.

One of the most dangerous effects of high moisture in the home is mold. Molds and mildew (two words for the same thing), are simple plants of the group known as fungi. They grow on the surfaces of objects when the relative humidity, or degree of moisture contained in the air, is high. Mold is commonly assumed to be found only in older homes, but it can be found wherever moisture accumulates, such as basements, kitchens, bathrooms, window sills, carpets, furniture against outside walls, wall cavities, unventilated storage areas, laundry rooms, or wherever leaks and flooding occur. One study showed 16 different kinds of mold in one finished basement.[139] As a general rule of thumb, if you can see it or smell it, you have mold.

And mold spreads! Microbes commonly grow within the ductwork of forced air heating systems, which can result in the spreading of mold and dust

What is Relative Humidity?

Relative humidity is a measure of the amount of water in the air compared with the amount of water the air can hold at the temperature it happens to be when you measure it. (We typically measure relative humidity in metric.) For example, at 30 degrees Celsius, the air can hold 30 grams of water vapor per cubic meter of air. At 10 degrees C, the air can hold 9 grams of water vapor per cubic meter of air. These are basic physical facts at sea level pressure. Let's say at 3 P.M. the air's temperature is 30 degrees and the air has 9 grams of water vapor per cubic meter of air (you can measure water vapor in your home using a device called a hygrometer). To calculate relative humidity, we divide 9 (water vapor in air) by 30 (temperature) and multiply by 100 (percentage rate) to get 30 percent relative humidity. If it were 9 degrees outside, we'd get a relative humidity of 100 percent, because 9 (water vapor in air) divided by 9 (temperature) multiplied by 100 *is* 100 — the air now has as much vapor as it can hold without condensing into liquid water. In your home, it is best to keep relative humidity below 50 percent.

throughout the house. Unless kept spotlessly clean, toilets and many modern appliances that use water reservoirs, such as vaporizers and humidifiers, can breed microbes. Many standard construction materials are susceptible to water damage and fungal growth: they can become breeding grounds for mold and bacteria within a few days. Even when molds are contained inside walls or other building cavities such as attics or crawlspaces, the slightest air current can send fungal spores swirling through the air where they are easily inhaled.

Once inhaled, mold produces allergic reactions, hypersensitivity, and infectious diseases. Certain fungi found indoors produce mycotoxins, which can be carcinogenic, teratogenic (induces birth defects), immuno-suppressive (reduces immune system performance) or oxygenic (poisons tissues). In addition to health issues, mold discolors surfaces, causes odor problems, deterioration of building materials, and homeowners can incur large bills for structural damage caused by water or water vapor trapped behind walls.

You may have mold If...

- You see mold or discoloration, ranging from white to orange and from green to brown or black.
- You smell a musty odor.
- There is condensation on your windows.
- Building materials, such as drywall and plaster or plywood, look cracked or discolored in areas where previous water damage occurred.
- You see loosened drywall tape.
- You see rotting material, such as warping wood.

Getting Rid of Mold

Once you have identified mold, you can get rid of it by following these steps:

- Wash the area with soap and water.
- Disinfect the area with a solution made up of ten percent household bleach and a little detergent (the detergent will help with the dirt and oil on the surface and act as a surfactant to help thoroughly wet all surfaces).
- Clean and disinfect area quickly — mold grows within one to two days.
- Let cleaned area dry overnight.

- Remove materials affected by mold, especially porous materials such as sheetrock, carpeting, and plywood.
- Bag and discard the materials at the work area rather than potentially spreading contaminants throughout the home.
- Wear gloves and high quality respiratory protection when cleaning areas affected by mold growth and when removing damaged materials.
- Provide continuous and controlled ventilation in the work area, with slightly more negative pressure in the contaminated area so that air flows from clean to dirty areas

Although the above cleaning methods will temporarily get rid of mold of your home, there are two ways to prevent mold from forming again. First, keep the interior surfaces of exterior walls and other building assemblies from becoming too cold (think of condensation on a cold soda can). Air conditioners tend to cool particular spots on a wall, resulting in mold growth where humid air enters the wall cavity. Impermeable wall surfaces, such as vinyl wallpaper, can exacerbate the problem by further trapping natural moisture flow. Therefore, you should prevent the overcooling of rooms, increase the permeability of interior finish materials in humid climates, and relocate ducts and diffusers to elevate the temperature of the surface. Adding insulation to walls, ceilings, ducts, or cold water pipes also raises the temperature of the interior surface and prevents condensation. Air-vapor barriers can be used to keep wall cavities warmer and drier, and two-stud corners reduce heat loss and condensation at corners.

The other way to control mold growth is by limiting interior moisture levels. Install a balanced, whole-house ventilation system that controls moisture by bringing in drier outside air where and when possible. Controlled ventilation, such as venting moist air from bathrooms, clothes dryers, and kitchen stoves, and space heaters to the outside can significantly reduce mold issues in these areas. Also, it helps to store firewood outdoors — storage of one cord of green firewood indoors over the winter can produce the same amount of moisture as does a family of four through respiration!

Your air conditioner acts as a dehumidifier after it runs for about eight minutes. Use this to control relative humidity on an ongoing basis, Install your hygrometer right next to your thermostat to constantly monitor your home.

Some people use dehumidification to remove moisture from a space; the dehumidifier warms the moisture-laden air to reduce its ability to hold moisture, thereby forcing moisture to condense. This can be helpful in the short term, but you should also control moisture sources to limit interior moisture levels. In other words, fix basement plumbing and leaks, avoid porous absorbent material like cardboard or newspaper in the basement, use and remove carpets and rugs from cold floors, and remove obstacles obstructing air flow in damp areas that can give molds a place to grow.

Consider using a hygrometer to monitor humidity levels, to help you maintain a healthy 30 to 55 percent humidity. If you find moisture levels are too low, you can counter the low humidity by reducing ventilation or you can use a humidifier, cleaning it often and monitoring the humidity level.

Now that you understand how your home's renovation impacts world energy, world resources, and the health of you and your family, you have a solid background for changing the world, one room at a time....

CHAPTER 6

... One Room at a Time

So you want an efficient, resourceful, healthy kitchen that will change the world? We'll help you. We've divided this chapter into a room-by-room survey to give you an overview of the stages of the building process for each particular remodeling project — so you can make connections to our more in-depth coverage of specific green features in later chapters.

Keep in mind that, underlying this simple format, we are still considering the house as a whole. In other words, the addition of a large bathtub might add 1,200 pounds to the load capacity of the floor, which will require additional structural support — and this in turn will have an effect on such things as adjoining walls or pipes that service the entire house.

The descriptions of remodeling for each room are by no means complete: there are always more innovative ideas that can be applied to specific situations and, as technology improves, new ways to save energy and resources, improve air quality, and so on. Use the room descriptions as a guide for discussion with your architect and contractor, and refer to the checklist at the end of each description to locate more information on specific features. Always keep an open mind to new ideas — and choose whatever works best for your renovation project.

The Bathroom

Most people renovate their bathroom to upgrade old fixtures and, in the course of remodeling, find they need to repair water damage. As discussed in Chapter 5, excess moisture not only causes house materials to rot, but can also result in serious health issues. We'll show you ways to control bathroom moisture that will result in improved comfort, better air quality, and lower utility bills. We will also discuss other important bathroom remodeling issues, including low-impact construction and water- and electricity-saving strategies.

Fig. 6.1: Green bathroom options, remodeled for efficient use of space could include:

- *high-performance, low-flow shower head with chlorine filter*
- *compact flourescent bulbs*
- *lighting controls*
- *windows that open*
- *landscaping for shade*
- *greater natural daylight*
- *upgraded single pane windows*
- *water filters*
- *low-flow faucets*
- *insulated plumbing & pipes*
- *solvent-free adhesives*
- *low-flow or greywater flushing toilet*

Credit: Jill Haras & Kim Master.

Job Site and Landscaping

As you plan for your bathroom remodel, try to build with as little impact as possible. This means recycling job site waste and salvaging reusable materials. That avocado green sink may look hideous to you, but could be someone else's treasure. In fact, maybe you want an avocado green sink that someone donated or could use the sink you are replacing in a second bathroom — be resourceful when looking for replacement bathroom features! Old lumber, door and window casings, and baseboards can all be reused when they are removed carefully. When reusing old materials is not possible, some companies can recycle them. For example, porcelain toilets and other ceramics can be recycled into concrete, and metal bathtubs can be used for making steel. Contact your local waste management authority for recycling information specific to your neighborhood.

Structural Framing

If you want to change the size or shape of your bathroom, this will probably entail taking down walls. Some walls you can demolish yourself, depending on the type of wall. Load-bearing walls — the walls that literally hold up the house — can be altered, but this requires new structural support in the form of a beam supported by posts on either side of the room. In addition, temporary wood frame walls will be needed to support the structure above until the new wall is in place. This work can be intensive, and is best completed by a professional. However, you can easily cut into or tear down non-bearing walls yourself without damaging the adjacent structures. You still need to be on the lookout for plumbing leaks — even if you have a professional work on the project, he could inadvertently break a pipe and cause a leak, or cut live electrical wires. Look for obstacles ahead of time to avoid unnecessary expenses and hardship.

Opening walls in your bathroom is a great opportunity to redo bad wiring or plumbing. You can also replace any rotted or damaged wood, repair leaks that have gone unnoticed, check to see if you have adequate insulation in the exterior walls, and caulk cracks in the structure so that air doesn't leak through.

Plumbing

First, look at what kind of pipes you have. If your home was built before 1950, you may have lead pipes that should be replaced! Otherwise, cast iron, steel or plastic drainpipes are adequate, as are copper, brass, and steel. Polyvinyl chloride (PVC) pipes can also be used; although the PVC manufacturing process produces toxins, the pipes are durable and affordable, and won't need to be replaced for a long time. Second, if you've opened up bathroom walls, consider moving the plumbing from outside walls to interior walls where heat from your home will keep the pipes warmer. Insulating the hot water pipes will further reduce heat loss through the piping.

To save water, add an "on-demand" hot water circulation pump that can send hot water to fixtures in seconds without wasting water while you wait for it to get hot. Some jurisdictions allow households to use greywater (all waste water except toilet water) for outdoor watering use or as toilet-flushing water. Although you need to be especially careful what you put down the drains, this will dramatically reduce water use around your home. If this is not currently allowed in your neighborhood or is too costly, you may consider "pre-plumbing" while the walls are open during renovation — installing the plumbing necessary for later greywater use to make this retrofit easier and less costly in the future.

When buying a new toilet, look for models with less than 1.6 gpf (gallons per flush). Likewise, look for low-flow bathroom faucets and showerheads that produce an adequate flow, but use 60 percent less water by combining the water with air pressure. This not only decreases your water consumption, but also lowers your water bill and reduces your impact on the local water treatment system. If you buy a large whirlpool tub, make sure you have it installed before new framing makes it difficult to fit the tub in the bathroom. You may have to add an additional water heater to fill the tub entirely. This is a great time to install a tankless unit. (See Chapter 17 for more on tankless water heaters.)

Additionally, faucets and showers can be equipped with filters to make your water safer to drink and less irritating for sensitive individuals to bathe in. Carbon filters are an inexpensive option for most people who desire cleaner-tasting water. Chlorine can be an issue if you suffer from eczema or have sensitive skin, in which case you may want to purchase an additional carbon filter for your showerhead.

Electrical

Bathrooms may be small, but they need good overall lighting and task lighting for washing, shaving, and applying makeup. Incandescent bulbs are standard for home lighting but they waste a lot of energy. Fluorescent tubes and compact fluorescent bulbs are significantly more energy-efficient for general and vanity lighting. Fluorescent lighting has improved vastly over the past decade — it is now available in flattering light (as opposed to the older cool, bluish color), and with dimming options that enable you reduce the light energy output. Compact fluorescent bulbs are similarly effective for task lighting. Although fluorescent bulbs cost more initially, you'll definitely save money on your electric bills and replacement bulb costs over time.

Another option for recessed or indirect lights is halogen lighting. These produce as much light as incandescent lights but use half the power. Use halogens for mood lighting and to light specific features, like glass block. While you are in the walls it is a good opportunity to wire for music for that perfect night with candles and bubble bath....

Keep in mind that lights over showers or other wet areas need a waterproof housing. Also, it saves energy to wire different lights to different switches so that you can turn on only the lights you need in the bathroom at a given time.

Radiant electrical panels are wonderful for when you're sitting in the tub or while you're drying off. They can be mounted on a wall or ceiling, and they heat you rather than the air. They can be put on a timer or controlled with a switch (see "Install Zoned, Hydronic, Radiant Heating" in Chapter 16).

If you are doing serious rewiring, put in a separate circuit for electrical devices. A hair dryer alone can use up most of a circuit's capacity. Add extra "his and hers" plugs.

Insulation

As we mentioned briefly in the above framing section, renovating your bathroom is a great time to upgrade your insulation, because the walls are already open. The bathroom is where you want to stay the warmest. If possible, add new insulation in the form of rigid foam board on the inside of the framing. Insulating just to code doesn't mean that it will keep you comfortable when you are wet in cold winter months; how much insulation you need also depends on your climate. Check the existing insulation to see how thick it is, and if it is evenly distributed so that heat does not escape through gaps. Also check for water damage or compaction, which can reduce the effectiveness of insulation. (Water damage can take the form of blackened wood — a sign of mold or rot — or obvious water marks.) When installing new insulation, fill the area around the tub: this helps keep drafts down and the tub warm. The most efficient, resourceful, and safe types of insulation include recycled-content fiberglass, and cellulose insulation. Make sure the insulation stays dry after it is installed; water vapor inside the wall can condense when it meets cool outer layers, causing rot and harmful mold growth. A vapor barrier installed on the warm side of the insulation can also prevent moisture from getting into the floor, ceiling, or walls.

Always use advanced infiltration reduction practices to further save energy in leak-prone areas. This includes using a low-toxic sealant around plumbing fixtures, such as the hole in the subfloor around the tub drain, and use rubber gaskets behind electrical outlets. The trap under the tub should be sealed well too, since the plumbing pipes go all the way down to the ground, and it is a typical place for drafts to occur.

Take precautions when removing paint from windows!
According to the EPA, painted window sashes and frames in homes built before 1978 may contain lead-based paint. This is a special concern because the friction of opening and closing windows can release lead dust into the home. (See "Lead" in Appendix 1.)

Energy

Windows in bathrooms are important for many reasons. First, bathrooms without any windows at all are illegal in many jurisdictions because the windows provide necessary ventilation. Second, a large window can make the room brighter, resulting in proven mood-uplifting effects.[1] Third, leaky windows can waste a lot of energy, therefore efficient double-paned windows with reflective or low-emissivity film can effectively keep unwanted heat or cold outside your bathroom. Try not to put the window directly over the bath or in the shower area. This can can cause condensation problems when the steam hits the cool window surface. Vinyl or fiberglass window frames will also minimize the potential for condensation damage, and are a better option than wood or aluminum. Other options include well-insulated skylights or a smaller, more efficient solar tube that effectively brings light in through a reflective pipe.

Heating, Ventilation, and Air Conditioning (HVAC)

Fixing up an existing bathroom generally does not require changing air registers, radiators, or baseboard heaters — usually it can simply be "retuned" by a heating contractor. A new bathroom, however, will require an extra heat register or radiator installed by a professional, usually with no extra strain on the heating system. Make sure the workers use duct mastic on duct joints rather than less-effective duct tape. The new ductwork should be installed in conditioned (or insulated) walls rather than in walls facing the outside of your home, where energy can more easily escape from the pipes.

Ventilation requirements are usually stipulated in the building code for each jurisdiction. Often the code will state that you need to have either an operable window or a fan to get rid of unwanted moisture caused by showering and bathing. We recommend that a fan be installed in any case — you don't want to open a window in the winter! The fan should be energy-efficient and should be vented to the outside. Most people prefer quiet fans under two "sones" — when they are quiet they tend to get used more often. If extensive HVAC work is being considered, keep in mind that a heat recovery ventilator (HRV) can transfer heat from outgoing bathroom exhaust air to warm the incoming fresh air for the entire house.

Water Heating

If the current water heater is old or if you'll be using a lot more water because of a new whirlpool bath, you'll need an upgrade, such as a more efficient water heater with an energy factor (EF) of .60 or higher. Tankless water heaters are a more expensive option, but they save water, energy, and petroleum resources by heating water as it is needed. Installing a heat trap will help minimize water heat loss by preventing thermosiphoning. (See "Upgrading Your Water Heater" in Chapter 17.) If you decide to keep your existing water heater, consider an insulating jacket that works like a winter coat to keep heat in. It costs only about $10 to $20 and reduces heat lost through the tank by 25 to 40 percent.[2]

Interior Materials/Finishes

Bathroom finishes should be attractive, water-resistant, and healthy. Some people use wallpaper, but we do not recommend this option because moisture can get behind the paper (especially vinyl wallpaper), and cause mold growth. Also, avoid using particleboard and medium-density fiberboard (MDF) in cabinets and substrates under counter tops. Choose exterior-grade plywood or a formaldehyde-free alternative (see "Formaldehyde" in Chapter 5). If you do use particleboard or MDF, use a low-toxic sealant, such as water-based, low-VOC paint to keep the unwanted vapors trapped in the material.

Solid wood can be a healthy alternative. The best woods to use are Forest Stewardship Council (FSC)-certified with a low-VOC, water-based wood finish. Sustainably harvested FSC wood is also available for trim material. If you decide to paint the finish, use a hard wearing, washable, low- or no-VOC and formaldehyde-free paint that minimizes indoor air pollution.

Ceramic tiles are a popular finish because they are water-resistant, washable, and don't need a painted finish. Look for a low-toxic grout to fill the areas between the tiles. Natural stone, like marble or slate, is a more expensive option, but is extremely water-resistant and is a healthy, beautiful element to add around showers, tubs, or other areas of the room. Like ceramic tile, stone is a durable material that will save resources and energy because it will rarely need replacing. For all bathroom materials, select low-VOC adhesives and sealers when they are required to minimize toxins in your indoor environment.

Flooring, like other bathroom finishes, should be durable, water-resistant, and washable. We do not recommend carpeting — it traps dirt and moisture that leads

Volatile Organic Compounds (VOCs)

VOCs are a class of chemical compounds that can cause nausea, tremors, headaches, and, some doctors believe, longer-lasting harm. VOCs can be emitted by oil-based paints, solvent-based finishes, and other products on or in construction materials.

to unhealthy mold growth. Use small, washable rugs for softer surfaces in specific areas of the bathroom. The best flooring options are hard or smooth surfaces, recycled-content ceramic tile, and natural linoleum. Ceramic floor tiles, larger and thicker than walls tiles, are easily cleaned and long-lived. Some floor tiles are made from stone. Dense or vitreous tiles are the most moisture-resistant. Tiles should be installed with low-toxic grout and adhesive when possible.

Natural linoleum is the preferred alternative to vinyl. Although similar in appearance to vinyl, natural linoleum is made from natural materials including linseed oil, jute, and wood dust. It is durable, available in a variety of patterns and colors, and can be installed without toxic adhesives.

The Kitchen

Redoing the kitchen is the number one renovation project. We spend a lot time there, cooking, eating, or just lounging around, so it might as well be the best room in the house. Fixing up the kitchen also has a tremendous impact on the resale value of your home, and can be a great investment. Like the other rooms in this chapter, we will divide our discussion of the kitchen into the generic categories basic to remodeling projects (site, plumbing, electrical, etc.) and emphasize an important upgrade for the kitchen when considering energy use — appliances.

Job Site and Landscaping

It is always a good idea to try to reuse materials in the kitchen, from countertops and ceramic dishware to flooring and appliances. Check your regional waste management authority to find out who can use the materials you don't want. If a material can be reused, look into recycling it; construction materials, glass, fluorescent light bulbs, telephones, sinks, and appliances can all be recycled.

As you're designing your kitchen, think about creating a space that won't have to be remodeled again, thereby saving energy and resources in the future. You should be able to move around three key cooking areas — the stove, the sink, and the refrigerator — easily. Consider adding features that allow you to do more than

Bathroom Checklist

Job Site and Landscaping (see Chapter 7)
- [ ] Reuse construction and deconstruction waste.
- [ ] Recycle job site waste.

Structural Framing (See Chapter 9)
- [ ] Use advanced framing techniques (AFT).

Plumbing (See Chapter 12)
- [ ] Insulate pipes.
- [ ] Remove plumbing from outside walls.
- [ ] Install on-demand hot water circulation pump.
- [ ] Install low-flush toilets.
- [ ] Consider greywater flushing.
- [ ] Install high-performance showerheads.
- [ ] Install low-flow faucets.
- [ ] Investigate your water supply.
- [ ] Install chlorine filters on shower heads
- [ ] Install activated carbon filters.
- [ ] Install water distillers.
- [ ] Install a reverse osmosis system.

Electrical (See Chapter 13)
- [ ] Install compact fluorescent light (CFL) bulbs.
- [ ] Install halogen lighting.
- [ ] Install lighting controls.
- [ ] Install sealed or airtight IC recessed lighting.

Insulation (See Chapter 14)
- [ ] Use formaldehyde-free, recycled-content fiberglass insulation.
- [ ] Use cellulose insulation.

- [ ] Caulk, seal, and weatherstrip.
- [ ] Increase insulation thickness.

Solar Energy (See Chapter 15)
- [ ] Install double-paned windows.
- [ ] Install low-e (low-emissivity) windows.
- [ ] Incorporate natural light.

Heating, Ventilation, and Air Conditioning (HVAC) (See Chapter 16)
- [ ] Use duct mastic or Aeroseal™ instead of duct tape.
- [ ] Install ductwork within the conditioned space.
- [ ] Install operable windows for natural ventilation.
- [ ] Install a bathroom exhaust fan.
- [ ] Install a heat recovery ventilator (HRV).

Water Heating (See Chapter 17)
- [ ] Consider a tankless water heater.
- [ ] Use the smallest water heater possible.
- [ ] Install hot water jacket insulation.
- [ ] Install heat traps.

Interior Materials/Finishes (See Chapter 19)
- [ ] Use formaldehyde-free materials.
- [ ] Seal all exposed particleboard or MDF.
- [ ] Use rapidly renewable flooring materials.
- [ ] Use recycled-content tile.
- [ ] Replace vinyl flooring with natural linoleum.
- [ ] Use low- or no-VOC and formaldehyde-free paint.
- [ ] Use solvent-free adhesives.

Fig. 6.2: Green kitchen options, remodeled for efficient use of space could include:

- *upgrade single-pane windows*
- *windows that open*
- *window coverings for shade*
- *landscaping for shade*
- *light pipes*
- *range hood venting to outside*
- *water filters*
- *low VOC, water-based finishes*
- *formaldehyde-free materials*
- *sealed particleboard & MDF*
- *Energy Star™ appliances*
- *low water-use dishwasher*
- *best-available technology refrigeration*
- *recycled-content tiles*

Credit: Jill Haras & Kim Master.

just cook and eat — like adding a small desk in an unused space or creating an area for entertaining. Also, installing a recycling center into the cabinetry makes recycling kitchen waste more convenient.

Structural Framing

For information regarding structural changes you may need to make to the kitchen, please refer to Chapter 9, Structural Framing.

Plumbing

As long as you do not move or add plumbing fixtures, you can use existing hookups for a new sink or dishwasher. More drastic changes may require new plumbing. Try to avoid moving your sink, because this can significantly increase the cost of the job. Again, if your house was built before 1950, the pipes may be lead, in which case they need to be replaced to protect your health (see "Lead" in Chapter 5, and "Lead" in Appendix 1.)

For minimal cost, you can insulate your hot water pipes so that energy does not escape as you are running water. To further save water, install a low-flow faucet that uses less water but with seemingly no effect on water pressure because the spray is mixed with air.

Water quality can be improved simply by adding a carbon filter to your tap. Each water filtration method filters distinct pollutants from the water, so you should find out what's in your water from your regional water authority before you purchase a filter. (See "Water Quality" in Chapter 5).

Electrical

In general, people prefer bright kitchens. While light colors and big windows help, good general lighting and task lighting will provide for the best cooking, eating, and socializing environment, any time of day or night. The most energy-efficient lights are fluorescent and halogen lights. Energy-efficient dimmers are available that can create an intimate dining area. You can also save energy by wiring lights to different switches, allowing you to illuminate specific areas of the kitchen. For safety and energy-saving purposes, recessed lights should be labeled "IC-AT," meaning they're designed for direct insulation contact and an airtight housing.

In terms of circuitry, new appliances may require new circuits and/or upgrading of existing circuitry. Many old kitchens can overload existing circuits, causing a fire hazard. Stoves need a heavy-duty circuit, and other appliances also need their own circuits. New circuits in the kitchen are a good investment since there will always be a new gadget to plug in.

Insulation

If you've opened up walls in your kitchen for the renovation, this is a cost-effective time to upgrade your insulation. Uneven insulation distribution and blackened wood or other signs of moisture damage are indications that your insulation needs to be replaced. Otherwise, you might just consider adding more insulation to improve the comfort and overall energy-efficiency of your home. Green insulation options include recycled

Kitchen lingo.

Backsplash: The part of the wall surface behind the sink or countertop. As it is splashed frequently, it should be finished with a waterproof surface.

Kickspace: The recessed space at the bottom of a base cabinet. The kickspace allows you to stand close to the cabinet without stubbing your toes.

Soffit: The space above the upper cabinets. This can be left open for storage or display, covered with molding, or boxed in with drywall. Sometimes it is used to hide the vent duct running from your range hood.

Vinyl Windows: A Green Product?

We have discussed how "bad" vinyl and PVC are for your health (see Vinyl Chloride in Chapter 5). So why do we recommend it for windows? There are pros and cons to almost every material. Wood is expensive and uses wood resources. Vinyl is a low-cost option that has good temperature and sound insulation qualities, thereby helping you to save energy that would otherwise have to come from polluting, non-renewable energy sources. Many people are concerned about health risks associated with vinyl, but hard vinyl used in windows does not offgas vinyl chloride as much as soft vinyl (such as the kind in vinyl flooring).

content cellulose, cotton batts made from trimmings in jeans factories, and recycled-content fiberglass. Advanced infiltration reduction practices will also help you save energy. This involves looking out for leaks in doors, windows, plumbing, ducting, and electrical wire, and penetrations through exterior walls, floors, ceilings, and soffits over cabinets. Leaks should be sealed with a low-toxic caulk, sealant, or expansive foam.

Solar Energy

Windows can make the kitchen significantly brighter and will help to open the room up. Purchase the most energy-efficient windows you can afford for optimum energy-savings and comfort. At a minimum, they should be low-emissivity (low-E) windows that block out unwanted heat during the summer, while allowing in plenty of sunlight, and should have low-conductivity frames like vinyl or wood. You may want to shade windows with overhangs or landscape features to prevent your kitchen from overheating in the summer.

Heating, Ventilation, and Air Conditioning (HVAC)

Unless you are adding major space to the kitchen, you won't need extra heating for this room. However, you may want to relocate a register in the kickspace under your kitchen counter to fit a new design. If you are putting in new flooring or there is an unfinished basement below, you might also consider radiant floor heating, which uses water in tubes under the floor to deliver heat to the room. Not only does this make typically cold kitchen floors warmer, but it heats objects instead of air — efficiently making you more comfortable. As with all rooms in the house, ducts should be sealed with duct mastic for greatest efficiency, because duct tape deteriorates over time.

Given all the moisture created while cooking, ventilation should be a top priority in the kitchen to avoid mold, odors, and harmful combustion gases caused by cooking. Operable windows are always a good idea, but an impractical solution for many homeowners in the winter. A range hood directly over the stove effectively carries pollutions and odors outside through an air duct to the outside. Vent hoods are mounted on a cabinet above the stove, while canopy hoods hang down from the ceiling over stoves on an island. The noise of fans is measured in sones: the lower the sone (typically below 4.5) the better.

Appliances

The major appliances in your kitchen are the stove or range, the refrigerator, and the dishwasher. Stoves and ranges do not have great variations in efficiency, which is why there is no EnergyStar® rating for them; however, another major issue with them is indoor air quality. Cooking releases particulates, odors and moisure. (Gas stoves can also emit harmful combustion gases; see "Combustion Gases" in Chapter 5.) To keep these things to a minimum, be sure to ventilate properly. The basic "recirculating" range hood does little more than blow everything back into the house; instead, install the hood close to the cooktop (25-30"), and vent it to the outside with as short a duct as possible to reduce the amount of fan power needed. A basic 190-360 cfm range hood fan should be adequate.

Refrigerator efficiency has improved significantly over the past decade, but it is still the biggest energy user in the kitchen. Don't buy a bigger fridge than you need, as this will only increase its energy consumption. Likewise, keeping your old one for extra food storage will only increase your energy bills unnecessarily — it is best to buy a slightly larger new refrigerator that accommodates all your needs. Keep in mind that extra features, like water and ice dispensers, will make your refrigerator less efficient. Likewise, refrigerators with the freezer compartment to the side of the refrigerator are less efficient than models with the freezer door above or below the refrigerator door. Avoid locating your refrigerator close to heat sources like the stove or microwave that will reduce its cooling efficiency.

Most of the energy required for dishwashers is used to heat the water. Therefore, look for models that heat less water or include a booster heater that allows you to turn your main water heater down to save energy. A heat "on/off" setting allows you to turn off the heat and air-dry dishes to further save energy.

Also look for water saving features. Extra-insulated models will operate more quietly.

Interior Materials/Finishes

Although many materials and finishes in a renovated kitchen can potentially cause serious health problems, there are ways to avoid this. For example, the particleboard and medium density fiberboard (MDF) used to make kitchen cabinets and counters contain urea formaldehyde glue which can cause a range of health issues (see "Formaldehyde" in Chapter 5). The toxins can be sealed into the materials with several coats of a water-based sealant, preferably at the factory, before the material gets into your home. Better, you can purchase exterior-grade plywood that contains less toxic phenol formaldehyde, or formaldehyde-free medium density fiberboard (MDF).

Your best — but likely most expensive — option is custom cabinets made from solid wood, and solid surface countertops, which are durable and healthy. Finger-jointed wood is an excellent option for trim because it uses smaller diameter wood by joining the pieces like intertwined fingers, creating a stronger piece of wood which might otherwise have been cut from a larger tree. Also, use low- or no-VOC, formaldehyde-free paints, low-VOC wood finishes, and solvent-free adhesives where they are applicable. In general, paints are better for indoor air quality than wood finishes, but remember that the shinier and darker the paint, the more likely it is to offgas toxins into your living space. Calculate finish needs beforehand to avoid wasted leftovers; store, donate, or recycle finishes when leftovers are unavoidable.

Kitchen floors should be durable and washable, because chances are things will spill on the floor while people are cooking or eating. (My dogs also do a good job of making their own mud artwork on the kitchen floor.) The best options are wood, ceramic tile, exposed concrete, or natural linoleum. Wood should be FSC-certified or come from a natural, rapidly-renewable resources like bamboo, to avoid cutting down old-growth forests that take hundreds of years to regenerate. Wood is relatively soft and comfortable, but may be damaged with long-term exposure to water; for example, oak turns black with repeated exposure to water. Always use low-toxic adhesives and water-based wood finishes to minimize indoor air pollution.

Recycled-content ceramic tile, stone tile, or exposed concrete are durable surfaces. Concrete can be mixed with various pigments to create beautiful, warm

Kitchen Checklist

Job Site and Landscaping (See Chapter 7)
- [ ] Remodel for efficient use of space.
- [ ] Reuse construction and deconstruction waste.
- [ ] Recycle job site waste.

Plumbing (See Chapter 12)
- [ ] Insulate pipes.
- [ ] Install on-demand hot water circulation pump.
- [ ] Investigate your water supply.
- [ ] Install a whole-house water filtration system.
- [ ] Install activated carbon filters.
- [ ] Install water distillers.
- [ ] Install a reverse osmosis system.

Electrical (See Chapter 13)
- [ ] Install light pipes.
- [ ] Install compact fluorescent light (CFL) bulbs.
- [ ] Install halogen lighting.
- [ ] Install lighting controls.
- [ ] Install sealed or airtight IC recessed lighting.

Insulation (See Chapter 14)
- [ ] Use formaldehyde-free, recycled-content fiberglass insulation.
- [ ] Use cellulose insulation.
- [ ] Caulk, seal, and weatherstrip.

Solar Energy (See Chapter 15)
- [ ] Install double-paned windows.
- [ ] Install low-e (low-emissivity) windows.
- [ ] Install low-conductivity frames.
- [ ] Install window coverings and overhangs.
- [ ] Use landscaping to shade windows.

Heating, Ventilation, and Air Conditioning (HVAC) (See Chapter 16)
- [ ] Install zoned, hydronic, radiant heating.
- [ ] Use duct mastic or Aeroseal™ instead of duct tape.
- [ ] Install operable windows for natural ventilation.
- [ ] Vent kitchen range hood to the outside.

Appliances (See Chapter 18)
- [ ] Replace your standard dishwasher with a low water-use model.
- [ ] Look for EnergyStar® label on gas appliances.
- [ ] Buy best available technology refrigerator.

Interior Materials/Finishes (See Chapter 19)
- [ ] Use formaldehyde-free materials.
- [ ] Seal all exposed particleboard or MDF.
- [ ] Use FSC-certified wood flooring.
- [ ] Use rapidly renewable flooring materials.
- [ ] Use recycled-content tile.
- [ ] Replace vinyl flooring with natural linoleum.
- [ ] Use exposed concrete as finish floor.
- [ ] Calculate paint needs beforehand.
- [ ] Use low- or no-VOC and formaldehyde-free paint.
- [ ] Use low-VOC, water-based wood finishes.
- [ ] Deal properly with finish leftovers.
- [ ] Use solvent-free adhesives.

patterns. Although pigment adds to the cost of the concrete, it does not require any finishes or maintenance and therefore is a cost-effective option.

Natural linoleum — not to be mistaken with vinyl flooring — is made from natural materials like linseed oil, jute, cork dust, and wood dust. Unlike vinyl flooring, which offgases toxic vinyl components, this surface is relatively non-irritating (unless you are affected by linseed oil). Natural linoleum is durable, can be finished in solid colors or patterns, and can be installed without the need for adhesives.

Keep in mind that even though carpeting can be warm and cozy on your feet, it is not a recommended kitchen floor option because it traps dirt and moisture. Small, washable mats should be used instead.

Room Additions

The purpose of adding on an addition is to provide more space in you home. Often, it is to take that little box they called the master suite in the '70s and create a contemporary master suite and bath. Perhaps it is a new den so you can all be together after dinner. Whatever drives the project, aesthetics, comfort, health, and functionality are all part of the plan.

It is important that the addition work well with the existing house. There is nothing worse at the end of the job than to look at the house and see a strange protrusion sticking out of one side. Keeping the architecture consistent is important both from an aesthetic standpoint and to the value at the time of resale. Adding a room is an opportunity to improve the comfort and remedy many of the existing problems of the entire house.

Designing an addition is like designing a whole house, and it's an opportunity to think about your house as a system, with the addition a smaller system itself. Virtually everything involved in new construction is encountered in building an addition, especially if you are including a kitchen or a bathroom.

Beginning with the site of the addition and ending with its flooring, we'll show you how to make your additional room as energy-efficient, resourceful, and healthy as possible.

Job Site and Landscaping

If possible, when you first think about adding a room, consider exactly where it will go to optimize sunlight and wind exposure, and to minimize disruption to

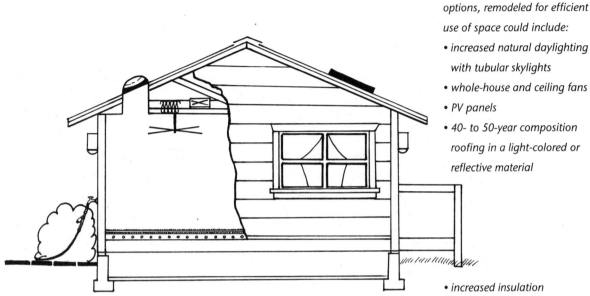

Fig. 6.3: Green room addition options, remodeled for efficient use of space could include:
- *increased natural daylighting with tubular skylights*
- *whole-house and ceiling fans*
- *PV panels*
- *40- to 50-year composition roofing in a light-colored or reflective material*

- *increased insulation*
- *compact fluorescent bulbs*
- *siding made from materials other than wood*
- *finger-jointed studs*
- *recycled-content decking*
- *FSC-certified wood*
- *limited use of wood treated with unsafe chemicals*
- *rapidly-renewable flooring*
- *recycled-content carpet & underlay*
- *zoned, hydronic radient heating*
- *outside drip irrigation*
- *protect native topsoil with native plants*

Credit: Jill Haras & Kim Master.

existing landscaping. You will likely need to demolish a wall or tear down part of a roof. Often contractors will reuse many of the deconstructed materials, or you may be able to utilize them yourself for the new space. If not, choose a contractor who is willing to make an effort to recycle construction materials. These days, everything from windows to wheelchair ramps can find a happy home outside of our overflowing landfills.

Protect native soil by removing it carefully and storing it until renovations are complete — this will save you the cost of buying new soil later. Avoid digging utilities in root zones, or try hand-digging utilities to protect existing vegetation. Mark fragile site features so that vehicles do not inadvertently run over delicate plantings or tree roots. Try to limit traffic in general, because cars and trucks tend to contaminate indoor air quality and compact the soil to a point where existing plant roots are damaged irreversibly. Remember, your yard is a micro-ecosystem, and damaging one tree could disrupt the stability of what you have nurtured over years.

To re-landscape disturbed areas, use native plants that require minimal water or fertilizers. If you do water, consider a drip-irrigation system that directly waters plant roots, wasting significantly less water than a sprinkler system, where most water evaporates. Also consider permeable surfaces, like gravel driveways and walkways, that prevent drainage problems caused by impermeable surfaces such as conventional asphalt.

Foundation

The foundation will be there for as long as the house is lived in. It pays to think about the permanence, the perfection, and the protection a foundation provides. You will probably never dig it up again, so do what you can from the start. A perfectly dimensional foundation layout (square, plumb, and level) is essential so that the above structure fits together and functions well. Use recycled-content fly ash concrete — it will improve the foundation's durability and reduce the potential for moisture migration through the concrete. In addition, consider energy-efficient insulated concrete forms. Insulate the foundation at this early building stage while it is easy with foam panels on the outside of the walls or slab so you don't have to deal with cold floors later. Perhaps most importantly, install drainage around the footing to keep moisture away from the house and prevent future mold problems.

Structural Framing

Use Forest Stewardship Council (FSC)-certified, engineered, or reclaimed wood to reduce the need to cut down old-growth forests. Consider structural insulated panels (SIPs) for the exterior walls, floors, and the roof. These are constructed of a rigid foam core sandwiched between two panels of oriented strand board (OSB). SIPs are more energy-efficient, provide excellent sound-proofing, reduce infiltration, are erected quickly (saving labor costs), and save wood relative to conventional frame construction.

Advanced framing techniques, such as using two-stud corners instead of conventional four-stud corners (studs are vertical framing members) will save wood studs and allow for extra insulation to improve your home's energy efficiency. Finger-jointed studs also enable you to save wood material, as they are constructed by joining smaller pieces of wood (that would otherwise be wasted), in a manner that resembles interlocking fingers.

Whatever choice you make, do it intentionally. Each step of construction determines how comfortable you will be, how much you pay in future energy bills, and how you personally steward the larger environment. The structure of the addition makes an indelible impact on the rest of the building, and you will only make these decisions once.

Exterior Siding

Obviously you will want to match the new addition's siding with the existing siding, but sometimes you can use a material (like stone), that will look stunning and blend well with an existing wood structure. Or you may consider redoing the siding of the entire house to make it more durable and attractive. There are numerous options for exterior siding that eliminate the need for old-growth wood, including regionally produced brick, indigenous stone, natural stucco, or cementitious siding. These materials are also low-maintenance, impact-resistant, and fireproof. Additionally, fiber-cement siding is more durable than wood (warranted to last 50 years). It looks like wood but won't warp, twist, melt, or burn. It is also moisture- and termite-resistant, inhibits fungal growth, holds paint very well, and is easy to install and finish.

While you are working on the exterior of your home, consider building a deck — an easy way to add enjoyable living space. Locate it on the south side if possible. You will gain extra months of use by letting the sun temper the space and melt the snow. (North decks are cold and icy in winter.) Use sustainable decking materials, such as FSC-certified or reclaimed wood for the structural components. Ensure that treated wood is not CCA-treated lumber that contains chromium or arsenic. Also consider recycled composite wood and plastic decking for the surface. Building a patio made of indigenous stone or brick will last a lifetime and add value. It is important to slope the earth away from the house to keep water from pooling beside the foundation wall or under the deck. Plant trees for a bit of shade, and if there is a tree exactly where you want your patio or deck, build around it to keep the house cooler in summer.

Roofing

The main function of a roof is to keep the house dry and protected from the elements. All too often people compromise on roofing materials. This is where you want the best you can afford and you want a roofer that really understands how the existing roof and the addition roof intersect. Make sure that the roofs provide positive drainage, and be sure to install flashing carefully to avoid future leaks.

All roofing shingles require significant amounts of energy during manufacture, so look for 40- to 50-year composition roofing you won't have to replace. Recycled-content asphalt, slate, concrete tile, lead-free metal, recycled-content plastic, and photovoltaic roofing tiles are other options worth looking into. For all roofing materials, avoid adhesives and calculate how much material you'll need, in order to be able to assess contractors' estimates accurately. Light-colored roofing and radiant roof barriers installed on the underside of your roof sheathing are recommended to reflect unwanted heat away from your home.

Plumbing

For information regarding plumbing changes you may make to the addition, please refer to Chapter 12, Plumbing.

Electrical

Design electrical lighting with natural light in mind. For example, if you incorporate clerestories or light shelves into your addition, you'll likely need only minimal task lighting during the day. Compact fluorescent lights are the best option for electrical lighting, including task lighting. They last 10 times as long as incandescent lights, and a 15-watt fluorescent light can give off as much light as a 60-watt incandescent. The light is more flattering than older fluorescent lights, and they're now available with dimming options so that you can change the mood of your room while reducing energy output. Compact fluorescent bulbs are also available for outdoor lighting fixtures. It is important that outdoor lighting fixtures do not create light pollution that disrupts wildlife, neighbors, and possibly even your health. To this end, avoid use of floodlights, specify "full-cutoff" luminaries, and focus lights downward. Also consider timers, motion detectors, or photocells to ensure lights are on only when needed.

Insulation

The building code for your jurisdiction determines the *minimum* required level of insulation, but you really want as much insulation as you can get into the walls and ceiling. We have had artificially low energy prices for decades, and that is about to come to an end. Insulation will reduce your need for heating and air conditioning, making your home more comfortable and affordable. In general, framing determines how well the house will be insulated — if you build with 2" × 4" walls, you can wrap one inch of rigid foam around the exterior to improve your comfort level; if you frame with 2" × 6" walls, you can get even more insulation into the walls. The ceiling is the least expensive place to install additional insulation.

It is important to seal and insulate exterior walls with advanced infiltration reduction practices. This entails paying close attention to where the new addition joins the existing house. Often, this area isn't sealed properly, so it is open to the attic at the top and filled with wires and pipes that were shoved in during remodeling. Advanced sealing entails using spray foam or caulking at all intersections and penetrations so no air can enter the house. Even though it is standard practice to seal around doors and windows, make sure you can't see daylight around the framing and the window. Gaps like these are small details that all too often get overlooked and covered over; they will then create drafts for all the years you live there.

During the renovation, when walls are already open and trucks with insulation materials are on your property, is the best time to insulate the other walls of the house. Blow loose fill into attic or walls cavities. Consider recycled-content fiberglass, rock wool, or cellulose. Dense pack fiberglass or cellulose provides air sealing as well as insulation benefits. There will be small holes in either the exterior siding or in the drywall inside that you will have to fill, patch, and paint.

Solar Energy

Purchase the most energy-efficient windows you can afford for optimum energy-savings and comfort. At a minimum, use low-emissivity (low-E) windows in low-conductivity frames like wood or vinyl to save energy. A skylight is another option that significantly brightens a room; make sure it is well-insulated and low-E if possible so that it doesn't become an energy-wasting feature. Skylights on the south side gain heat in the winter, while skylights on the north side lose heat; all skylights have the potential for overheating in the summer when the sun is high in the sky. A solar tube, essentially a pipe lined with a reflective surface, is a more efficient option — it lights a room more effectively with less heat loss.

Passive solar heating and natural cooling help create a comfortable temperature in your home, reducing the need for mechanical heating or cooling. Orienting the addition to maximize sunlight and prevailing breezes, as well as installing reflective or low solar heat gain windows on east and west windows can help maximize comfort inside your home. You might also consider constructing some form of "thermal mass." A thermal mass is a wall made of stone, brick, or other dense material in south-facing rooms. It can be an attractive feature while soaking up sunlight from windows to store the heat in your home when the sun has set. Window awnings and trees, conversely, can block unwanted sunlight in the summer. One homeowner I worked for wondered why the new furnace never went on in his new addition — it simply wasn't needed because good insulation and the energy-efficient windows provided enough passive solar heat for the space!

Active solar energy uses sunlight to generate heat in solar panels to heat water. Photovoltaic panels generate electricity. Solar energy is a great investment, but for those who can't afford the upfront cost, pre-plumbing the new addition for solar water heating or installing easily removable roofing material to make way for PV later are options that minimize the cost and labor required when adding solar to your home in the future.

Heating, Ventilation, and Air Conditioning (HVAC)

Whether you replace your existing HVAC system with a larger one or put a separate system in the new space depends on other changes you have made to the house. If you have insulated the entire house, you may be able to use the existing system. Even if you must replace an old and inefficient system, you will likely be able to buy a smaller, less expensive system because your home is better able to keep heat in or out. Similarly, you may be able to purchase a separate split system heat pump that will suffice for an efficient addition. Installing new ductwork in the conditioned space (rather than on outside walls) and applying duct mastic to duct joints will also improve furnace efficiency.

If you are going to replace the system, at a minimum, install a forced air furnace or sealed combustion furnace with a 90 percent or greater AFUE (Annual Fuel Utilization Efficiency). Another option is radiant floor heating which uses water in tubes under the floor to deliver heat to the room. Zoned radiant heating enables you to heat certain rooms more than others, helping you save energy if other rooms in the house tend to be colder than your new, efficient addition.

For cooling the new addition, first consider operable windows and ceiling fans. Fans run on 98 percent less electricity than air conditioners, and substantially reduce energy costs. Also consider evaporative coolers that work by blowing house air over a damp pad or by spraying a mist of water into the house air. The dry air evaporates moisture and cools off. This is the same process as a breeze that makes you feel cooler when you get out of a swimming pool, and it works best in dry, hot climates. Since it is often hard to duct a single space, these coolers can be perfect. The wicks (moisture absorbing element in the cooler) must be inspected several times a year for mold. If you must use air conditioning, room air conditioners use less energy than central air conditioning. For more ways to avoid air conditioning or to use your AC unit more efficiently, refer to Chapter 16.

Nursery Percolation

Chemical Cleaners + Volatile Organic Compounds + Polyvinyl Chloride + Phthalates + Formaldehyde = Toxic Chamber

As parents, we want to provide a healthy, safe and secure environment for our children. However, we may not be able to provide a 100 percent chemical-free environment: since World War II, at least 75,000 new synthetic compounds have been developed and released into the environment; fewer than half of these have been tested for potential toxicity to humans, and still fewer have been assessed for their particular toxicity to children. However, with a little care, a lot of label-reading and an awareness of what we are putting into our children's spaces, we can offer them a healthier future.

Table 6.1 on pages 156 and 157 shows the effect of toxins that carry a high-exposure burden and which are often released during a typical nursery renovation. It includes some of the things you can do to make this room healthy for your new child.

Connie Menuey McCullah, Odin's Hammer Construction, Berkley, CA

Interior Materials/Finishes

You can make your home healthier by using no-VOC, formaldehyde-free paint; low-VOC water-based wood finishes; solvent-free adhesives; and formaldehyde-free materials in place of more toxic materials like particleboard and medium density fiberboard (see "Health" in Chapter 5).

One of the most popular flooring options for bedrooms is carpeting, but this can also be the least healthy. Not only do the fibers trap pollutants, which a vacuum cannot pick up, but the carpet also often offgases formaldehyde and toxins from the adhesive backings. If you must use carpeting, ensure that it carries the Carpet and Rug Institute Green Label. That will ensure that it has minimal offgasing. Consider recycled-content carpet tiles that you can replace individually, if they are damaged.

A better option is wood that has not harmed old-growth forests; such as FSC-certified wood, reclaimed wood, or rapidly renewable wood like bamboo or cork. Wood is highly durable, washable and attractive in any room.

My friend Connie McCullah is a green building professional and co-owner of Odin's Hammer, a residential remodeling company in Berkley, California. As a woman concerned about her children and the planet, she specializes in implementing safer alternative materials and methods of design and building to improve indoor air quality and energy efficiency. She has compiled the

Table 6.1: Nursery Room Toxins

Activity or Product	Product Ingredients	Exposure Type	Exposure Burden	Notes	Safer Alternatives
Cleaners	Blue Cleaners – ammonia Bleaches – chlorine	Offgasing[1]	High	May cause eye, nose, throat, or skin irritation. Use gloves during application.	Lemon juice, white vinegar, baking soda. Never mix different conventional products — you end up producing a more toxic brew!
Paints	Washable or enamel paints: – VOCs[2]	Offgasing	High	Eye, nose, and throat irritation; headaches and nausea.	Use no- or low-VOC paint.
Floor or wall coverings	Vinyl floor, wall paper, PVC[3], chemically treated carpets	Offgasing	High	May cause kidney or liver damage.	Use no- or low-VOC paint and solid wood flooring.

Notes:

[1] *Offgasing*: The release of gases or vapors into the air.

[2] *VOCs (volatile organic compounds)*: A class of chemical compounds that can cause nausea, tremors, headaches, and, some doctors believe, longer-lasting harm. VOCs can be emitted by oil-based paints, solvent-based finishes, and other products on or in construction materials.

[3] *Polyvinyl chloride (PVC)*: Thermoplastic polymer of vinyl chloride. Rigid material with good electrical properties and flame and chemical resistance. PVC is a known human carcinogen. Due to the environmental releases during manufacture and during fires, it is banned in many parts of Europe. Used in soft flexible films, including flooring, and in molded rigid products like pipes, fibers, upholstery, and siding. Identified by a "Y" inside a recycling triangle found on packaging. Greenpeace has developed an online resource listing PVC alternatives.

information in Table 6.1, above and opposite, on the kind of toxins that may be released when remodeling a room that is intended to serve as nursery.

Activity or Product	Product Ingredients	Exposure Type	Exposure Burden	Notes	Safer Alternatives
Mattresses, crib bumpers	PVCs, formaldehyde	Offgasing	High	EPA classifies formaldehyde as a Class B carcinogen.	Unbleached cotton or fabric with padding.
Cribs, dressers, etc.	Particleboard, formaldehyde, plastic, PVC	Offgasing	Very high to High	PVC often has a pungent, disagreeable odor; often a masking agent is used to cover this smell	Formaldehyed-free medium density fiberboard or exterior grade plywood, or used furniture that has had time to offgas.
Clothing, linens, curtains	Flame retarding or fire resistant chemicals	Offgasing, absorption	High	Eye, nose, and skin irritations.	Unbleached, chemical-free cotton clothing, like linen, hemp, silk wool, cotton. Wash new clothes in 4–5 washings with a nonphosphorus soap to break down chemicals. Purchase used clothing that has had time to offgas.
Toys, soft and flexible	PVC, phthalates	Ingestion, offgasing	Very high	May cause cancer, alterations in sexual development.	Wooden toys or toys stuffed with natural fibers. If using pacifiers, nipples, or plastic bottles and bags, do not heat over a flame or in the microwave. The plastic will breakdown and leach into the liquid. Liquids should be warmed before filling containers.

Source: Connie McCullah, personal conversation. Oden's Hammer Construction, Berkley, CA.

Room Addition Checklist

Job Site and Landscaping (See Chapter 7)

☐ Limit vehicular traffic.

☐ Mark fragile site features.

☐ Hand-dig utilities in root zones.

☐ Protect native topsoil.

☐ Reuse construction and deconstruction waste.

☐ Recycle job site waste.

☐ Install drip irrigation.

☐ Incorporate permeable paving.

Foundation (See Chapter 8)

☐ Use concrete containing recycled waste.

☐ Install insulated concrete forms (ICFs).

☐ Use recycled-content rubble for backfill drainage.

Structural Framing (See Chapter 9)

☐ Use advanced framing techniques (AFT).

☐ Use reclaimed lumber.

☐ Use FSC-certified wood.

☐ Use engineered lumber.

☐ Use finger-jointed studs.

☐ Use SIPs for walls and roofs.

Exterior Siding (See Chapter 10)

☐ Use alternatives to wood.

☐ Use fiber-cement exterior siding.

☐ Use finger-jointed trim.

☐ Install drainage planes.

☐ Build patios with brick or local stone.

☐ Use recycled-content decking.

☐ Use FSC-certified wood decking.

☐ Limit use of unsafe treatments.

Roofing (See Chapter 11)

☐ Calculate how much roofing you'll need.

☐ Avoid adhesives.

☐ Provide a light-colored or reflective roof.

☐ Install radiant barriers.

☐ Install 40–50 year roofing.

Electrical (See Chapter 13)

☐ Install clerestory windows.

☐ Install light shelves.

☐ Install CFL bulbs.

☐ Install lighting controls.

☐ Minimize light pollution.

Insulation (See Chapter 14)

☐ Use formaldehyde-free, recycled-content fiberglass insulation.

☐ Use cellulose insulation.

☐ Use mineral/rock wool insulation.

☐ Caulk, seal, and weatherstrip.

Solar Energy (See Chapter 15)

☐ Install double-paned windows.

☐ Install low-E windows.

☐ Install low-conductivity frames.

☐ Incorporate natural light.

☐ Incorporate and distribute thermal mass.

☐ Install window coverings and overhangs.

☐ Install a solar water heater.

☐ Install grid-connected PV panels.

☐ Install off-the-grid PV panels.

☐ Install solar hot water pre-plumbing.

Room Addition Checklist – continued:

Heating, Ventilation, and Air Conditioning (HVAC)
(See Chapter 16)

☐ Install a 90 percent or greater AFUE forced air furnace.

☐ Install sealed combustion furnaces.

☐ Install zoned, hydronic, radiant heating.

☐ Properly size your HVAC system.

☐ Use duct mastic or Aeroseal™ instead of duct tape.

☐ Install ductwork within the conditioned space.

☐ Install operable windows and fans.

☐ Install evaporative cooling units.

☐ Use your AC wisely.

Interior Materials/Finishes (See Chapter 19)

☐ Use formaldehyde-free materials.

☐ Select FSC-certified wood flooring.

☐ Use rapidly renewable flooring materials.

☐ Install recycled-content carpet and underlayment.

☐ Use low- or no-VOC and formaldehyde-free paint.

☐ Use low-VOC, water-based wood finishes.

☐ Use solvent-free adhesives.

The Office

Largely as a result of advances in telecommunication, more than 27 million people in the US are working out of their homes. The home office should be as comfortable and energy-efficient as possible to keep you productivity high and your utility bills down. This section will focus on energy-efficient equipment and lighting — but the planning, insulation, solar energy, and interior finishes mentioned previously are also important aspects to consider when remodeling a home office.

Job Site and Landscaping

As you plan for your new office space, consider what you will do with your obsolete electronic equipment. An estimated 30 million computers are thrown out in the US every year; of those, only about 14 percent are recycled.[3]

"Electronic equipment is one of the largest known sources of heavy metals, toxic materials, and organic pollutants in municipal trash waste," said Leslie Byster, of Silicon Valley Toxics Coalition, a nonprofit group in California that studies computer industry waste. "The only household item I can think of that is worse would be pesticides."[4]

Fig. 6.4: Green office installation options, remodeled for efficient use of space could include:

(see below)

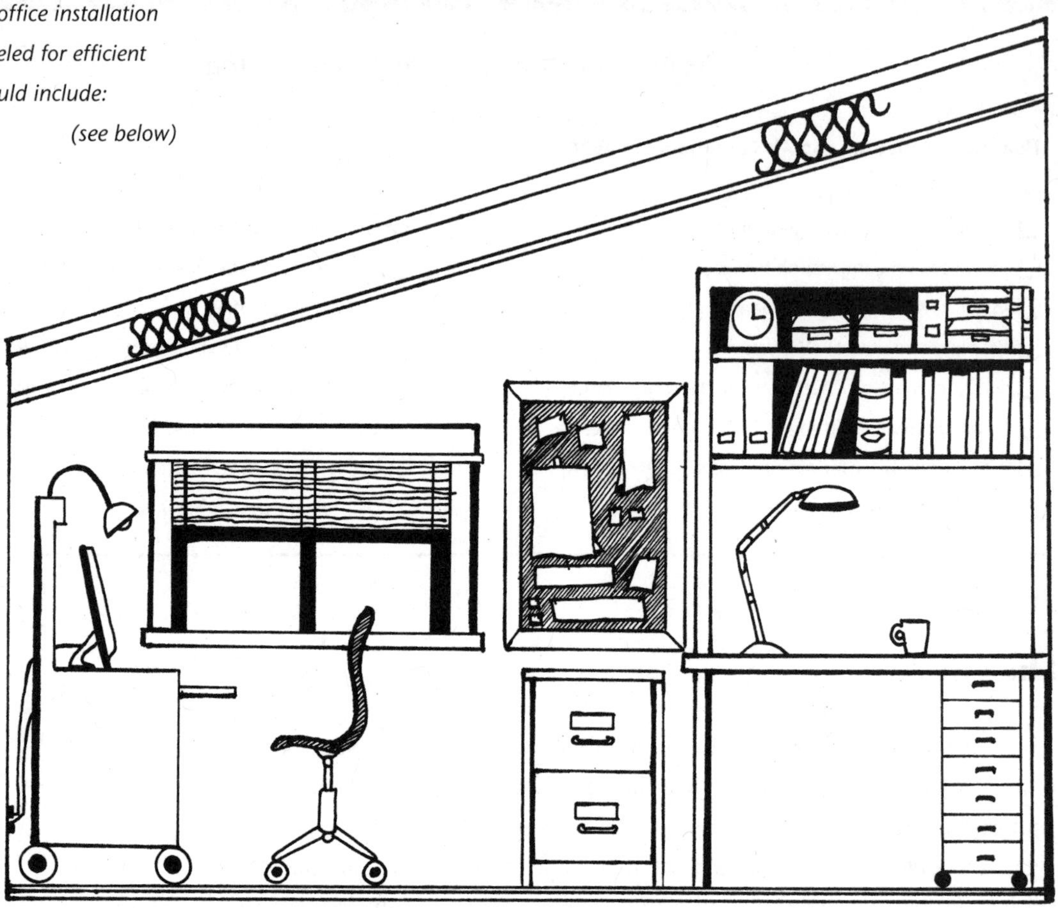

• windows glazed according to their aspect
• window coverings & overhangs
• landscaping to shade windows
• increased roof insulation
• compact flourescent bulbs/lights
• recycled-content carpet & underlay
• low-VOC, water-based paint & finishes on walls & furniture
• unplugging equipment when not in use

Credit: Jill Haras & Kim Master.

The lead in the cathode ray tube of a monitor is especially dangerous. While the government has banned lead in paint and gasoline because it can cause brain damage in children, there is still an average of five pounds of the metal in each tube.[5] In response to the environmental threat, California and Massachusetts passed laws in 2001 that require recycling of old monitors. Three other states have similar plans in the works. Encourage electronic equipment recycling in your jurisdiction and support retailers such as Best Buy and Staples that have started holding special collection days where people can bring in old electronics for recycling.

Structural Framing

For information regarding structural changes you may need to make to the office, please refer to Chapter 9, Structural Framing.

Plumbing

For information regarding plumbing changes you may make to the office, please refer to Chapter 12, Plumbing.

Electrical

Laser printers use one-third of their printing power when they are on standby. Turn off your laser printer when you're not printing!

Source: "No Regrets Remodeling," Home Energy Magazine, *1997, p. 169.*

Good lighting is the best way to increase your productivity. Natural sunlight should suffice during the day. One great option is a light pipe — a long tube lined with a reflective surface to help channel significant amounts of light into your office without overheating the room. When daylight is not available, fluorescent lights will provide the most light for the least amount of energy. Inexpensive straight-tube fluorescents can be used in a variety of attractive ceiling fixtures for general lighting, and can also be placed in soffits directly over the desk for task lighting. Get the more efficient T-8 lamps with electronic ballasts for these fixtures.

Low-voltage halogen track lighting is an attractive option that illuminates pictures or credentials on a wall and it is a better choice than standard incandescent spotlights. New compact fluorescent track lights are also available, and are even more efficient than halogens. You can also get compact fluorescent wall sconces to highlight walls indirectly. Install separate switches for general lighting, task lighting, and highlighting so that you do not have to waste energy by having all the lights on at once.

As you buy electronic equipment, look for EnergyStar® labels that indicate the device will have lower operating costs than other equipment. Also, many electronic devices continue to use electricity even when switched off, so physically unplug these appliances or turn off their power strips. This will also protect your equipment against voltage surges.

Insulation

Working from home can be a challenge when there are teenagers blasting music from their bedrooms, neighbors using their lawnmowers, and dogs barking. Insulating the interior walls between your office and the rest of the house will help you keep your peace of mind so you can get your work done more efficiently and truly enjoy "family time" at the end of the day. Insulation will also help the room maintain a more comfortable temperature, saving you money on energy bills.

Avoid recessed can light fixtures that allow unwanted hot or cool air into your working space.

Solar Energy

If you've ever worked in a windowless cubicle, you know how important windows are to your productivity (and sanity). Look for low-E, low solar-heat-gain glass, especially if sunlight will shine directly on your electronics. A lower cost option is to use landscaping and window overhangs to shade windows. Orient desk surfaces and computer screens to avoid sunlight glare.

A solar space heater is a great way to heat a small space like an office. Active solar air heating systems have solar collectors that use the sun to heat air which is then used to provide space heat. They do not require a heat storage component, and they can complement most existing heating systems. Solar space heaters are the least expensive and simplest solar technology to install. And, just as important, solar power will reduce the amount of pollution and greenhouse gases that are emitted to the atmosphere to heat your home.

Heating, Ventilation and Air Conditioning (HVAC)

Incandescent lights give off 90 percent of their energy as heat rather than as light. Computers, printers, copiers, and fax machines also create substantial waste heat. This may make your office and even other parts of your home uncomfortable in summer, and the excess heat can create problems with sensitive electronics like hard drives. In the summer, you may suddenly need air conditioning; in the winter, heating in your home may become uneven because of the excess heat in the office.

Your best solution is to eliminate sources of unwanted heat in the first place. Energy-efficient office equipment will also minimize waste heat. Operable windows and ceiling fans will help ventilate and cool your home with little energy expense. (See HVAC section under "Room Additions".)

Interior Materials/Finishes

Select light colors for walls, ceilings and floors to reflect light deep into the room. Wood can be a light, attractive finish. Try to buy sustainable flooring like bamboo or FSC-certified wood. If you choose to install carpeting, use a formalde-hyde-free, recycled-content carpet and backing. Carpet tiles are now available, which allow you to simply replace small squares of carpet that get stained or damaged. Always use paints and wood stains that contain little or no toxic VOCs.

Office Checklist

Job Site and Landscaping (See Chapter 7)

❏ Recycle job site waste.

Electrical (See Chapter 13)

❏ Install light pipes.

❏ Install compact fluorescent light (CFL) bulbs.

❏ Install halogen lighting.

Insulation (See Chapter 14)

❏ Increase insulation thickness.

❏ Avoid recessed can light fixtures.

Solar Energy (See Chapter 15)

❏ Install double-paned windows.

❏ Install low-E windows.

❏ Install glazings tuned to orientation.

❏ Install window coverings and overhangs.

❏ Use landscaping to shade windows.

❏ Install a solar space heater system.

Heating, Ventilation and Air Conditioning (HVAC)
(See Chapter 16)

❏ Install operable windows for natural ventilation.

❏ Eliminate sources of unwanted heat.

Interior Materials/Finishes (See Chapter 19)

❏ Select FSC-certified wood flooring.

❏ Use rapidly renewable flooring materials.

❏ Install recycled-content carpet and underlayment.

❏ Use low- or no-VOC and formaldehyde-free paint.

❏ Use low-VOC, water-based wood finishes.

The Basement

Fixing up the basement is the easiest and most cost-effective way to create more space in your home. The basement can be transformed into a recreation area, rental suite, or home office. However, beware of a wet basement! If you can't afford to completely remedy basement moisture problems, do not attempt to renovate this area. The finish materials will simply rot and cause odor and health problems. However, we will discuss how to remedy moisture issues and build a healthy, comfortable, efficient basement environment from foundation to ceiling.

Job Site and Landscaping

The most common problem in basements is moisture. Some water problems can be fixed simply by addressing the house site. For example, you can extend outside gutter drainage pipes farther away from the house, add gravel-filled drains to deflect water

Fig. 6.5: Green basement installation options, remodeled for efficient use of space could include:

• *FSC-certified wood*
• *increased natural daylighting*
• *increased insulation thickness*
• *relocation of plumbing from exterior walls & insulated pipes*
• *exposed concrete as a finished floor*
• *radon mitigation*
• *HEPA filter*
• *ductwork installed or relocated within conditioned space & sealed with duct mastic instead of duct tape*

Credit: Jill Haras & Kim Master.

away from the house at the bottom of the drainage pipe, incorporate a rooftop water catchment system with gutters, or slope the ground away from the house so that water does not run into the basement. More serious problems will require professional help to excavate around the house. Don't get duped into thinking you can paint the basement walls with "water proof" paint to fix the problems. Professionals will provide a back hoe for digging and mechanisms to move water away from the house.

As you design the interior, try to incorporate storage space in corners, under stairs, or in odd spaces. For example, an unused corner could become a walk-in closet or storage room. Create multipurpose rooms: a playroom today might make a great den or exercise room in the future. Save yourself anticipated costs by planning ahead!

Foundation

One clear sign of moisture is a crack in the foundation wall or floor. Some can be easily repaired with mortar or special hydraulic cement. More subtle water migration through walls can be sealed; water condensation can be remedied by making the wall warmer or the air drier. Serious moisture problems can cause not only larger cracks and structural damage, but also mold. You need to eliminate mold carefully, because exposure to some molds can be debilitating or even fatal. Call a professional or use a full-face respirator for severe mold issues.

To get rid of mold, first wet the area with bleach for 15 minutes. Second, wash the moldy area with one part bleach, four parts water, and a bit of non-ammonia detergent. Lastly, rinse the surface and dry it quickly. This should remove the mold and retard its growth so that you can then focus on fixing the source of dampness that caused the mold. You may have to repeat the cleaning process.

Another foundation issue is radon gas. It seeps into your home from soil deposits and can cause serious health problems, including lung cancer. Radon test kits are typically available in hardware stores, online, or through mail-order. Sealing cracks and seams in the floors and walls, as well as electrical and plumbing pipes will help keep

Is the basement dampness caused by water migration or by water condensation?
Caulk a film of polyethylene to a dry section of the basement wall for one to two days. Water on the bottom of the sheet indicates moisture; water on the top indicates condensation.

radon out. You can also seal the foundation with a vapor barrier and install a gas trap to prevent air from entering your home through a floor drain or sump. If the problem is not fixed by these measures, get professional help to install vents that exhaust gases from your home (see "Radon" in Chapter 5).

Structural Framing

There are several framing obstacles that may come up as you renovate the basement, including steel posts, furnaces, or laundry tubs. You can try to make the post part of a dividing wall or storage unit. Furnaces are more difficult because they are big and connect to a network of ducts and pipes. You can either replace the furnace with a smaller, more efficient unit, or you could get a technician to rewire and re-duct the distribution system. Laundry tubs you can move if they are connected to the main drain. However, you may have to work around floor drain tubs because floor drains should not be covered up in case of floods.

Using the basement may include fixing the staircase. Making it wider and less steep with a hand railing is a safe improvement that makes it more accessible. A stair exit to the outdoors is a more costly improvement, but will significantly improve the area if you plan to rent it out; it may be necessary in order to comply with building code requirements for fire exits. To make sure you provide adequate drainage and foundation support, seek professional advice.

Plumbing

Plumbing fixtures can be installed in a basement, but the feasibility of this varies from house to house. The main waste drain typically goes through the basement, so you can break up the concrete around it in order to get a new waste line to attach below grade. Water supplies are often overhead but may not be in the same location as the waste line. You can easily run hot and cold water lines to new sinks and toilets — just make sure each fixture is vented by joining it to a pipe that allows air to escape to the outside. However, if the main drain is above the fixtures, you may need a professional to help you run connections downhill to the drain or to create a special system to pump waste up to the drain.

Since most basements (and hence, bathrooms in basements) are small, compact fixtures such as corner showers and round-fronted toilets will save space. Toilets and faucets should be low-flow models to save water. For safer drinking water, consider installing carbon filters or a whole house filtration system. Investigate the types of pollutants in your water before purchasing a filtration system, because each filter cleans different pollutants from the water.

Always insulate hot water pipes to save energy. If you are opening up walls, this may be a good time to move pipes to interior walls, where they will lose less energy to the outdoors.

Electrical

If your basement renovation is fairly complex, and includes, say, a kitchen, an electrician will need to upgrade your service. Avoid putting all the lights on one circuit, so that the basement will not go completely dark if you blow a fuse. In wet areas, like bathrooms, you will need what is called a "ground fault interrupter circuit breaker" to prevent the user from getting a shock if a short circuit occurs.

Good lighting is essential in the basement because it is typically a dark area. IC-recessed lighting with compact fluorescent bulbs is an efficient lighting option that saves ceiling space. Fluorescent tubes and halogen lights provide flattering light that is significantly less expensive to operate and maintain than incandescent lights, despite initially higher costs. Pay extra attention to dark stairways, and include switches at the top and bottom of the staircase for added safety and convenience.

Insulation

If you repair the foundation from the outside, consider insulating the foundation wall with a closed-cell rigid foam insulation, such as extruded poly-styrene. Rigid foam insulates well and can be reused if it is ever removed. Install the insulation over a low-toxic dampproofing layer. Extend the insulation above where the floor hits the foundation to minimize air leakage, and add a drainage mat over the insulation so that water drains away. Add flashings, or flat sheets of metal, to protect the top edge of the insulation from water penetration. Interior wall insulation is easier and cheaper to install, if you don't have to excavate outside, but it reduces your living space, it won't protect the concrete wall waterproofing membrane, and it may become saturated with moisture if there is a failure.

Soundproofing may be a good idea to keep noise from your rowdy downstairs tenants or ping-pong fanatic kids from disrupting your peace of mind upstairs (or vice versa). To minimize noise transmission between floors, seal cracks and holes around pipes and ducts where they penetrate the ceiling. Also make sure the basement door is airtight. Then, install recycled-content cellulose between ceiling joists. Next, install drywall or tile using resilient channels that effectively minimize contact between the joists and ceiling materials, or lay recycled-content insulation blankets on top of a suspended ceiling. Lastly, look for ceiling panels or tiles with a high noise-reduction rating.

Solar Energy

You can get more natural light in your basement by upgrading your windows. First, enlarge the well around the windows. You can put light-colored rocks at the bottom and paint the sides of the well white to reflect more light into the basement. If you'd like even more light, enlarge the window itself; it's easiest to extend the window downward — just make sure there are no pipes or ducts in the wall. You can also cut new windows into the walls if necessary. Awnings can increase the energy efficiency of windows, and greenhouse windows project forward to bring in more light from all sides of the window. Keep in mind that fire codes require that at least one window on each floor must act as a fire escape — so this window should be large and easy to open for someone to get out in case of emergency. Check your local codes for exact requirements. It is important to purchase the most energy-efficient windows you can afford to replace the typical metal-cased, single-glazed windows that are typical in foundation walls. The old windows will be a constant source of cold and drafts. Use low-e windows with low-conductivity frames to maximize comfort and minimize energy bills.

There are also newer technologies on the market, including solar tubes and fiber optic lighting that can bring natural light in from several floors above the basement. With solar tubes, light passes through reflective pipes that are incredibly energy-efficient and effective at lighting large areas despite their small size. A chase can be run through rooms above, such as a closet, to get the light to the basement. With fiber optic lighting, light passes through glass fiber into a mechanism where it is then distributed anywhere in the house, including the basement.

Heating, Ventilation, and Air Conditioning (HVAC)

Heating a basement is generally easy, since it is usually a small area; one heating duct or spot heater is often sufficient. Larger areas may require a small system with outlets or vents to circulate the heat throughout the basement rooms. Make sure that you install the new ductwork in conditioned spaces (i.e., not on the exterior walls), and use duct mastic on duct joints to effectively seal them.

New basement kitchens or bathrooms will require ventilation systems that exhaust air to the outside of the house. For the basement in general, you will also need ventilation. You can run a duct with a small fan up the wall to pull air from the floor above into an air register low on the basement wall. The warm air will then rise through the room and shouldn't affect household airflow.

Interior Materials/Finishes

You determine how healthy your basement will be by how you finish it. Many materials emit toxins into your indoor air that can lead to serious medical conditions. Wallpaper is a popular wall covering, but its vinyl coating causes irritating odors. Additionally, moisture can get behind the paper, where mold enjoys feasting on the gluten-based paste. Your best option is to use a low-VOC, formaldehyde-free paint over drywall, or a low-VOC finish over FSC-certified, reclaimed, or solid wood paneling. Conventional wood or simulated wood paneling should be avoided because it offgasses formaldehyde (see "Formaldehyde" in Chapter 5).

Particleboard or medium-density fiberboard (MDF) cabinetry will release formaldehyde into your air. You can seal these materials with a couple of coats of a water-based acrylic sealer to minimize emissions.

The most economical and resourceful flooring option is to leave the concrete slab unfinished. It can be stained to give it color and polished to give it a finished look. Other flooring may be necessary, however, to make the room more comfortable. If you are going to build a new floor, include a moisture barrier layer underneath the framing sleepers (like floor joists laid on the concrete that serve as a nail base for the new subfloor), as well as rigid insulation panels between the sleepers and beneath the exterior grade plywood subfloor and your finished surface.

Be on the look out for asbestos!

Asbestos is particularly dangerous on surfaces where it is loose or disintegrating. It should be removed by a professional unless it is already sealed or in solid form.

For finished surfaces, choose natural linoleum over vinyl; the vinyl offgasses harmful chemical, whereas natural linoleum is usually less irritating. Healthier, more durable options include ceramic tiles and well-sealed FSC-certified wood floors. Use low-toxic, water-based adhesives rather than unhealthy solvent-based products. Recycled-content carpet is acceptable in areas that are free of moisture, but in general they provide a warm and cozy home for mold spores, dirt, and pesticides tracked in from your neighbor's yard.

Changing the world one room at a time means thinking about every choice you make. As you can see, the environmental consequences are endless. Taking the time to consider your options is beneficial to you, your children, and the planet. In Part III, we give you more detailed information on particular green features to guide you through these choices at each stage of the remodeling process.

Basement Checklist

Job Site and Landscaping (See Chapter 7)

☐ Remodel for efficient use of space.

☐ Use rooftop water catchment systems.

☐ Consider gravel-filled "Dutch drains."

Foundation (See Chapter 8)

☐ Incorporate radon mitigation.

Structural Framing (See Chapter 9)

☐ Use reclaimed lumber.

☐ Use engineered lumber.

Plumbing (See Chapter 12)

☐ Insulate pipes.

☐ Install low-flush toilets.

☐ Install low-flow faucets.

☐ Investigate your water supply.

☐ Install a whole-house water filtration system.

☐ Install activated carbon filters.

☐ Remove plumbing from outside walls and insulate pipes.

Electrical (See Chapter 13)

☐ Install compact fluorescent light (CFL) bulbs.

☐ Install halogen lighting.

☐ Install sealed or airtight recessed lighting.

Insulation (See Chapter 14)

☐ Use cellulose insulation.

☐ Caulk, seal, and weatherstrip.

☐ Increase insulation thickness.

Solar Energy (See Chapter 15)

☐ Install low-e windows.

☐ Install low-conductivity frames.

☐ Incorporate natural light.

☐ Consider fiber optic or solar tube daylighting.

Heating, Ventilation, and Air Conditioning (HVAC) (See Chapter 16)

☐ Use duct mastic or Aeroseal™ instead of duct tape.

☐ Install ductwork within the conditioned space.

Indoor Air Quality/Finishes (See Chapter 19)

☐ Use formaldehyde-free materials.

☐ Seal all exposed particleboard or MDF.

☐ Install recycled-content carpet and underlay.

☐ Use recycled-content tile.

☐ Replace vinyl flooring with natural linoleum.

☐ Use exposed concrete as the finish floor.

☐ Use low- or no-VOC and formaldehyde-free paint.

☐ Use low-VOC, water-based wood finishes.

☐ Use solvent-free adhesive.

Part III

Everything You Need to Know
(and More)

As we mentioned at the beginning of this book, the intent is not to insist that you use every green feature listed here. Some, as detailed in Chapter 3, are more expensive than others. Others, as you may learn from your architect and building contractor, may not even be feasible for your project, given varying regional climates and code regulations. In Chapter 5, we described how certain aspects of building may have more of an impact than others on energy, on natural resources, and on our health. And as shown in Chapter 6, some home features, such as plumbing, are obviously more relevant to bathrooms than bedrooms. Our intention is to give you all the green options available so that you may pick and choose what's most important to you. In the following chapters, you will find more in-depth information regarding the green features mentioned in the previous room by room chapter. You can also use the index to flip directly to the pages of interest.

To assist you in finding your way through the detailed descriptions, we have inserted symbols that indicate specific benefits of the green feature under discussion.

$ = cost effectiveness

 = energy efficiency

 = resource conservation

 = health benefit

CHAPTER 7

Job Site and Landscaping

As you begin looking at how you want to remodel your house, it is important to think about the construction process itself. How will construction impact your property, your health, and the environment? There is a lot to consider — from vehicle traffic, to dust, to waste. In order to minimize construction site disturbance, health risks during renovation, construction waste, and storm water issues, we discuss these topics alongside design early in the renovation planning stage. Once you break ground for an addition, it may be too late to consider the following green remodeling strategies.

Minimize Construction Site Disturbance

$, 🌲 – Limit Vehicular Traffic

- Construction trucks on your property compact the soil, harming existing tree roots and making it more difficult for future plants to flourish in these areas. You can limit damage to your property and save future costs for landscaping by clearly delineating vehicle restrictions before construction begins.
- **Recommendation**: While you are remodeling, limit vehicular traffic to designated areas, restrict parking of construction vehicles, and make arrangements for particularly heavy vehicles (such as concrete trucks or cranes) to back in or out of the job site so that space for turning around is not required. Ensure that concrete trucks are washed out in slab or pavement sub-base areas, not in areas with more fragile vegetation.

$ – Mark Fragile Site Features

- During renovations, construction vehicles often needlessly damage existing leach fields, buried utility lines, and trees. You can save the costs of repair if you locate and identify these features before construction begins.

- **Recommendation**: Flag the locations of walkways, leach fields, and buried utility lines with stakes. Circle past the drip line of important trees and lower branches with orange plastic construction fencing.

$, 🌲 – Reduce Excessive Sunlight Exposure

- When you clear land for an addition, trees are often exposed to sunlight for the first time and can dry out quickly. Reducing excessive sunlight exposure, or "sun shock," protects your trees and saves you the cost of buying replacement trees and then waiting for mature growth. Large trees around a new house can also boost your property's value.
- **Recommendation**: Consider thinning deciduous trees during the winter months when they will be acclimated to higher sunlight levels. Also water trees during dry periods and ask an arborist if wrapping trunks is appropriate to reduce sun exposure.

$, 🌲 – Hand-dig Utilities in Root Zones

- Utilities for new heating or plumbing systems are typically dug with machinery that can damage nearby trees and soil ecosystems. Although hand-digging utilities can be labor intensive, you will protect tree roots and recoup costs by spending less on plantings following construction. Saving older trees can greatly reduce cooling loads on your home and preserve its market value.
- **Recommendation**: Avoid digging utilities in root zones. When this is unavoidable, consider hand-digging utilities.

$, 🌲 – Protect Native Topsoil

- Major renovations can create erosion problems on your property that are costly to fix and can waste quality soil. Any chemical treatments on your lawn or landscape might also pollute local water supplies as topsoil runs off your property.
- **Recommendation**: To protect essential sediment nutrients and save the cost of buying new soil, store native topsoil. To reduce erosion, regrade and plant disturbed areas as soon as possible after renovation work. When

finish grading cannot be done quickly, piled soil and disturbed areas should be protected with straw, filter fabric, or temporary seeding.

Minimize Renovation Construction Health Risks

♡ – Minimize Exposure to Vehicle Exhaust

- During renovations, there are often multiple trucks near your home for weeks or even months. The exhaust from these trucks contains carbon dioxide, carbon monoxide, and other substances that are unhealthy to breathe.
- **Recommendation**: Protect your family's health by keeping exhaust from construction equipment away from your home as much as possible. Ask workers to shut down diesels rather than letting them idle. Although you should air out the interior of your home during renovations, close your windows temporarily while trucks and equipment are in use.

♡ – Don't Let Workers Track Contaminants into the House

- Even if you have been told that your workers use only the safest products for remodeling your home, workers are likely to have unhealthy toxins on their shoes from other job sites. Even in your own home, toxins can be found everywhere from lead in peeling paint to pesticides from soils around the foundation. Once in your home, these pollutants become trapped in carpets where they might remain and offgas for years (even after vacuuming), or result in unhealthy mold spores. Wiping your feet on a doormat reduces the amount of lead in a typical carpet by a factor of 6.[1] With a high-efficiency particulate arresting (HEPA) vacuum cleaner, the amount of lead and other toxic substances in your carpet can be reduced by a factor of 10 (or, in some cases, 100).[2]
- **Recommendation**: Have construction workers take off their shoes when they enter your home, or at least wipe their feet on a stiff-bristle doormat. Use an effective vacuum cleaner that comes equipped with a rotating brush and, preferably, a dust sensor and a high-efficiency particulate arresting (HEPA) filter.

♡ – Protect Your Home from Dust

- There is no avoiding the fact that renovations create lots of dust. Dust (mites) can cause hay fever or asthma.[3] Dust can also cause children to become hyperactive, inattentive, and impulsive.[4] Keeping your air free of dust will minimize the associated health risks. In addition, these precautions will keep your heating, ventilation, and air conditioning (HVAC) system running smoothly, protect your furniture from dust, and save you hours of cleaning.

- **Recommendations**: Store removed HVAC equipment in a clean, dry place and cover with plastic. Seal all the registers in the rooms under construction. Use temporary filtration media on all furnace intakes to ensure clean combustion and safe operation. Try to seal off the area(s) being renovated from the rest of the house, or consider living away from home during the most intrusive construction. Better yet, what a great excuse to go on vacation!

♡ – Positively Pressurize the House

- When there is negative pressure in a home, air from outside is being "sucked" in and air on the inside is unable to escape the house (see Chapter 4: Building Science Basics). This is especially dangerous during construction, when indoor air is potentially laden with dust, toxic chemicals from synthetic construction materials, such as finishes and adhesives. Pressurization also helps keep radon from entering the house.

- **Recommendation**: To reduce the impact of construction and to protect your family from breathing harmful toxins, keep your windows in the unaffected space open as much as possible to introduce fresh air. If this is not an option, such as in winter, have your contractor negatively pressurize his or her work area by installing a fan that exhausts the air in the construction zone out of your living space.

$, ♡ – Protect Materials from Moisture Exposure

- There is a growing number of studies that link allergies, immuno-depression, and illness to the amount and type of mold or fungal growth in a home.[5] By keeping building materials dry, you can prevent unhealthy and unsightly mold growth in your home, as well as saving the cost of removing mold in the future.

• **Recommendation**: Mold needs moisture to grow; therefore, it is essential to keep all building materials dry during renovations. Don't have materials delivered too early to the site. If this is impossible, try to store materials in the garage or a covered area.

> *I believe in taking a jewel box approach to architecture where, instead of making a huge building that is cheaply made and cheaply finished, make a small building that is well-crafted and has wonderful custom details.*
>
> **Cate Ledger, Architect, Oakland, CA**

Reduce Construction Waste

⚘ – Remodel for Efficient Use of Space

• The average US house size in 1999 was 2,250 square feet, up from 1,100 square feet in the 1940s and 1950s.[6] Larger houses tend to have taller ceilings and more features — and therefore consume more construction and maintenance resources that end up as waste at the landfill, while many areas are running out of landfill space. Remodeling for smaller, more efficient spaces reduces the amount of resources used and construction waste generated. In addition, these smaller homes save money — enabling homeowners to focus on more expensive quality features.

• **Recommendation**: Use space efficiently to minimize wasted resources.

$, ⚘ – Reuse Construction and Deconstruction Waste

• Building materials and products, including concrete, asphalt, wood, glass, brick, metal, roofing, insulation, doors, windows and frames, tubs and sinks, cabinets, fixtures, and flooring, are often wasted when homes are remodeled. In fact, the US leads the world in municipal waste production, generating 200 million tons a year, enough to fill a convoy of garbage trucks stretching 8 times around the globe.[7] Anytime you can reuse waste from your remodeling project or from someone else's construction or deconstruction project, you help to conserve resources, save landfill space, decrease disposal costs, and avoid the costs of new materials.

Increasing the amount of materials recycling in the US to only 60 percent could save an amount of energy equivalent to 315 million barrels of oil per year.[9]

• **Recommendation**: Set aside space to separate materials for reuse. For example, extra bricks or stone might be used for a patio or walkway. Or call a licensed contractor who offers dismantling services to salvage materials for reuse. Usable items can also be dropped off at used building material stores. (Keep in mind that all donations are tax deductible, if the reuse store is operated by a nonprofit organization.) Conversely, check these stores and around your area for materials you might be able to reuse from homes being deconstructed or remodeled. Decking that doesn't fit on a particular home (and is about to go to a landfill), might be perfect for your new deck. Flooring, doors and windows, tubs and sinks, cabinets, and fixtures can also be salvaged easily.

$, 🌲 – Recycle Job Site Waste

• The National Association of Home Builders estimates that a typical homebuilder in the US pays $511 per house for construction disposal, of which 90 percent or more is recyclable.[8] Recycling significantly reduces your tipping fees: if you recycled 90 percent of your renovation waste, you'd pay only $51 instead of $511! Recyclable materials include wood, metals, concrete, dirt, some plastics, shingles, and cardboard.
• **Recommendation**: Identify the types and quantities of materials generated at the job site. Contact local recycling facilities and haulers to identify terms and conditions required for recycling materials. Allocate space for recycling bins and containers.

Implement Resourceful Landscaping

🌲 – Design Resource-efficient Landscapes or Gardens

• When you remodel your home, it is a great time to rethink landscaping areas that have been disturbed by construction. Conventional landscapes such as Kentucky bluegrass have high water and chemical inputs and are often over-planted or planted without regard for climate and soil conditions. This results in excess water and fuel consumption, water pollution, and waste generation. Sustainable landscape techniques are in harmony with the local ecosystem and help conserve water, reduce use of chemical fertilizers and pesticides, create healthier soil and plants, and increase biodiversity in landscape areas. In a state-of-the-art assessment, it was proven that xeriscapes cost half as much as standard irrigated landscapes and could almost eliminate water use in drought-prone yards.[10]

What is xeriscaping?

Xeriscaping is landscaping that minimizes water demand. It is based on intelligent planning and design, proper soil preparation, use of drought-tolerant plants, reductions in the amount of lawn space dedicated to grass, water harvesting (diverting rainwater to plants), efficient irrigation design, and mulching to reduce water loss from the soil. Many xeriscapes include beautiful flowers, shrubs, and trees.

Source: Daniel D. Chiras, The Natural House, *Chelsea Green, 2003, p. 413*

- **Recommendation**: Xeriscape and use native plants when you re-landscape after remodeling.

– Landscape to Block Winds

- Older homes exchange all of the inside air with outside air as often as once or twice per hour, whereas new homes average about half an air change per hour. The older and leakier your home, the more important it is to reduce the wind velocity to save energy and heating costs. A mature windbreak comprised of several rows of tall evergreens can reduce wind velocity by up to 50 percent.[11] According to a 2001 report at Virginia Tech, a 50 percent wind reduction can reduce heating fuel consumption 20 to 40 percent[12] in leaky homes. Even in calmer areas, wind barriers can reduce fuel use by 10 percent or more.[13]
- **Recommendation**: The most effective windbreaks are trees or shrubs that have low crowns and dense foliage; as they grow, they will block winds even better. Windbreak tree placement varies with location, but typically means placing them on the north, northeast, or northwest facades.

– Use Landscaping to Shade Windows

- Refer to "Use Landscaping to Shade Windows" in the Passive Solar Energy section of Chapter 15.

Fig 7.1: Using landscaping to shield against the elements: 1. trees close to the house on east and west aspects protect against summer sun; 2. trees on the south aspect should be deciduous to permit winter sun while sheilding summer sun; 3. coniferous wind breaks protect house from cold winter winds. Credit: Jill Haras & Kim Master.

Fig. 7.2: Replenish groundwater with attractive permeable pavements.

🌲 – Install Drip Irrigation

- Water sprinklers waste significant amounts of water through evaporation. They exacerbate water runoff and often result in over-spray onto windows, walks, and streets. When you replace these traditional sprinkler systems with drip irrigation systems, which deliver precise amounts of water slowly and evenly at the plant's roots where the plant uptakes water, you reduce the water waste and weeds, and leaf burn caused by sunlight on moist leaves.
- **Recommendation**: Replace standard sprinkler systems with drip irrigation systems for all residential applications except turf.

Minimize Storm Water Run-off

These days, storm water flow rates are significantly higher for a given storm than they would be naturally, because more and more people are using impermeable paving methods. Storm water picks up polluted materials like pesticides from around your home, flows quickly over impervious surfaces, and contaminates water supplies. In addition, improper drainage can cause moisture damage to your home and increase the risk of flooding. You can prevent storm water damage to the environment and your home as you remodel by considering the following recommendations.

🌲 – Incorporate Permeable Paving

- Permeable paving allows water to seep into the soil, reducing the volume of polluted storm water that flows into local water supplies. It also replenishes soil moisture and local aquifers, reduces irrigation requirements, and lowers flooding risks.
- **Recommendation**: For walkways and paths, use gap-spaced unit pavers (such as stones with sand in between them), decomposed granite, or gravel. For driveways, consider gravel or grass-stabilization systems.

🌲 – Protect Existing Plants; Plant New Vegetation

- Vegetation will increase the amount of water that can seep into the soil, thus significantly decreasing or even eliminating problems such as flooding and runoff caused by excess storm water.
- **Recommendation**: Existing vegetation should be protected and, where necessary, new vegetation planted to retain moisture in the soil. In general, woodlands with dense underbrush provide higher moisture permeation than turf.

$, 🌲 – Use Rooftop Water Catchment Systems

- Rooftop water catchment systems can significantly reduce both runoff and the use of potable water for landscape irrigation. Rainwater can be channeled through gutters and downspouts to an aboveground cistern or underground gravel dry well.
- **Recommendation**: Install a rooftop water catchment system wherever there is guttered roof runoff and room for a cistern. Simple measures may be required to comply with local building codes or to keep mosquitoes and other insects away.

Fig. 7.3: In this rooftop water catchment system, water flows from the existing rooftop gutters into a storage bin for easy access. Credit: Green Culture™.

Job Site and Landscaping Checklist

Minimize Construction Site Disturbance

☐ Limit vehicular traffic.

☐ Mark fragile site features.

☐ Reduce excessive sunlight exposure.

☐ Hand-dig utilities in root zones.

☐ Protect native topsoil.

Minimize Renovation Construction Health Risks

☐ Minimize exposure to vehicle exhaust.

☐ Don't let workers track contaminants into the house.

☐ Protect your home from dust.

☐ Positively pressurize the house.

☐ Protect construction materials from moisture exposure.

Reduce Construction Waste

☐ Remodel for efficient use of space.

☐ Reuse construction and deconstruction waste.

☐ Recycle job site waste.

Implement Resourceful Landscaping

☐ Design resource-efficient landscapes or gardens.

☐ Landscape to block winds.

☐ Use landscaping to shade windows.

☐ Install drip irrigation.

Minimize Storm Water Run-off

☐ Incorporate permeable paving.

☐ Protect existing plants; plant new vegetation.

☐ Use rooftop water catchment systems.

CHAPTER 8

Foundation

Foundations are typically one of three types: full basement, crawlspace, or slab-on-grade. *Basements* extend the farthest underground, and are typically used in areas where there are deep frost lines and low water tables. A home addition will often have a *crawlspace* — essentially, a basement with shorter walls one to three feet deep. *Slab-on-grade* is just that: a cement slab on top of the earth. Most foundations are made of cement, although there are various mixtures you can add to the cement (such as fly ash) to reduce the overall cement content (and the impact on our natural resources), and to strengthen the cement.

Insulating Foundations

Insulating foundations is important — an uninsulated foundation may account for up to 50 percent of the heat lost from an otherwise tightly sealed, well-insulated house in a cold climate. Proper installation of foundation insulation material is necessary to avoid the condensation, material damage, and structural decay caused by the difference in temperature between the house interior and the adjacent earth. Poor design may also aggravate radon infiltration and insect infestation.

Slabs or crawlspaces can be insulated on the interior or exterior, and there are various advantages and disadvantages to each option (see Table 8.1, over page). Whereas other rigid foam materials absorb water, extrudable polystyrene (XPS) is ideal for exterior insulation because water cannot penetrate the material. If a foundation is insulated on the interior, batt or wet-spray is traditionally used, and this is then covered with drywall. Rigid foam insulation boards such as extruded polystyrene (XPS), expanded polystyrene (EPS) or polyisocyanurate insulation boards might also be used on the interior. (See Chapter 14 to compare the environmental impacts of these insulations.) Most insulation boards are then covered with drywall.

Table 8.1: Interior Versus Exterior Insulation

Interior Insulation

Advantages	• It is simpler to install on existing foundation walls.
	• Material costs may be low since you can use almost any insulation material.
Disadvantages	• Many types of insulation require separation from habitable spaces by a fire-resistant material, since they are often extremely flammable and will release toxic gases if ignited.
	• It reduces usable interior space when retrofitted.
	• It fails to protect the waterproofing membrane.
	• It may become saturated by moisture.

Exterior Insulation

Advantages	• It minimizes heat loss through the foundation.
	• It protects waterproofing membrane.
	• It can serve as a capillary break to block moisture infiltration.
	• It prevents freeze-thaw cycle damage to foundation.
	• It reduces interior moisture.
	• It does not reduce usable interior space when retrofitted.
Disadvantages	• Installation is more difficult than interior insulation in retrofits.
	• Material cost is higher.
	• Some exterior insulation materials are susceptible to insect infestation.

Exterior foundation insulation uses only XPS, also known as closed cell insulation, directly on the outside of the basement walls. The insulation is then covered with a material like a cementitious board or stucco finish. You might also use a foam-form foundation system in which polystyrene foundation forms are set on conventional footings, much like building a child's Lego wall. Concrete is poured into the core of the forms, where it cures to form the structure and thermal components of the basement wall.

We will describe the various types of insulation and their respective environmental and health risks more completely in Chapter 14.

The Foundation as a System

The following green alternatives deal primarily with the foundation as a whole system, and with its structural materials. They are intended to help you choose among the many systems available, and to ensure that you understand the impact of your choice.

$, ⌒, 🌲 – Excavate Frost Protected Shallow Foundations (FPSF)

• The US consumes over 100 million metric tons of cement every year.[1] However, you can minimize the amount of cement used for your addition when you reduce your excavation, foundation wall depths, and slab width. Choosing a FPSF typically results in a reduction in concrete use of up to 50 percent. In 1999, the National Association of Home Builders (NAHB) estimated there were 3,000 FPSFs in the US with builders reporting about a 40 percent savings in foundation costs.[2] Moreover, excavating to 16 inches rather than 48 inches will result in significantly less soil removal and piling on the site, thereby reducing soil compaction and vehicle disturbance.[3] Finally, the NAHB Research Center claims that the required insulation levels for a frost protected shallow foundation make your home more energy efficient, exceeding existing code requirements for foundation insulation.[4] By placing insulation horizontally two to four feet from the foundation, heat loss can help keep the ground from freezing around the foundation, reducing freeze-thaw cycle damage.

• **Recommendation**: Excavate the foundation to 16 inches rather than the 36–48 inches typical for a cold climate application. Ask your building code official about appropriate depth of the foundation for your specific project.

$, 🌲 – Use Concrete Containing Recycled Waste (i.e., slag, bottom ash, fly ash)

• The manufacture of cement is contributing more to global warming than most other building materials: The CO_2 emissions produced for US cement consumption is equivalent to the CO_2 produced by 22 million passenger cars.[5] Recycled fly ash, a coal-fired power plant waste product, can replace up to 50 percent of Portland cement used in conventional concrete — thereby decreasing the overall environmental impact of cement production — and will also increase the strength, water resistance, and durability of the concrete. Other industrial and agricultural waste products, including ground blast furnace slag and rice hull ash, can also be used to replace some of the Portland cement in concrete.

- **Recommendation**: Typically, 15 to 50 percent of cement can be replaced with fly ash or other industrial waste products in concrete mixes. Note that it must be cured longer than standard concrete.

⌂, ⛰ – Use Autoclaved Cellular Concrete (ACC)

- ACC is a lightweight precast concrete product, usually manufactured as blocks or panels, which can reduce the total amount of concrete used and the negative environmental impacts associated with concrete. The ACC process uses aluminum to aerate and foam the concrete, which is then steam-cured in an autoclave to create a high strength-to-weight ratio material. ACC insulates much better than concrete and has very good sound absorbing characteristics.
- **Recommendation**: Check with local codes for structural applicability, and if possible, choose ACC in place of concrete. (It can be assembled with regular masonry methods and tools.) ACC is typically more available in the southern US and California.

$, ⌂, ⛰ – Use Insulated Concrete Forms (ICFs)

- Lightweight interlocking rigid foam blocks or panels hold concrete in place during curing and remain in place afterwards to serve as thermal insulation for concrete foundations and/or above-grade walls. Foam form systems have higher thermal efficiencies than solid concrete foundations or walls, and can reduce the total amount of concrete used, yielding material cost savings. Unlike untreated lumber, ICFs are not subject to rot and result in a higher strength wall than standard cast-in-place concrete. Most ICFs have metal or plastic ties that will hold a screw so that you don't have to frame a wall to install the drywall, thereby saving resources and money.
- **Recommendation**: Use rigid foam forming systems wherever an insulated foundation is desirable. Sizes and styles vary by manufacturer, but generally, rebar is placed in the hollow foam cores and concrete (preferably containing slag or fly ash) is poured into the cores to create a load-bearing structure.

Fig. 8.2: Insulated concrete forms (ICFs) save concrete materials, energy, and money. Credit: David Johnston.

$, ⌂, ⚘ – Reuse Form Boards

- Forms are used whenever a slab is poured, and form boards are used to mold the foundation. The boards are often two-by-tens or larger solid-lumber made from old-growth trees. Reuse of forms saves money and conserves resources, as large dimension, solid-sawn lumber is becoming increasingly expensive and scarce.
- **Recommendation**: By carefully removing and separating the forms, they can be reused several times. Use non-toxic form release treatment to make it easier to remove the form, minimizing the risk of splitting the form during removal.

⚘ – Use Metal Forms

- Metal forms, like wood forms, are used to mold the foundation. They come in all sizes and shapes, and produce a smooth finished surface on the concrete. Metal forms can be reused almost indefinitely, thereby reducing wood use and the cost of buying new forms.
- **Recommendations**: Metal forms can be used in most applications where wood forms are used.

⚘, ♡ – Use Biodegradable, Non-petroleum Form-release Agents

- Concrete form-release agents are products used to remove the cement structure from the form used to mold it. They are typically made from petroleum byproducts, and can be a major source of VOCs and soil contamination. Biodegradable or rubber-based, non-petroleum form-release agents meet or exceed the EPA's VOC and toxicity regulations. They are better for your health and provide a smoother finished surface, with fewer "bug" holes than conventional form release agents.
- **Recommendation**: Use biodegradable, non-petroleum form-release agents wherever concrete form agents are specified. Water-based products must be protected from freezing temperatures during storage and application.[6]

⚘, ♡ – Use Non-asphalt-based Biodegradable Damp Proofing

- Damp proofing materials protect your foundation from water intrusion and decay, but the petroleum in asphalt damp proofing can leak into the soil and ground water. Synthetic rubber and cement-based damp proofing products are available that do not contaminate soil and ground water.

- **Recommendations**: Use synthetic rubber and cement-based damp proofing instead of traditional damp proofing products.

🔎, ⛰, ♡ – Install Non-vented Crawlspaces

- Non-vented crawlspaces require less insulation than vented crawlspaces because there is less air infiltration and energy loss. In the summer, the non-vented crawlspace stays cooler, thereby helping to keep the adjacent living space cooler. In winter, the non-vented crawlspace will be drier, reducing the potential for mold. Since the crawlspace is not a living space, ventilation is an unnecessary and expensive feature.

Insulation R-values.

R-values, used to rate insulation, are a measurement of the insulation's resistance to heat flow. The higher the R-value, the better the insulation.

- **Recommendation**: Check with your building department to find out if they will approve non-vented crawlspaces. Many codes require low-volume mechanical ventilation. Install polyethylene sheeting with overlapping seams on the crawlspace floor and up the foundation wall to keep moisture out of the crawlspace.

🔎 – Insulate Foundation Before Backfilling

- All foundations, including slab floors, can be insulated to minimize heat loss from the floors and basement, a major source of energy loss and higher utility bills.
- **Recommendation**: Insulate the foundation with at least R-4 insulation.

⛰ – Use Recycled Content Rubble for Backfill Drainage

- Concrete and rubble can be crushed and used for backfill and drainage purposes at the base of foundations, saving money and natural resources.
- **Recommendation**: Use recycled materials for backfill. Make sure the foundation drain is on top of the footing, underneath the siding, and at the outside perimeter edge of the flashing.

♡ – Incorporate Radon Mitigation

- Radon gas is one of the most dangerous contaminants you may find in the air in your home. The EPA has estimated that exposure to radon may be the

Foundation Checklist

❏ Excavate frost protected shallow foundations.

❏ Use concrete containing recycled waste (i.e., slag, bottom ash, fly ash).

❏ Use autoclaved cellular concrete (ACC).

❏ Use insulated concrete forms (ICFs).

❏ Reuse form boards.

❏ Use aluminum forms.

❏ Use biodegradable, non-petroleum form-release agents.

❏ Use non-asphalt-based biodegradable damp proofing.

❏ Install non-vented crawlspaces.

❏ Insulate foundation before backfilling.

❏ Use recycled content rubble for backfill drainage.

❏ Incorporate radon mitigation.

second leading cause of lung cancer, after cigarette smoking. In the US, the average indoor radon level is 1.3 picocuries per liter (pCi/L), while average outdoor level is only 0.4 pCi/L. It is estimated that in the US, 6 percent of all houses are above the threshold value of 4pCi/L; reducing levels in these homes would reduce the incidence of radon-related lung cancer by one third.[7] The EPA and the Surgeon General recommend that all home owners test their radon levels and take action to increase ventilation if tests result in a radon level above 4pCi/L. (See "Radioactive Contaminants" in Chapter 5).

• **Recommendation**: To reduce concentrations, place an airtight membrane under carpets or provide some form of sub-slab ventilation. Cover any exposed earth with a polyethylene air barrier and seal all cracks and joints in the foundation wall and floor slab with caulking or foam. Install a self-priming drain or gas trap in the floor drains leading to a sump or to drainage tiles. Remove radon from well water using activated charcoal filters or aeration units. If radon is suspected, lay perforated PVC pipe under the foundation floor for future retrofit of an above-ground radon mitigation system. For further information about radon, go to the EPA's webpage at: <www.epa.gov/iaq/radon/>.

CHAPTER 9

Structural Framing

TRADITIONALLY, FRAMING ACCOUNTS FOR ABOUT 15 percent of the total cost of home construction. Advanced framing techniques (AFT) uses engineering principles to minimize material usage while meeting model building code structural performance requirements. Using these framing techniques to reduce the amount of framing material by up to 20 percent allows you to better utilize resources while saving material and labor costs.

The following advanced framing techniques can be applied to many renovation projects:

- Increase wall stud spacing to 24 inches on center (o.c.), rather than the standard 16 inches. This uses fewer framing members, therefore saving materials and labor costs.
- Space floor joists and roof rafters at 19.2 or 24 inches instead of 16 inches to save floor joist material (more space means less material) and labor.
- Use inline framing, in which floor, wall, and roof framing members are in line with one another and loads are transferred directly to the foundation (sometimes called "inline" or "stack" framing). This technique enables you to use the engineered strength of the material to transfer point loads, thereby eliminating the need for large headers in load-bearing walls. Check with local codes to make sure this is permitted on your project.
- Design homes on two-foot modules, thereby reducing cutoff waste and installation labor, since most panel products come in even-incremented dimensions such as four by eight feet.
- Use two-stud corner framing instead of three-stud, and replace two- and three-stud backing for drywall attachments at wall "T" intersections with inexpensive drywall clips or scrap lumber. Three-stud corners are the traditional standard due to relative ease of applying drywall, but by using two-stud corners with drywall-clips to hold the drywall, you use less wood

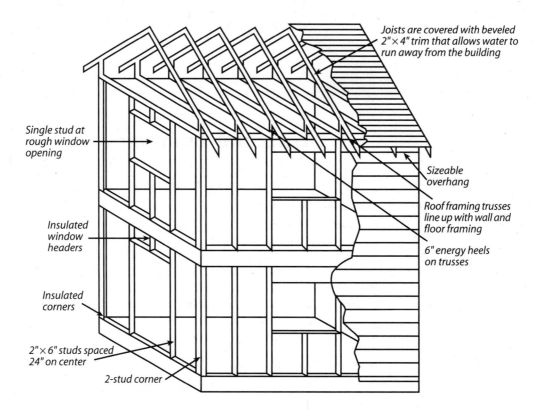

Joists are covered with beveled
2"× 4" trim that allows water to
run away from the building

Single stud at
rough window
opening

Sizeable
overhang

Roof framing trusses
line up with wall and
floor framing

Insulated
window
headers

6" energy heels
on trusses

Insulated
corners

2"× 6" studs spaced
24" on center

2-stud corner

*Fig. 9.1: Advanced framing
techniques.*

*Credit: Jill Haras & Kim Master,
redrawn by Jeremy Drought.*

while effectively supporting the drywall. This also gives additional room for insulation, raising the whole-wall R-value.

- Carefully size headers used to support the load above an opening. Builders typically use two-by-tens simply because it is the industry standard, but smaller sizes are adequate and save material.
- Eliminate headers in non-load-bearing walls. They are not needed, because a header's sole job is to protect the opening from the load above.
- Substitute 2" × 6" studs (24" o.c.) for conventional 2" × 4" (16" o.c.) framing. By using 2" × 6" studs, you save time and labor costs (offsetting slightly higher per-item material cost), since they are 24 inches as opposed to 16 inches apart. The increased spacing allows for additional insulation in wall cavities, which improves thermal resistance and saves you money on energy bills.
- Space roof rafters at 19.2 inches or 24 inches to save on roofing materials.
- Use trusses instead of stick-framed roofs. With trusses, builders can use 2" × 4" lumber. This saves the larger dimensional wood from old-growth forests that is typically required for stick-framed roofing rafters.

- Install energy heels of 6 inches or more on trusses to allow for increased ventilation and insulation.
- Provide a sizable overhang to shade your home and prevent precipitation from damaging the siding of your home.

Knowledge of these advanced framing techniques is key to saving money and resources on your project. Additionally, intelligent material choices detailed in the following section will further save precious wood reesources.

🌲 – Use Reclaimed Lumber

- As demands on forest resources have increased, non-forest sources of wood have grown in importance. High-quality reclaimed wood can be salvaged from places such as buildings slated for demolition or rivers, where sinker logs sank decades ago during river-based log drives. Reclaimed wood is often available in species, coloration, and quality not found in today's forests. It is not tied to recent timber harvesting, it reuses materials, and it can reduce the construction and demolition load on landfills.
- **Recommendations**: Use reclaimed lumber (in place of new material) for nonstructural applications. For structural applications, look for reclaimed lumber that is engineer-stamped and graded. The supplier may require you to pay for an engineer's grading and stamping. In some cases, reclaimed wood suppliers have only limited quantities of woods with matching coloration or weatherization patterns. Ample lead time and accurate materials estimates can help ensure availability of the desired wood.

🌲 – Use Forest Stewardship Council (FSC)-certified Wood

- The FSC is an international nonprofit organization that accredits certifiers and promotes standards for sustainable forestry certification worldwide. The FSC's principles include management for biological diversity, long-term forest health, and the long-term economic well-being of local communities. FSC certification guarantees that forests are managed in a way that assures the long-term availability of precious woods and does not clearcut large sections of old-growth forests. (See "Certified Wood" in Chapter 5).
- **Recommendation**: Use FSC-certified wood whenever new wood framing is required. You may have to order the lumber if it is not stocked by your

Fig. 9.2a: Standard roof truss.

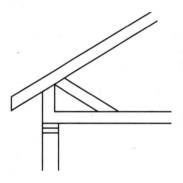

Fig. 9.2b: Raised heel truss ("energy heel")

Fig 9.2 a & b: Standard roof truss and raised heel truss.
Credit: Jill Haras & Kim Master, redrawn by Jeremy Drought.

Fig. 9.3a & b: There is a wide variety of engineered wood products available for framing, including wood I-Joists (9.3a), laminated veneer lumber, glue laminated lumber, oriented strand board, and parallel strand lumber (9.3b). Credit: David Johnston.

local lumberyard. Allow sufficient time for delivery before the framing starts.

☖ – Use Engineered Lumber

• Solid-sawn lumber in sizes of 2" × 10" or greater typically comes from old-growth forests, because these are the only trees that can produce such large-dimension wood. However, engineered lumber products can now be produced efficiently from small-diameter, fast-growing plantation trees. The small pieces of wood are combined with adhesives under heat and pressure to produce practical and economical alternatives to plywood and solid wood framing in sheathing, headers, beams and joists.

Inevitably, the compressive manufacturing process involved in engineered wood is expensive: more than 1 cubic foot of younger lower quality raw wood must go into each cubic foot of engineered wood products, along with a good deal of energy, typically derived from wood wastes. However, the benefits outweigh the costs; engineered buildings are built more quickly and easily. They offer consistent performance, predictable quality and dimension, and superior structural integrity with minimal construction waste. Engineered lumber uses wood fiber more efficiently than conventional lumber — some mills can use 95 percent of the tree![1] Moreover, reducing demand for large-dimension lumber decreases pressure to cut down old-growth forests.

• **Recommendation**: For the following list of engineered wood applications, avoid using products with urea formaldehyde glues. Phenol formaldehyde (the typical glue used for exterior products) does not offgas as much as urea formaldehyde, but it is still not an ideal product. Water-resistant, formaldehyde-free methylene-diisocyanate (MDI) adhesives are best. (For more information, see Chapter 19, "Interior Materials and Finishes.")

• **Use Glulam beams**. These glue together a number of layers of solid wood members to replace 4" × 6" or larger solid beams. You can also achieve better results by sandwiching carbon-fiber, steel, or other strong materials in between layers of wood. Glulam beams combined with carbon fiber can save two-thirds of wood previously required, cut total costs, and make light, airy beams attractive for large structures.[2]

- **Use Oriented Strand Board (OSB) for Subfloor and Sheathing**. OSB is manufactured from fast growing farm trees, whereas plywood is made by shaving sheets of wood from large, old-growth logs. OSB comes in sheets and is used for sheathing and subfloors.

- **Use Wood I-Joists for floors and ceilings**. Wood I-joists are engineered to use only the wood fiber necessary for the structural function required. They typically use OSB for the web and either laminated veneer lumber or solid-sawn lumber for the chords (top and bottom pieces). Wood I-joists use 50 percent less wood fiber to perform the same structural function as similar-sized solid-sawn lumber, and will never twist, warp, or split. They are stronger and lighter than 2" × 10" or 2" × 12" lumber and they can span greater distances.

- **Use Engineered Studs**. Engineered wall framing systems use wood studs made by pressing together small-diameter, low-grade hardwoods. They are also four times as strong and more predictable than commodity-grade studs. Additionally, they are straighter and free from knots, defects, or other irregularities that can compromise quality construction. Engineered studs can be used wherever conventional studs are typically used in vertical applications, particularly when straight walls are critical, as in kitchens where cabinets will be hung.

- **Use Finger-jointed Studs.** Finger-jointed studs are typically manufactured from any short-length, stud-grade lumber, such as Engleman spruce, lodge pole pine, or Douglas fir. The pieces of wood interlock in a way that looks like fingers interlocking, thereby forming a longer piece of wood made up of shorter pieces. Finger-jointed studs are straighter, stronger, and more durable than solid-sawn studs, which can warp and twist. As a result, these studs eliminate crooked walls, and reduce material waste. Finger-jointed studs save the time otherwise spent sorting studs for acceptable pieces and "shimming" walls so the drywall is straight. Finger-jointed studs are cost-competitive and can recover 500 to 700 board feet of good dimensional lumber from each ton of what was previously wood waste.[3] Use finger-jointed studs in place of conventional studs in vertical applications only. Use of finger-jointed studs may need to be submitted to your local structural engineer for approval.

Fig 9.4: Finger-jointed studs eliminate crooked walls and reduce material waste. Credit: David Johnston.

Structural Framing Checklist

☐ Use advanced framing techniques (AFT).

☐ Use reclaimed lumber.

☐ Use FSC-certified wood.

☐ Use engineered lumber.

☐ Use structural insulated panels (SIPs).

℘, 🌲 – Use Structural Insulated Panels (SIPs)

- SIPs are constructed of a rigid foam core sandwiched between two sheets of OSB. SIPs are more energy efficient, provide better soundproofing, and reduce infiltration relative to frame const-ruction. They save wood and can be erected more quickly. Although panelized walls (and roof panels) are more expensive initially than buying raw materials for conventional construction, the resulting savings in site labor, material waste and cleanup fees, and shorter construction timeframes offset much, if not all, of the costs of the panels. Likewise, they may cost more initially due to the insulation component, but again that cost is largely offset when compared to adding the insulation on site. Keep in mind that SIPs are not easier to disassemble, reuse or recycle.

- **Recommendations**: Use SIPs for structural exterior walls, roofs, and floors in place of stick framing. Generally, almost any addition plan can be modified to be built with SIPs.

SIPs tips.

- Select a company that optimizes panel utilization by using, for example, computer assisted drawing (CAD).
- Store panels under cover, out of the sun and off the ground.
- Use panel scraps for constructing headers, filler sections above windows, and for other uses.
- Use adequately thick panels to ensure energy efficiency.
- Specify panels that are provided with special foam-sealing channels — or another comparable system — for sealing between panels during erection to avoid moisture and air migration.
- To further seal panels, tape interior and exterior panel joints with quality construction tape.
- Use panels with foam treated against insect infestation, especially in the southeastern US.
- Always provide mechanical ventilation in SIP houses due to tight construction.
- Plumbing and electrical runs need to be predetermined so the manufacturer can accommodate these needs by forming chases inside the foam to run wire or pipes.

CHAPTER 10

Exterior Finish

Siding

A BUILDING'S EXTERIOR WALL SURFACE is one of its most visible and defining features. Ideally, you want to choose an aesthetically appealing siding material for your home renovation — one that can stave off years of harsh weather, that will require minimal repainting and waterproofing, and that will optimize or avoid the use of wood. When you re-side your home, it is also a good time to consider wrapping the house with rigid foam to increase the wall insulation (see Chapter 14).

Wood is a popular siding that is readily available in most regions; unfortunately, it has become a problematic siding solution due to decades of overharvesting and the subsequent declining quality of materials. Vinyl siding (made primarily from polyvinyl chloride, or PVC) is wood siding's chief competitor. PVC is currently under attack by many environmental groups, including Greenpeace, because of the persistent organic pollutants associated with PVC production. It is also incredibly difficult to recycle used vinyl siding into other products.

Fig 10.1: Innovative reuse of materials, such as this Volvo rear window as a sun screen, can make the exterior beautiful and unique. Credit: Ethan Kaplan, Leger Wanaselja Architecture, San Francisco, CA.

🌲 – Install Drainage Planes

- Without proper rain dispersion elements, water can easily be pushed by wind to where it is in contact with moisture-sensitive building materials. This can cause mold and decay, and can potentially lead to structural failure. Drainage planes will ensure the longevity of your home's materials (see Chapter 4, Building Science Basics).
- **Recommendation**: Install a drainage plane such as #30 felt paper or a housewrap over the sheathing and under all siding. Make sure the drainage plane is "shingled" like a roof so all water drains down the wall and away from the house.

Don't get ripped off!

Siding contractors generate more complaints to the Better Business Bureau than all other building contractors.

- Ask for proof that the contractor is properly licensed in your state.
- Get proof that the contractor is insured for workers' compensation and liability.
- Search your local Better Business Bureau website for complaints, or check with your local consumer affairs department.
- Get at least three written estimates. Be wary of especially low ones; the contractor may be underestimating time or materials, or cutting corners.
- Get a list of references and call those past clients with detailed questions about their experiences with the contractor.
- Make sure the contract provides for a reasonable deposit (like 10%), with balance payable on satisfactory completion. Have an attorney review it.
- Keep a copy of the product warranty; you'll need it if there's a defect or product recall down the road.
- Keep siding wrappers, and keep extra siding handy for repairs.

Source: "Home Improvement," Consumer Reports, August 2003, pp 18 to 32, 38 to 46.

– Use Alternatives to Wood

- Consider using recycled plastic lumber trim, regionally produced brick, indigenous stone, natural stucco, cementitious siding, or molded cementitious "stone." These attractive finishes use less wood, and in the case of recycled content, make use of otherwise wasted materials. They are low maintenance, impact resistant, and fireproof.
- **Recommendations**: Use the above-mentioned materials in place of exterior wood finishes. Siding may require periodic refinishing if painted, but much less frequently than with wood.

$, – Use Fiber-cement Exterior Siding

- Fiber-cement siding is composed of cement, sand and cellulose fibers. It's available in 4- to 12-inch lap planks or 4' × 8' sheets.

It is textured to look like wood lap siding or stucco finish for sheets. Cement siding is more durable than wood (warranted to last 50 years). It looks like wood but won't warp, twist, melt, or burn; it is resistant to moisture and termites; it inhibits fungal growth, holds paint very well, and is easy to install and finish. Using fiber-cement siding reduces the demand for old-growth redwood or cedar siding. It may also reduce homeowner's insurance rates, especially in fire-prone areas.

- **Recommendation**: Replace conventional wood siding with fiber-cement siding. This product can be cut with a diamond-tipped saw blade, snapper shears, or with a guillotine cutter. Dust protection and control are required when cutting with a circular saw. Available primed and painted.

⚘ – Use FSC-Certified Solid Wood Siding

- Wood is renewable, locally available in many regions, relatively low in embodied energy, and ultimately biodegradable. Unfortunately, the most desirable and durable siding materials have traditionally come from old-growth trees, including mature cedar, redwood, and cypress. The ecosystems are often poorly managed and take hundreds of years to rejuvenate. When you buy certified FSC wood, you ensure the wood was sustainably harvested. (See Chapter 5 for more information about FSC.)
- **Recommendation**: Use FSC solid wood siding in place of traditional wood siding. It can be finished or left natural. FSC siding is available primed or painted. Be aware that wood siding can require frequent maintenance, generate significant waste during renovations, and can be short-lived.

$, ♡ – Install Window and Door Flashing

- Window and door penetrations through the exterior envelope are common places for water to leak into framing cavities and create mold. Specialized flashing materials and procedures, in addition to conventional house wrap, can prevent moisture from penetrating the building. Installing flashing, doors, and windows appropriately is an inexpensive way to prevent mold and decay and to ensure a healthy home environment.
- **Recommendation**: Install specialized flashing materials in addition to conventional house wrap. Check all flashing to ensure proper overlap of assembly. As in Figure 10.2, barrier strips are attached with nails and overlapped with successive barrier strips and weather-resistive barriers in shingle fashion. Do not depend on tapes or glues, as they may fail over time.

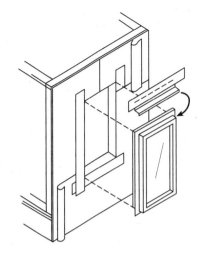

Fig. 10.2: Flashing Detail.
Credit: David Johnston, redrawn by Jeremy Drought.

Siding helpful hints.

- Check the old siding. If the siding is sound, new siding can go over it, especially if the original siding is plywood. Rotted wood siding should be replaced; check the wall behind it for damage and mold.
- If old siding is removed, have a professional apply a code-compliant drainage plane beneath the new siding, as well as beneath flashing around windows, doors, and other openings.
- Check the windows. Replace all rotted windows and frames.
- Save leftovers. Use spare panels for repairs. Keep labels and receipts in case there's a product recall or you have to collect on the warranty.

Source: "Home Improvement," Consumer Reports, August 2003, pp. 18 to 32, 36 to 48.

Decking

Building a deck is one of the easier ways to make your house more enjoyable. It is an inexpensive summer renovation project that is relatively easy to do yourself. However, as with any renovation project, there are ways to use resources more efficiently and to choose materials that do not negatively affect your health or the planet.

🌲 – Build with Brick or Local Stone

- Patios are generally more environmentally friendly than decks because they require fewer materials and resources to construct. When built out of bricks or local stone, patios are more durable, more weather resistant, and typically require less maintenance than decks. Many decks are made of old-growth trees like redwood and cedar, which take hundreds of years to regenerate. Pressure-treated wood is another common decking material, most of which is made with potentially carcinogenic chemicals (See "Wood Treatments" in Chapter 5). These chemicals leach into the soil, affect groundwater, and have led to an increased prevalence of arsenic poisoning, especially in young children.
- **Recommendation**: Build patios made of brick or local stone, instead of decks.

$, 🌲, ♡ – Use Recycled-content Decking

- There are two types of recycled-content decking: plastic lumber and composite lumber. Plastic lumber contains only recycled plastic resins, while composite lumber is made by combining recycled wood fiber and recycled plastic resins that are then formed into deck boards. The durability of these materials is greater than that of wood, providing cost-savings to the homeowner over the life of the product. They will not rot, crack, or splinter, they do not require staining, and they are not treated with potentially toxic chemicals. Recycled-content decking eliminates the need to use old growth trees for your deck.
- **Recommendation**: Use recycled-content decking in all deck applications, excluding structural applications. Both composite and plastic lumber can be used in place of old-growth redwood, cedar, or pressure-treated pine.

These products accept screws and nails and cut like wood. Follow manufacturer recommendations closely regarding the amount of expansion that will occur when using plastic lumber. Beware of plastic decking that uses virgin materials. It's best for the environment to use recycled materials, so look in the manufacturer's literature for recycled content.

🌲 – Use FSC-certified Wood Decking

- It is important to buy FSC-certified wood for your decks, which proves that the wood came from well-managed forests, ensuring the long-term availability of precious woods without damaging old-growth forests. (See "Certified Wood" in Chapter 5.)
- **Recommendation**: Use FSC-certified lumber to ensure your decking comes from forests managed in an environmentally and socially responsible manner. FSC-certified lumber can be used for all exterior decking applications or as structural deck members.

Wood Treatments

🌲,♡ – Use Naturally Decay-resistant Wood

- Along with cedar and redwood, the following woods are considered resistant or very resistant to decay: bald cypress (old-growth), catalpa, black cherry, chestnut, Arizona cypress, junipers, black locust, mesquite, red mulberry, burr oak, chestnut oak, gambrel oak, Oregon white oak, post oak, white oak, osage orange, sassafras, black walnut, and Pacific yew. Naturally decay-resistant wood will reduce the need to apply toxic wood treatments that have known adverse health effects. FSC-certification ensures the wood has been harvested in a way that will minimize harm to forest ecosystems.
- **Recommendation**: Use naturally decay-resistant, FSC-certified wood in place of wood that requires toxic treatments.

Beware of CCA Treated Lumber

Since the 1940s, lumber producers and manufacturers have used a chemical compound mixture containing inorganic arsenic, copper, and chromium (hexavalent chromium, the kind we're talking about, is toxic; trivalent chromium is a food supplement), called chromated copper arsenate (CCA), as a wood preservative. Manufacturers inject CCA into wood by a process that uses high pressure to saturate wood products with chemicals, creating what the public knows as "pressure-treated lumber." CCA protects the wood from dry rot, fungi, molds, termites, and other pests that can threaten the integrity of the wood. However, the EPA, the Centers for Disease Control (CDC), and the World Health Organization (WHO) classify a primary component of CCA — arsenic — as a known human carcinogen. Estimates suggest that between 75 and 90 percent of all the arsenic used in the United States is used for wood preservation; the rest is used for rat poison. Virtually all lumber used in outdoor construction is pressure-treated, with the majority of lumber treated with CCA. Once wood is treated, it is not recyclable. In some cases, the wood is considered toxic waste. Manufacturing of CCA lumber was phased out in the end of 2003. However, CCA lumber may still be sold after that date.

♡ – Limit Use of Unsafe Treatments

- A healthier alternative to chromated copper arsenate (CCA) wood treatment is alkaline copper quaternary (ACQ). ACQ is applied just like CCA, but it is made of 100 percent recycled copper as its main component and contains no harmful chromium or arsenic ingredients. Other less toxic chemical treatments include borates, ACA (ammoniacal copper arsenate), ACZA (ammoniacal copper zinc arsenate), ACC (acid copper chromate), and CZC (chromated zinc chloride).

- **Recommendation**: If you decide to use treated wood, use alternatives to CCA, such as ACQ. To optimize durability, choose treated wood with a water repellent incorporated into the wood. Periodically apply a sealer to protect the wood against UV degradation and surface checking. Although not widely used, borate wood treatment is also an acceptable alternative for all types of wood — engineered, sheathing, or dimensional — in situations where the wood will not likely get wet. It is non-toxic for handling, cutting, and disposal. Additionally, borates are effective against fungus, preventive against insects, and suitable for brush application, spray application, or dipping. The lesser known water-borne preservatives (ACA, ACZA, ACC, CZC) can be used for hard-to-penetrate woods.

Exterior Finish Checklist

Siding

☐ Install drainage planes.

☐ Use alternatives to wood.

☐ Use fiber-cement exterior siding.

☐ Use FSC-certified solid wood siding.

☐ Use finger-jointed trim.

☐ Install window and door flashing.

Decking

☐ Build patios instead of decks.

☐ Use recycled-content decking.

☐ Use FSC-certified wood decking.

Wood Treatments

☐ Use naturally decay-resistant wood.

☐ Limit use of unsafe treatments.

CHAPTER 11

Roofing

THE TYPE OF ROOFING YOU CHOOSE can have a serious impact on your health and on the environment. Sloped roofing materials, such as asphalt-based rolled roofing and shingles, will offgas toxins when heated by the sun. Flat roofing materials, such as tar and gravel, will also continually offgas when heated by the sun, emitting known carcinogens such as VOCs from asphalt, including benzene, polynuclear aromatics, toluene, and xylene. Although roofing materials are located outside the living space, odors can enter the home through doors, windows, and vents. Since most people are not in a position to move out for several weeks when their roof requires replacement, they will be exposed to high levels of toxic fumes every time the roof is replaced or repaired.[1]

Seventy-eight-percent of total annual roofing dollars spent in the US is spent on re-roofing.[2] According to some studies, re-roofing can be necessary for common roofing types after only 12 years,[3] especially in higher altitude climates, which have greater exposure to ultraviolet rays. This is not only an expensive hassle, but it also sends used roofing materials to landfills, where the polluting contents continue to offgas and leach into the soil and groundwater.

Unfortunately, re-roofing your house is often a necessity. You'll know when you need a new one: the roof leaks, the shingles start to curl, or icicles will hang from the roof. Even if you're lucky enough not to have these problems, a new roof with good air sealing, insulation, and ventilation will save energy and make your house more comfortable. Many unhealthy and persistent mold and mildew infestations begin with an undetected roof leak. No type of roofing installation is foolproof, but the use of high quality roofing materials and skilled installers will reduce the risk of leakage. Moreover, roofing replacements give you a chance to work on areas of the house that are usually impossible to get to. Here are some green alternatives to traditional roofing that help make re-roofing your home a healthy long-term investment for you and our environment:

$ – Consider the System as a Whole

- In new construction, choice of roofing materials should be integral with other decisions about the building. The roof is your main defense against water leakage in your house. Good roofing is often degraded when budgets are tight and compromises are made. You want a roof that will last at least as long as your mortgage because long-lasting roofing protects your home and your investment.
- **Recommendation**: Early in the decision-making process, figure out your roofing and how it will integrate with the rest of the house. Make sure the roof framing structure is designed to bear the weight of the product you choose.

$ – Calculate How Much Roofing You'll Need

- Roofs are measured in 100-square-foot areas, or "squares." Three bundles of three-tab shingles typically equal one square; laminates come in four bundles per square. It's a good idea to calculate beforehand how much roofing material you'll need so that you'll have an easier time comparing bids from contractors.
- **Recommendation**: To gauge how much roofing material you'll need, multiply the overall length and width of each roof section to determine its area. Add ten percent to allow for waste, then divide by 100 to determine how many squares you'll need. If the roof is new or you're having the old shingles removed, you'll need an underlayment (roofing felt) to create a moisture barrier for the wood sheathing and rafters underneath. The sheathing may have to be replaced in severely damaged areas. You may also have to install an "ice-and-water shield" along the eaves and the valleys where two wings of the roof intersect. New drip edges and metal flashing are often needed around pipes, chimneys, and the like.

♡ – Avoid Adhesives

- Adhesives used in roofing applications can emit harmful VOCs. Solvent-based adhesives affect the respiratory and central nervous systems, as well as organs like your liver and kidney. They are especially toxic during application and curing (drying) periods. Water-based adhesives are better because they only release water vapor as they dry, but they are not necessarily 100 percent safe. Other components in water-based adhesives, such as resins, biocides, other solvents, and even some natural ingredients, can be irritants or toxins.[4] Mechanical fastening eliminates the need for VOC-offgassing adhesive materials. In addition, the membrane can be easily removed and recycled when it fails or when the roof needs to be modified.[5]

- **Recommendation**: Use mechanical fastening, a range of processes that utilizes a variety of fasteners including nails, nuts and bolts, screws, and rivets, to assemble materials without heating or adhesives. If mechanical fastening is unavailable, use water-based adhesives.

⌒ – Provide a Light-colored or Reflective Roof

- Dark roofing materials absorb heat and make the house warmer in the summer, whereas light-colored roofing reflects heat away from the building. Higher solar reflectance can be achieved with lighter colored roofing materials such as shingles, tiles, and white-reflective membranes and coatings on flat roofs. Unless the building has a highly insulated roof, it is generally advisable to provide a reflective roof surface. Reflective roofs help reduce "the heat island effect," a phenomena in which heat-absorbing buildings can increase the outside air temperature in urban areas by two to eight degrees Fahrenheit. Light-colored roofing also reduces heat buildup through the roof. As a result, if you have air conditioning ducts running through the attic, the ducts will stay cooler. Considering that air conditioning is the most energy-intensive appliance in the house, the energy savings are significant! Light-colored roofing can also last longer because it does not thermally expand and contract as much as darker colors.

- **Recommendation**: Use a roofing material with as high a reflectance as possible.

⌒ – Install Radiant Barriers

- Reflect heat away from your home by installing a radiant barrier on the underside of your roof. A radiant barrier is simply a sheet of aluminum foil with paper backing, or metalized mylar sheet material. When installed correctly, a radiant barrier can reduce heat gains through your ceiling by about 95 percent.[6] They are particularly helpful if you have air conditioning ducts running through your attic, because the lower attic temperature keeps the ducts cooler. In fact, some homeowners save more from cooler ducts than from the benefits of keeping the ceiling cool.

Table 11.1: Reflectance of Roof Materials

Material	Solar Reflectance (%)	Roof Temperature over Air Temperature (°F)
Bright, white coating (e.g., ceramic) on smooth surface	80%	15°F
White membrane	70–80	15–35
White metal	60–70	25–36
Bright white coating (e.g., ceramic) on rough surface	60	36
Bright aluminum coating	55	51
Premium white shingle	35	60
Generic white shingle	25	70
Light brown/gray shingle	20	75
Dark red tile	18–33	62–77
Dark shingle	8–19	76–87
Black shingles or material	5	90

Source: Home Energy Magazine. *No Regrets Remodeling. Energy Auditor and Retrofitter, 1997. p. 202.*

- **Recommendation**: Install radiant barriers by stapling them to the attic rafters — do not staple them to attic floor joists where dust collects on them more quickly. Alternatively, use laminated foil-backed OSB for new roof construction, reflective side down.

♁, ♤ – Use Recycled-content Asphalt Shingles

- Asphalt shingles are among the most disposed-of building materials. Recycled-content asphalt shingles, on the other hand, contain recycled waste paper and/or use reclaimed material slag in their aggregate surface, thereby reducing the waste of raw materials during the roofing process.
- **Recommendation**: Install recycled-content asphalt shingles rather than traditional asphalt shingles.

⚘ – Install Slate Roofing

- Slate is minimally processed cut or split rock. It is expensive, but it creates a very distinctive look and is incredibly durable. Not only does it have a good fire rating, but properly installed slate roofs can last 100 years or more with only minor maintenance. Slate is considered a safe material that can be easily reclaimed and reused on new building projects.

- **Recommendation**: Consider slate roofing if it is quarried near your home. Adequately strong roof structures must be built to withstand this heavy material. Keep in mind that slate requires skill to install and repair.

⚘ – Install Concrete Tile Roofing

- Concrete tile is a waterproof material that may be coated or glazed for special effects. Like all concrete products, concrete tiles contain energy-intensive Portland cement. However, we support concrete tiles as a good green alternative to asphalt shingles because of their durability, low-to-no toxicity, easy maintenance, and excellent fire rating.

- **Recommendation**: Consider concrete tile if it is manufactured in your region, since it is heavy and difficult to transport. Keep in mind that larger structural systems will be required to support its extreme weight. Concrete tiles depend on rapid water runoff and will not hold up under the standing water conditions common with low-sloped roofs. Also note that cement roofing may shatter in high hail-prone areas.

⚘, ♡ – Install Clay Roofing

- Clay tiles are made from slabs of clay and lined with a glazed or unglazed finish. This natural, traditional roofing material comes in attractive colors and textures that improve with age. Although clay is more expensive than many roofing materials, it is a durable, fireproof material with no negative health effects. New clay roof tiles can be made to match old tiles, making them the ideal choice for refurbishment as well as new roofing in areas where it is architecturally appropriate.

- **Recommendation**: Consider using clay if it is quarried in your region. If you are installing clay tiles in cold climates, note that low-water-absorption ratings minimize the potential for freeze-thaw damage. Clay tiles are heavy, therefore some rafters may need to be reinforced. Keep in mind that clay roofing may shatter in high hail-prone areas.

♢, ⚐ – Install Lead-free Metal Roofing

- Use metal roofing made from aluminum, copper, or steel. The components come in many sizes and shapes, including panels, shingles, shakes, and tiles. Some products contain up to 100 percent recycled material, and most products can easily be recycled. Metal roofing is easy to install, and is fireproof, lightweight, long-lasting and recyclable. It is available in a wide range of shapes, colors and patterns. Metal is the most favorable roofing material used in rainwater catchment systems. In northern climates, snow readily slides off metal roofs, thereby avoiding the damage caused by ice dams. Unlike all other roofing materials, metal roofs also provide added shear value or rigidity to the roof. Metal roofs do not radiate as much unwanted heat into the attic as asphalt shingles, because the metal is thin and does not have heat holding capacity. Using a white painted or galvanized finish will further reflect heat away from the roof and attic.

Preventing ice dams.

As snow melts and pools behind the ice, water may seep between shingles and leak into the house. But don't blame the roof if harmful ice dams form at the eaves! To prevent ice dams, you need adequate insulation in the walls and the ceiling below the attic to help contain the heat that causes the snow to melt in the first place. Seal holes, cracks, and crevices in the walls to keep warm air from rising to the attic. Also make sure the attic has adequate ventilation to draw warm air to the outside.

Source: "Home Improvement," Consumer Reports, August 2003, pp. 18 to 32, 38 to 46.

- **Recommendation**: Use high-quality unpainted coatings on metal roofs that are not subject to extreme moisture and corrosive salt air. Avoid roofing where lead is used in the alloy or coating. (See "Lead" in Chapter 5.) Use of different metals for roofing, flashing, and fastening is not recommended; in the presence of water, different metals in contact with each other are susceptible to galvanic corrosion. Try to use roofing profiles (fastening mechanisms) with hidden screws or clips to avoid penetrating the metal on the surface and inviting future leaks. Keep in mind that metal roofs can be somewhat loud during rainstorms, therefore occupants should be comfortable with this form of "white noise."

♨ – Install Fiber-cement Composite Roofing

- Fiber-cement is made of Portland cement, sand, clay, and wood fiber. The product typically carries a 50 year warranty. Fiber-cement composite roofing is durable, fireproof and recyclable.
- **Recommendations**: This roofing can be readily used on standard roof structures. Fiber-cement composite slates or shakes are not recommended in northern regions or at higher altitudes because they do not perform well in freeze-thaw climates or in hail-prone areas.

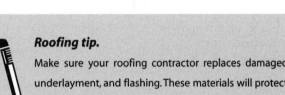

Fig 11.1: Fiber-cement roofing will outlast your mortgage. Credit: David Johnston.

♨ – Install Recycled-content Plastic/Rubber Shingles

- Plastic shingles are made from recycled materials like industrial rubber, used tires, and plastic — and the shingles themselves can be recycled at the end of their life. Although the long-term effects of UV light and the expansion and contraction of plastic materials remain unknown, many of these products have a 50-year warranty. They are also sound-absorbant and hail-proof.
- **Recommendation**: Use plastic shingles in place of traditional asphalt shingles. Some varieties can be installed just like clay or slate tiles; other types look like rough cedar shakes.

> **Roofing tip.**
> Make sure your roofing contractor replaces damaged sheathing, underlayment, and flashing. These materials will protect your home even more than the shingles. Also, be on the lookout for mold on the roof sheathing.

♀, ♨ – Install a Steep Slope Roof with Sizable Overhang

- Water will puddle and linger on poorly constructed flat roofing. As a result, flat roofs have a high leakage rate that may lead to devastating mold problems. A steep slope roof has a "pitched" roof line that prevents water from puddling on your roof. The roof overhang plays an important role in protecting the walls and foundations from water damage by directing water away from your home. Overhangs can be sized to suit the solar conditions in your region, providing shade in the summer while allowing maximum heat entry in the winter. Non-toxic roof materials are readily available and are standard products for sloped roof construction, whereas they are an exception in flat or low-sloped residential roof construction.

Fig. 11.2: Be sure to plant drought-tolerant local species on your roof, suited to your climate. Credit: Larry Master.

- **Recommendation**: Install a roof with a minimum 3/12 pitch. This fraction refers to the rise/run of the roof.

$, 🌲 – Harvest Rainwater

- Whenever it rains, naturally distilled water falls on rooftops, is guided into gutters, and is quickly sent into sewers to be combined with human and industrial waste and "taken away" at great expense. Without gutters it will fall off the roof, damage your walls and foundation, wash away your landscape and soil, and possibly run onto the street and cause flooding. However, rainwater can be diverted for use onsite before it is lost along with domestic waste. Rainwater is soft and pure and requires no treatment; rainwater recovery works well even in drought-prone areas and can often reduce your water bills. In addition, water-supply and storm water drains become unnecessary. In fire-prone areas, you might even position big containers at a height that can gravity-feed a hose, thereby reducing your fire insurance premiums.[7]
- **Recommendation**: Install cistern tanks that collect rainwater. This pure water can be used for landscape watering. You might also install simple rainwater "buckets" to collect rainwater from gutters for watering.

🌲 – Consider Green Roofs

- Green roofs, as the name implies, are planted with vegetation. Also known as "living roofs," these are protected-membrane roofs with soil and plantings (as well as insulation) installed above the membrane. These systems are encouraged and even subsidized in Europe because they reduce flooding risks and cooling needs.[8] They can detain over half the rainwater from a typical storm, reducing the often-high loads placed on sewer systems after a rainfall.[9] In addition, a green roof can be a wonderful architectural element that absorbs carbon dioxide (see "Global Climate Change" in Chapter 5), and helps to reduce building heat gain and urban heat islands.

> **Heat island effect.**
> A phenomena in which heat-absorbing buildings can increase the outside air temperature in urban areas by 2 to 8 degrees Fahrenheit (1 to 4 degrees Celsius).

- **Recommendation**: Install a green "living" roof in place of a traditional low-slope or flat roof. A green roof includes insulation, drainage, geotextile, soil, and vegetation layers. These multilayered green roof systems are thicker and heavier than conventional roofs and are often used for gardens, therefore the roof structure needs to be engineered to accommodate the increased weight of the roof and people walking on it.

⚘ – Install 40- to 50-year Composition Roofing

- Composition asphalt/fiberglass roof shingles typically come in 15- to 50-year life spans. Although the 15-year warranty roofs tend to be less expensive, they will also have to be replaced more often. This costs more in the long-run and adds to manufacturing and landfill pollution.
- **Recommendations**: Use 50-year composition roofing anytime your roof needs to be replaced.

$, ⭗, ⚘ – Consider PV Shingles and Roofing Tiles

- These new roofing systems are coated with a film that converts sunlight into electricity. The shingles or tiles snap together, and the current flows at the edge of the roof. The shingles look a lot like traditional roofing, and any good roofer can install them. After the roofer installs the tiles, an electrician connects the roof system with the electrical system. Each 100 square feet of PV roof generates about one kilowatt of electricity. PV can be used as a means to decrease reliance on conventional power plants that contribute to air pollution. PV can be cost effective in areas that require night lighting such as outdoor lights. (See "Photovoltaic (PV) Energy" in Chapter 15.)
- **Recommendation**: Use PV shingles instead of traditional roofing. Be sure that the roof is oriented to the south or southwest. The components for a utility-tied system typically include panels, a power relay center, and an inverter. An alternative installation would be self-contained systems (battery included) for outside lighting, security lighting, or walkway illumination.

Fig. 11.3: Special roof shingles can convert sunlight into electricity. Credit: Innovative Power Systems.

Roofing Checklist

☐ Consider the system as a whole.

☐ Calculate how much roofing you'll need.

☐ Avoid adhesives.

☐ Provide a light-colored or reflective roof.

☐ Install radiant barriers.

☐ Use recycled-content asphalt shingles.

☐ Install slate roofing.

☐ Install concrete tile roofing.

☐ Install clay roofing.

☐ Install lead-free metal roofing.

☐ Install fiber-cement composite roofing.

☐ Install recycled-content plastic/rubber shingles.

☐ Install a steep slope roof with sizable overhang.

☐ Harvest rainwater.

☐ Consider green roofs.

☐ Install 40–50-year composition roofing.

☐ Consider PV shingles and roofing tiles.

CHAPTER 12

Plumbing

Installing and Improving Your Plumbing System

I T's ALWAYS A GOOD IDEA TO IMPROVE YOUR PLUMBING and make your home more efficient. However, if you already have other home renovations planned that involve opening up walls, this might be an especially cost-effective time to insulate hot water pipes. With the exception of solar hot water pre-plumbing, our ideas in this section may not stand out as "green"— they are simply good building practices. Keep in mind green building *is* good building!

♫ – Insulate Pipes

- A good water heater won't do you much good if the water gets cold in the pipes. Uninsulated pipes — especially in exterior walls that are more exposed to cold outdoor temperatures — waste significant amounts of energy. Insulating your hot water pipes will reduce heat losses as the hot water flows to your faucet. It will also reduce standby losses, or energy lost through the walls and up the flue of the hot water tank.

- **Recommendation**: Any time is good to insulate pipes, but it's easiest to insulate when you are changing the plumbing, since pipes affixed to, or inside, walls and ceilings are harder to wrap with insulation. Insulate hot water pipes in all runs through unconditioned spaces, including basements, crawlspaces, and attics. Insulation is available in open cell foam tubes (R-3 to R-5), rigid foam (R-7) and fiberglass batts (R-2 to R-3). Since rigid foam has the highest R-value, or insulating value, it is typically the best option if are looking to save the most energy. Foam or fiberglass pipe sheaths usually come with a lengthwise slit, so you can just snap them around the pipe. Tape the ends to maintain a tight fit around the pipe.

℘ – Remove Plumbing from Outside Walls

- Pipes in exterior walls will cool more quickly than pipes in the interior walls, which tend to be significantly warmer. Pipes in interior walls have the advantage of the home's interior heat to keep them warm; this is an obvious energy-saving advantage over pipes in exterior walls that are more exposed to colder outdoor temperatures.

- **Recommendation**: Remove pipes from outer walls and install them in interior walls. This can be costly, but makes sense if you are in the process of tearing down walls anyway.

℘, 🌲 – Install On-demand Hot Water Circulation Pump

- An "on-demand" hot water circulation pump can send hot water to fixtures in seconds without wasting water while waiting for it to get hot. It uses a pump to rapidly move water from a water heater to fixtures, and stops when the water reaches a preset temperature. Note that this differs from a recirculating or continuous system — a more wasteful option that keeps hot water constantly circulating in the pipes.

- **Recommendation**: Install the pump at the furthest faucet from the water heater. Only one pump is needed to supply hot water to all the fixtures, and it can be easily installed without a plumber.

🌲 – Install Automated Home Leak-monitoring Techniques

- One tenth of US household water consumption is lost through leaks from toilet valves, dripping faucets, and aging pipes. Toilets are the biggest offenders, often wasting as much as 750 gallons a month; a leaky faucet can waste as much as 300 gallons per month.[1] Moreover, leaks can cause moisture build up and mold (see "Moisture and Mold" in Chapter 5). By eliminating water leakage, you save the cost of wasted water. In addition, efficiency enables individual septic systems to work better. One 12-house survey found that saving a quarter to half of household water use greatly reduced septic system malfunctions, made their treatment more effective, and resulted in lower long-term operating costs.[2]

- **Recommendation**: Install an automated leak-monitoring system that can often be integrated with cost-saving automatic meter-reading. Acute leaks trigger alarms to customers, utilities, or plumbers. Look online under "residential water leak detection" to find companies that offer these services.

$, ⚶ – Install Solar Hot Water Pre-plumbing

- Insulated copper pipes are installed from the attic to a hot water closet or mechanical room for future solar installation. This option allows an active solar system to be installed at a later date if desired.
- **Recommendation**: Provide south-facing roof area for the collectors. Also provide access for piping to connect to your mechanical room. The most cost-effective time to pre-plumb for solar is when you are remodeling interior spaces.

Toilets

A typical US single-family home uses about 70 gallons per person per day indoors,[3] 40 percent of which is used up by toilet flushing. It is the largest single use of water in the home:[4] one flush of a standard US toilet requires more water than most individuals, and many families, in the world use for all their needs in an entire day.[5] Here are some ways you can reduce the amount of water your toilet wastes.

$, ⚶ – Install Low-flush Toilets

- Older toilets waste 3 to 5 gallons per flush (gpf). US federal law now mandates that all new toilets use no more than 1.6 gpf. This ambitious standard has necessitated sophisticated redesign of toilet bowls, flushing systems, and tanks. Some compact models use a pressure system to achieve an effective flush; a larger water surface also helps. Some toilets have a dual-flush system with an additional 0.8 to 1.6 gallon low flush. Toilet designs continue to evolve, but many of the earlier problems with low-flush toilets (which often led to double flushing to make up for poor flushing performance), have been resolved.
- **Recommendation**: Replace older toilets with newer, low-flush models.

⚶ – Install a Composting Toilet System

- Rather than mixing waste with potable water and flushing it down the drain, as conventional toilets do, composting toilets convert human waste into nutrient-rich fertilizer for non-food plants. Composting toilets result in dramatic reductions in water use, reduced groundwater pollution or sewage treatment impacts, and the recycling of nutrients. Some composting

Fig 12.1: In this model composting system designed for families, waste from two toilets on the upper floor flows into the downstairs composting drum. Compost will need to be removed only once a year. Credit: Sun-Mar.

chambers can be used with microflush toilets, but most are non-flush. This option is most applicable in areas where septic systems are very expensive to install. Greywater treatment and disposal still must be addressed.

- **Recommendation**: Consider installing a composting toilet system to cut down on water use.

🌲 – Consider Greywater Flushing

- Greywater is waste water from sinks, showers, and washing machines that is not contaminated by human waste. It is a ubiquitous but normally wasted water resource. However, it can be used for toilet flushing or subsurface irrigation. Typical recovery and reuse rates average about 50 gallons per house per day, cutting total water use about in half, and saving even more in multifamily homes where greywater is used to flush toilets.[6]
- **Recommendation**: Check with your local building department for requirements; in many towns this may not be acceptable by local plumbing codes. Greywater plumbing separates the waste pipes from sinks, showers, and washing machines from the toilet waste. The greywater drains are then run to a holding tank similar to a septic tank, which in turn is used to water plants, lawns, and gardens.

Plumbing Fixtures

Water is drawn from groundwater and surface water sources in the US at unsustainable rates. In the US, we currently withdraw 340 billion gallons of fresh water per day from streams, reservoirs, and wells.[7] Withdrawals from aquifers sometimes exceed annual recharge rates: almost 14 million acre-feet of water are taken out of the Ogallala aquifer a year and not replaced — as much water as flows in the Colorado River in a "good" year![8] The following plumbing recommendations significantly improve water-use efficiency.

🌲 – Install High-performance Showerheads

- In the U.S., showers account for 18 percent of indoor water use and 37 percent of a home's hot water use.[9] A family of four each showering five minutes a day will use about 700 gallons per week — a three-year drinking supply for one person in the US[10] Since 1992, the legal maximum for new showerheads has been 2.5 gallons per minute (gpm), but this lowered figure still reflects inefficient water use. A high-performance showerhead uses 1–1.5 gpm — up to 60 percent less water than a traditional showerhead[11] — and will pay for itself in mere months from water-heating energy savings alone. Some showerheads have only a single orifice made of slippery plastic that prevents them from clogging in "hard" water that has a high mineral content (especially calcium), while features such as faucet aerators allow you to enjoy the pressure of a standard-flow showerhead without the excessive water use.[12]

- **Recommendation**: Use a high performance showerhead in place of a traditional showerhead. For those who enjoy pressure-pounding showers, "faucet aerator" varieties exist with mixing chambers that emit a powerfully wetting and massaging combination of air and water. Some showerheads even include clip-on taximeters that measure water flow, temperature, and dollars used during the shower.

$, ⛉ – Install Faucet Flow Reducers

- Flow reducers fit into the aerator at the tip of the faucet and reduce the rate of water flow through the faucet by as much as 40 percent with little noticeable effect.[13]
- **Recommendation**: Use flow reducers on all faucets.

$, ⛉, ♡ – Install a Pedal Faucet Controller

- A pedal faucet controller allows you set the temperature with your hands and then step on pedals to start the water flow. When the pedal is released, the flow stops, allowing for easier control of water flow. (If you need to leave the water running, you engage a lock-on button at the front of the pedal and the faucet works normally.) Since this is operated using your feet only, you reduce the risk of passing on germs by touching faucets with unwashed hands. In a test performed by Pedal Valves, Inc. at a medical center kitchen, pedal valves installed at three sinks saved a combined 285 gallons per day of hot water and 122 gallons per day cold water. Based on an installed cost of $500, payback is around 16 months, or several years in residential settings.[14]
- **Recommendation**: Install pedal faucet controllers in place of a traditional faucets.

Fig 12.2: Choose water filters over bottled water.

Credit: Paragon Water Supplies.

Water Filtration

Before you buy a water filter, find out what pollutants you need to remove from your water. Each water purification method removes different pollutants, so you need to buy the right equipment for your water supply. Start by calling city hall, your local water district office, or local department of health to find out what's in your water. Considering the current condition of our water supplies (see "Water Resources" in Chapter 5), there's probably a lot of harmful contents you want to get rid of!

Evaluating the myriad filtration products out there is a nightmare, so we've tried to make it simple by explaining what each filter does and highlighting the more popular carbon filters, distillers, and reverse osmosis systems as bulleted green alternatives. You may not be able to find what you want at the hardware store; you'll probably have to order what you need by looking online or by asking people at your natural food store if they know of local dealers. Look for an NSF International Seal (formerly the National Sanitation Foundation, <www.nsf.org>). NSF standards actually exceed EPA standards, which allows you to purchase their water filtration products with confidence.

♡ – Investigate Your Water Supply

- Find out if there's a problem with your water supply by obtaining a copy of your water quality report, which water utilities are required to send to customers annually. This report will help you determine what type of water filter to buy, if indeed you need one. If your water comes from a private well, the EPA recommends that well owners test their water annually for nitrate and coliform bacteria, and more frequently for radon or pesticides, if those are a problem in your area.
- **Recommendation:** To get a copy of your water utility report, request it from your local water supplier or from the EPA's Safe Drinking Water Hotline 1-800-426-4791 or online, <www.epa.gov/safewater/dwinfo. htm>. You can check with your local health department to find out which contaminants are common where you live. National Testing Labs can do the testing by mail. Check online, <www.watercheck.com>, or call 1-800-458-3330.

Table 12.1: Water Filters

Type of Filter	What It Does
Carbon filter	Carbon-activated filters adsorb lead, chlorine byproducts, and some organic chemicals, as well as odor and tastes. They won't remove heavy metals, pesticides, nitrites, bacteria or microbes, but they are the least expensive filter type and are sufficient for most needs.
Ceramic filter	Combined with carbon filters, these will remove bacteria, cysts, asbestos, and sediments.
Distillers	Distillers boil water into steam, then condense it back into water in a separate chamber, leaving behind particles and dissolved solids. Since water is heated, distillers kill microbes. They eliminate many other pollutants, including arsenic, but not VOCs and chlorine, which are usually removed by an accompanying carbon filter.
Copper/zinc alloy systems (or KDF)	Installed in shower heads, these remove chlorine, chloroform, and heavy metals, reducing contaminants that can be inhaled and absorbed through the skin in water and steam.
Reverse osmosis	Typically expensive and difficult to install. Membrane acts like an extremely fine filter to eliminate industrial chemicals, heavy metals, nitrates and asbestos (but not chlorine byproducts, radon or certain pesticides). The contaminated water is put on one side of the membrane and pressure is applied to stop, and then reverse, the osmotic process. These systems waste enormous amounts of water and flush contaminants back into the water supply. It generally take a lot of pressure and is fairly slow, but it works.
Ultraviolet light purifiers	Kills some bacteria, but is intended as supplemental to filtration — not a way to eliminate illness-causing pathogens.

Source: P.W. McRandle. Water Filters [online]. [cited July 25, 2003]. Green Guide, 2003.
<www.thegreenguide.com/reports/product.mhtml?id=23>.

♡ – **Install Chlorine Filters on Shower Heads**

• Chlorine is absorbed six times faster through the skin than through the digestive system. It has been shown that chlorine absorption can have

Filters are better than bottled water!

According to the World Wildlife Fund (WWF) there are more standards regulating tap water quality than that of bottled water in the US. Clever marketing in industrialized countries has convinced many consumers that bottled water is healthier, but the WWF study could not find any support for this claim. In fact, most bottled water is tap water. Filtering or boiling drinking water would eliminate the need for fleets of trucks that distribute bottled water while creating traffic congestion and air pollution. Moreover, we would save landfill space by keeping millions of plastic bottles out of our waste stream.

Source: Lester R. Brown, Eco-Economy, 2001, p. 142.

adverse health effects on some people, and especially on children.[15] Water filters on showerheads reduce chemicals and particulates from the water stream. Showerhead filters are especially beneficial to people with sensitive skin or eczema.

- **Recommendations**: Install the water filter between the pipe and the existing showerhead.

♡ – Install a Whole-house Water Filtration System

- Whole-house filtration systems eliminate chlorine and many other chemicals, particulates, and micro-organisms. This type of filtration is a good idea if you have well water; sediment buildup from wells has a tendency to decrease the life span of hot water heaters and plumbing fixtures.
- **Recommendation**: Install a whole-house filtration system between the cold water line and the main drinking water faucets in the house. Do not run hose bibs and toilet lines through a water filter.

– Install Activated Carbon Filters

- Activated carbon is the least expensive water purification method: it absorbs lead, chlorine byproducts, and some organic chemicals, as well as odors and tastes. Carbon filters work by absorbing pollutant molecules in their honeycomb-like pores. Until the filter becomes saturated, it can remove volatile compounds, or nonparticulate substances that evaporate. Volatile compounds include chlorine and pesticides like DDT. (Refer to Chapter 5 to learn the risks associated with volatile compounds.)
- **Recommendation**: Install carbon filters on all faucets, including showerheads, in your home. Carbon blocks, rather than granules, are the preferred filter — the carbon lasts longer because there is more of it, and the compressed design inhibits bacterial growth. Carbon filters can

become saturated, and ineffective very easily; you need to be aware of when to change it. Changing filters also helps prevent bacteria growth.

Water Filter Alternatives for Polluted or Brackish Areas

♡ – Install Water Distillers

- Water distillers work by boiling water into steam, and then condensing it back into "pure" water in a separate chamber, leaving behind particles and dissolved solids. Distillers kill microbes, and also eliminate many other pollutants, including arsenic.
- **Recommendation**: A distiller is not completely automatic, although it's not a big deal to operate: you have to turn on the power supply and hook up hoses to the water source. For example, you can have the distiller mounted on your sink and simply snap the water hose to the faucet.

♡ – Install a Reverse Osmosis System

- In reverse osmosis, a membrane acts like an extremely fine filter to create drinkable water from contaminated water. The contaminated water is put on one side of the membrane and pressure is applied to stop, and then reverse, the osmotic process. The reverse osmosis system is typically expensive and difficult to install; it also flushes away a few gallons of contaminant-containing water for every gallon purified. However, these systems remove industrial chemicals, heavy metals, nitrates, and asbestos. An additional carbon filter may help remove chlorine byproducts, radon, or certain pesticides which the reverse osmosis system does not eliminate.
- **Recommendation**: A reverse osmosis unit typically sits under the sink: simply flip the lever on a specially installed tap, and clean water comes out. The membrane, the carbon filter, and the special particle pre-filter must be changed periodically.

Plumbing Checklist

Installing and Improving Your Plumbing System

Insulate Pipes

- ☐ Remove plumbing from outside walls.
- ☐ Install on-demand hot water circulation pump.
- ☐ Install automated leak-monitoring techniques.
- ☐ Install solar hot water pre-plumbing.

Toilets

- ☐ Install low-flush toilets.
- ☐ Install a composting toilet system.
- ☐ Consider greywater flushing.

Plumbing Fixtures

- ☐ Install high-performance showerheads.
- ☐ Install faucet flow reducers.
- ☐ Install a pedal faucet controller.

Water Filtration

- ☐ Investigate your water supply.
- ☐ Install chlorine filters on shower heads.
- ☐ Install a whole house water filtration system.
- ☐ Install activated carbon filters.

Expensive Filter Alternatives for Polluted or Brackish Areas

- ☐ Install water distillers.
- ☐ Install a reverse osmosis system.

CHAPTER 13

Electrical

U S CITIZENS SPEND ONE QUARTER OF THEIR ELECTRICITY BUDGET on lighting, which equates to more than $37 billion yearly.[1] A typical household spends about $90 per year, or 10 to 15 percent of its annual electricity bill, on lighting, most of which is wasted on obsolete equipment, inadequate maintenance, or inefficient use.[2]

Wasting electricity has global and personal health consequences. In burning fossil fuels such as coal to supply electricity to homes and workplaces, power plants discharge clouds of soot and other pollutants into the atmosphere. This includes mercury, a brain-damaging metal that can cause learning disabilities, and carbon dioxide (CO_2), the greenhouse gas primarily responsible for global climate change. Power plants release an average of 1.34 pounds of CO_2 into the environment for every kilowatt-hour of electricity used in a home or elsewhere.[3] Furthermore, electricity generated from nuclear power plants generates radioactive waste.[4]

The following energy-efficient lighting recommendations will help you avoid unnecessary energy costs and the indirect toll of energy use on the environment.

Daylighting

Daylighting involves the integration of daylight with electric lighting, overall building design, mechanical systems, and interior design. Daylighting is an ideal way to increase comfort and aesthetics in a home while reducing energy consumption and improving your family's sense of well-being. According to ESource, Inc., 40 to 60 percent of lighting energy savings can be achieved using daylighting strategies.[5] Common examples include light pipes, skylights, clerestory windows, and light shelves.

Case Study: The Advantages of Daylighting

The Rocky Mountain Institute (RMI) profiled eight buildings in 1994, showing that productivity improved from a few percent to more than 15 percent after daylighting was implemented. And studies commissioned by Pacific Gas and Electric in California found with 99.9 percent statistical certainty that daylighting can have real, measurable benefits on learning rates: students in classrooms with the most daylight had a 20 to 26 percent faster learning rate than students in classrooms with artificial light. Most importantly, we just feel better with natural light in our homes.

Some speculate that the productivity and learning benefits occur because better light improves one's mood and morale. It is also thought to improve health, behavior, and arousal levels (through suppression of melatonin). We know, for example, that there is a higher prevalence of Seasonal Affective Disorders (a depressive condition that improves upon exposure to natural light) in areas of the world where there is little natural light. Regardless of the biological processes involved, natural lighting has proven energy, productivity, and learning benefits.

Fig 13.1a & b: Before and after photos of a tubular skylight installation.
Credit: Tubular Skylight.

🔑,♡ – Install Light Pipes

• Light pipes, or solar tubes, provide natural light for the home's interior or in areas where other issues, such as privacy in a bathroom, restrict the use of traditional glazing. High quality units tend to lose little heat in the winter and gain little heat in the summer, due to insulation. The light output can be significant: a 13-inch diameter tube can light about 200 square feet of space.

• **Recommendation**: Light pipes run from the ceiling, through the attic space to the roof, where skylight placement is not always feasible or too expensive. Performance of light pipes can be improved by constructing the tubes of insulating material or by wrapping them with insulation.

🔑,♡ – Install Skylights

• Skylights are typically wide and rectangular in shape. They are installed on rooftops and are intended to bring light into the home. Although not as energy-efficient as light pipes, traditional skylights still eliminate the need for electric lighting during daylight hours and improve the mood and general well-being of home occupants.

• **Recommendation**: Never install single-glazed skylights; they lose significant amounts of energy through the glass in colder climates. In warm climates, skylights should be used sparingly, as they tend to heat up the home, creating more of a problem than the daylighting they offer is worth.

🔵,♡ – Install Clerestory Windows

- Clerestories are horizontal, narrow windows set high into walls, just below the eaves, or into cathedral ceilings. They permit low winter sun to enter, illuminating and heating much of the house. These windows are wide rather than tall, therefore they need only a small overhang to block summer heat gain. Additionally, clerestory windows encourage natural ventilation, as hot air rises to escape the house at the top.
- **Recommendation**: In the summer, clerestories should be open to let the hottest air flow out. To keep winter heat in, however, these windows should be made of well-insulated glass. Design into south-facing roofs in cold climates. Interior ceilings, walls, and other surfaces should be as light-colored as possible to reflect and scatter daylight throughout the space.

🔵,♡ – Install Light Shelves

- Light shelves help channel daylight more deeply into your home. The Lawrence Berkeley National Laboratory claims that well-designed light shelves can extend the perimeter daylight zone up to 2.5 times the window height, thereby reducing the need for electric lighting and improving the health of your family with natural daylighting.
- **Recommendation**: Install on south-facing windows. The surfaces should be highly reflective, such as white or silver. The bottom of the light shelf should be placed above head height. Configure full-height interior partitions, book shelves, and files so that they do not block daylight penetration.

Fig 13.2: In this diagram, light enters the building through the clerestory ("clear-story") windows, or windows specifically designed into a part of a building rising above the roofs or other parts. Light easily bounces off walls opposite the clerestory windows, thereby providing natural light to the building interior.
Credit: Jill Haras & Kim Master, redrawn by Jeremy Drought.

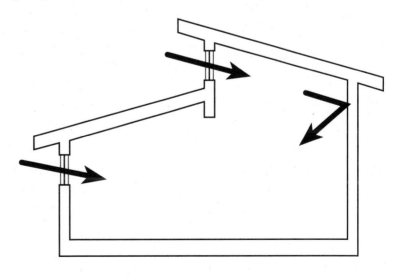

Green technology for the future: Fiber optic daylighting.
Fiber optic daylighting involves the transmission of light through glass, plastic, or gel fibers. The light then remains inside the mechanism and effectively distributes natural light anywhere in a building, unlike the limited vertical distances that skylights are able to illuminate. Fiber optic daylighting may eventually provide an important means of distributing daylight throughout the home.

Electrical Lighting

Electric lighting should be designed specifically to supplement daylight. For example, electric lighting should be dimmed or turned off when adequate daylighting is present. Simply turning off unneeded lights can reduce direct lighting energy consumption up to 45 percent.[6] The following recommendations will help you save energy and avoid the environmental degradation caused by energy use when electrical lighting is necessary.

$, ℘ – Install Compact Fluorescent Light (CFL) Bulbs

- Only 10 percent of the energy emitted from incandescent lights is in the form of light — the other 90 percent is heat.[7] For each incandescent you replace with a CFL, you will reduce CO_2 emissions from power plants by over 700 pounds over the life of the bulb.[8] The EPA estimates that if just one room in every US home were switched to CFLs, we'd reduce CO_2 emissions by over 1 trillion pounds. Although CFLs cost more initially, they can last up to 20 times longer than incandescents and use 75 percent less energy, thereby proving to be an extremely profitable investment.[9] One compact fluorescent bulb over its lifetime can save an amount of energy equivalent to driving a compact car from New York to Los Angeles.[10]
- **Recommendation**: Replace incandescent bulbs with CFLs that screw in like conventional bulbs. Table 13.1 opposite compares the costs of incandescent and CFL lighting.

Table 13.1: Compact Fluorescent Light Savings

	CFL (20W)	Incandescent (75W)
Bulb Cost (approx. 10,000 hours)	$17 (approx. 10,000 hours/1 bulb)	$9.75 (approx. 10,000 hours/approx. 13 bulbs @ $0.75 each)
Energy Cost (10,000 hours)	200 kWh/$16.98	750 KWh/$16.18*
TOTAL COST	$33.58 (at a rate of at $.0829 per kWh)	$71.93 (at a rate of at $.0829 per kWh)

Source: Sofia Perez. Light Bulb [online]. [Cited October 7, 2003]. Green Guide Institute, 2003.
<www.thegreenguide.conVreports/product.mhtml?id=8>.

🔎 – Install Halogen Lighting

• While compact fluorescent lighting remains the most energy efficient lighting option, halogen lighting is 50 percent more efficient than incandescent lighting[12] and is best used when short-distance, focused, high quality light is desired. While the first cost of halogen bulbs can be two to three times that of incandescent bulbs, they last much longer. Still, they are not the best option: fluorescents are nearly three times as efficient as halogens.[13] Moreover, halogens create a lot of heat and use transformers, which require energy even when they're turned off.

• **Recommendation**: Halogen lights are common in designer light fixtures and are generally considered an upgrade. Use in place of incandescent lighting, but keep in mind that using halogens for indirect lighting greatly decreases their efficiency.

🔎 – Install Lighting Controls

• Lighting controls use occupancy and light sensors and timers to turn lights off in unused areas or during times when lighting is not needed. Dimmers can also reduce the energy output of a light fixture when you adjust them to low light levels.

• **Recommendation**: Install lighting controls at specific locations or as a whole house system. Energy efficient compact fluorescent lights can be used in most applications.

Tips When Using CFL

- All light bulbs, and to a lesser extent fluorescent bulbs, contain small amounts of mercury. The amount of mercury is too small to be of danger to you in your home, but it still requires careful disposal. Call your sanitation department to find out the best way to dispose of used CFLs.
- Turn off the lights in any room you're not using, or consider installing timers, photo cells, or occupancy sensors to reduce the amount of time your lights are on.
- Use task lighting: instead of brightly lighting an entire room, focus the light where you need it. For example, use fluorescent under-cabinet lighting for kitchen sinks and countertops under cabinets.
- Consider three-way lamps; they make it easier to keep lighting levels low when brighter light is not necessary.
- Use four-foot fluorescent fixtures with reflective backing and electronic ballasts to help disperse light in your workroom, garage, and laundry areas.
- Consider using four-watt mini-fluorescent or electro-luminescent night lights. Both lights are much more efficient than their incandescent counterparts. The luminescent lights are also cool to the touch.
- Use CFLs in all the portable table and floor lamps in your home. Consider carefully the size and fit of these systems when you select them. Some home fixtures may not accommodate larger CFLs.
- When shopping for new light fixtures, consider buying dedicated compact fluorescent fixtures with built-in ballasts that use pin-based replacement bulbs. This makes it less expensive to replace the bulb.
- For spot lighting, consider CFLs with reflectors. The lamps range in wattage from 13–32 watts and provide a directed light using a reflector and lens system.
- If you live in a cold climate, be sure to buy a fluorescent lamp with a cold-weather ballast for outside lighting.
- Replace decorative outdoor gas lamps. Just eight gas lamps burning year round use as much natural gas as it takes to heat an average-size home during an entire winter.[11]
- Make sure you periodically wipe light fixture reflectors and diffusers to ensure that you're getting the maximum amount of light from the bulb.
- Pick lampshades and diffusers that let the most light through while still filtering glare.

𝒫 – Install Sealed or Airtight IC Recessed Lighting

- Typical recessed fixtures are like an open chimney: when not in use, cold air enters all winter and hot attic air enters the house all summer; when in use, the heat created by the bulb causes the room's conditioned air to rise through the fixture's hole and exhaust into the attic. Airtight fixtures avoid all of this loss of conditioned air. Recessed lighting is an ideal location for compact fluorescent lighting because these fixtures do not get as hot, and as result, they are not a fire hazard. Recessed lighting in insulated ceilings should be "IC-rated" fixtures for direct insulation contact, and they should be covered with insulation.

• **Recommendation**: Many recessed lighting fixtures still have holes in their housings. Purchase an "airtight" (AT) recessed can instead. Space between these housings and adjacent drywall should also be caulked to prevent additional airflow.

Myth: It is better to leave lights on than to turn them on and off repeatedly.

It used to be true that cycling lighting and appliances on and off wasted more energy than leaving the lights on. However, better designs have made this issue obsolete. Turn off the lights!

Avoid installation in cathedral ceilings or joist cavities because inadequate room for insulation can cause thermal problems.

– Minimize Light Pollution

• As a result of light pollution, two out of three Americans grow up without ever seeing the Milky Way.[14] Light pollution wreaks havoc in certain natural systems — from sea turtle nesting in Florida to migrating birds in Toronto and sycamore trees in urban parks. Some research even suggests human health is affected by lack of darkness. And of course, excessive outdoor lighting wastes energy and money and can annoy the neighbors! Once you turn outdoor lights off, the pollution ends. Those turtles, birds, the trees, and your insomniac neighbor will thank you for it. In addition, these strategies to reduce light pollution will help you light the perimeter of your home safely, efficiently, and at lower operating costs. And won't it be nice to see the stars?

• **Recommendation**: There are many ways to minimize light pollution around your home. First (and most obvious), minimize use of outdoor lighting. Avoid the use of floodlights and specify "full-cutoff" luminaries (fixtures) to avoid excessive glare. Focus lights downward. A good rule of thumb is to make sure that direct light shines a minimum of 20 degrees below a horizontal plane, and in no case above the horizontal plane. Avoid uplighting trees or architectural facades, as this results in unnecessary glare. Check to make sure that glare from your outdoor lights will not be a problem for neighbors, pedestrians, or motorists. Use timers, motion detectors, or photocells: these will ensure the lights are on for safety and security purposes when needed and off at all other times. And finally, use metal halide, compact fluorescent, and high pressure sodium lamps — the brightest, most energy efficient choices for outdoor lighting. Do not use low-efficiency mercury vapor lamps.

Electrical Checklist

Daylighting

❏ Install light pipes.

❏ Install skylights.

❏ Install clerestory windows.

❏ Install light shelves.

Electrical Lighting

❏ Install compact fluorescent light (CFL) bulbs.

❏ Install halogen lighting.

❏ Install lighting controls.

❏ Install sealed or airtight IC recessed lighting.

❏ Minimize light pollution.

CHAPTER 14

Insulation

REMODELING IS THE BEST CHANCE YOU'LL EVER HAVE to make your house more comfortable with better insulation. Improve all your insulation as long as the insulation contractor is in your house; the house will be more comfortable, you'll save energy and money, and you'll minimize noise from outside.

There are many types of insulation, and just as many ways to install it. We'll help you figure out what materials and techniques are safest and most effective. First, you should know how insulation works. Heat travels through the house in three ways:

- It is transported with air or water (convection). You can control convection by controlling air leakage.
- It radiates from a hotter surface to an unconnected cooler surface (radiation). You reduce unwanted solar radiation by shading windows and by using a radiant barrier on the underside of the roof.
- It is directly transferred through materials from molecule to molecule (conduction). To reduce conduction, you use insulation. (See Chapter 4, Building Science Basics.)

Heat moves from hotter areas to cooler areas. In other words, heat is conducted through walls, floors, and ceilings, from the hotter side to the cooler side. In the winter this would mean heat inside the house would want to move to unheated spaces (attics, garages, basements) and to the outside. In the summer, heat outdoors would want to move into indoor air-conditioned spaces. Your insulation goes in the thermal boundary, or the walls, ceilings, and floors, that separate the cold from hot areas. You want your insulation to slow this heat transfer as much as possible to keep your living space from getting uncomfortably hot or cold. How well your insulation does this is roughly measured by R-value or "thermal resistance" — the higher the R-value, the more effective the

Table 14.1: Insulation Comparisons

Insulation Type	R-Value per Inch	Where Applicable	Advantages
Batts:			
Fiberglass	2.9–3.8	All unfinished walls, floors, and attics.	Do-it-yourself.
Cotton	3.0–3.7	Fitted between frame studs, joists, and beams.	Suited for standard stud and joist spacing, if there are few obstructions.
Loose fill:			
Cellulose	3.1–3.7	Anywhere that frame is covered on both sides, such as walls or cathedral ceilings, attic floors and hard-to-reach places.	Easy to use for irregularly shaped areas and around obstructions. Dense pack provides air sealing as well as insulation.
(dense pack)	3.4–3.6		
Fiberglass	2.2–2.9		
(dense pack)	3.4–4.2		
Rock wool	2.2–2.9		
Sprayed insulation:			
Polyurethane foam	5.6–6.2	Walls, attics, floors.	Provides air sealing as well as insulation. Can provide complete coverage around obstructions.
Polyicynene foam	3.6–4.3		
Damp-spray cellulose	2.9–3.4		
Spray-in fiberglass	3.7–3.8		
Foam board:			
Expanded polystyrene	3.9–4.2	Basement masonry walls and floors.	High insulating value for relatively little thickness.
Extruded polystyrene	5.0		
Polyisocyanurate	5.6–7.0	Exterior walls under construction.	Covers wall framing, insulating studs, as well as cavities.
Polyurethane	5.6–7.0		
Phenolic (closed cell)	8.2	Exterior walls when adding siding.	
Phenolic (open cell)	4.4		

Source: Home Energy Magazine. "No Regrets Remodeling." Energy Auditor and Retrofitter, 1997. p. 202.

insulation. To figure out the most economic insulation level for your existing home, visit the US Department of Energy's Zip Code Insulation Calculator at <www.ornl.gov/~roofs/Zip/ZipHome.html>.

To help you determine what insulation is best for your remodeling job, Table 14.1 compares the various insulations, showing you how well they insulate, where they are most applicable, and their advantages in specific situations.

Most insulation products, by their function, have green qualities. In other words, the resource depletion, embodied energy, and other associated environmental costs of manufacturing insulation products are more than offset over time by the energy saving function of insulation. However, not all building products and techniques are equal in their environmental impact. We have compiled additional recommendations that will help you look for important green features, such as minimizing insulation leakage, indoor air quality impacts, and avoiding ozone-depleting blowing agents, while maximizing energy savings and recycled content.

⚘ – Avoid CFC- or HCFC-based Foam Insulation

- CFCs (chlorofluorocarbons) and HCFCs (hydrofluorocarbons) have high ozone-depleting potential. When insulating can be done without reducing overall energy performance, avoid HCFC-based insulation. Expanded poly-styrene (EPS) foam insulation has long been the rigid foam board of choice for those seeking to minimize the greenhouse gas and ozone-depletion effects of their rigid-insulation choices. Pentane gas, which is used to expand EPS, does contribute to smog, but modern plants can be built to recover 95 percent of pentane used in production, reducing this effect. Like EPS, several manufacturers have introduced foam insulations that are ozone-safe, including SuperGreen Foam™, Icynene®, Sealection©500, and Green Polyiso™.

- **Recommendation**: EPS, rigid fiberglass. and Green Polyiso™ can be substituted for extruded polystyrene (XPS) and polyisocyanurate. (Except for exterior foundation insulation, where XPS is the only suitable material for the moist conditions. See Chapter 7.) HFC-blown polyurethane (SuperGreen Foam™), CO_2-blown isocyanurate (Icynene®), or CO_2-

Fig 14.1a & b: Batts and blankets (above) are flexible, bound insulation made of cellulose, fiberglass, cotton, or rock wool. Sprayed insulations (below) are foam-in-place mixed, sprayed, or extruded into place using special equipment.
Credit: David Johnston.

When should I insulate?

- When you live in a house built before 1980.
- When the temperature in your house is uncomfortable.
- When you are already remodeling.
- When you pay higher energy bills than your neighbors in a similar-sized home.
- When you can hear the neighbors from inside your home.

blown polyurethane (Sealection©500) can be substituted for conventional HCFC-blown polyurethane.

🌲,♡ – Use Formaldehyde-free, Recycled-content Fiberglass Insulation

Most fiberglass insulation is made primarily from silica spun into glass fibers, and contains a phenol-formaldehyde binder. In the short term, the glass fibers that are released into the air may cause skin irritation and upper respiratory problems. In the long term, the fiberglass fibers are potentially carcinogenic. Some manufacturers are increasing the amount of phenol formaldehyde binder in the end product to minimize the amount of airborne fibers, but phenol formaldehyde is also thought to be a serious health hazard. You can avoid the health hazards associated with formaldehyde and fiberglass, as well as reduce the impact of mining for raw materials, by using formaldehyde-free, recycled-content fiberglass insulation.

• **Recommendation**: Formaldehyde-free fiberglass insulation is easy to find. Fiberglass insulations typically have 15 to 30 percent recycled glass content, including both post-industrial glass cullet and post-consumer bottle glass. Virtually all loose-blown fiberglass doesn't use a formaldehyde binder. Fiberglass insulation can be used for any typical insulation installation. As with most fiber insulation materials, you should install a continuous air barrier between the insulation and the living space to keep fibers out of the indoor air.

🌲,♡ – Use Mineral/Rock Wool Insulation

• Mineral wool insulation is made of molten slag, a waste product of steel production, or natural rocks (rock wool) such as basalt and diabase (or a mix of the two). Mineral/rock wool has a higher density than fiberglass, so it has better sound-blocking properties. It is also more fire-resistant than fiberglass.

• **Recommendation**: Mineral/rock wool can be blown in as loose fill or formed into rigid board stock.

🌲,♡ – Use Cellulose Insulation

• Cellulose is a highly effective insulation made out of recycled newspaper, with borate added as a fire and pest retardant. Damp-spray cellulose wall insulation is mixed with low-toxic binders to adhere to stud and joist cavity surfaces to at least an R-13 insulation level. Cellulose insulation has several environmental benefits over other types of insulation; in terms of resources, most cellulose insulation contains 75 to 80 percent post-consumer recycled newspaper and therefore diverts significant amount of materials from landfills. The energy performance of cellulose is excellent; it is comparable to high-density fiberglass batts, at roughly R = 3.7 /inch. Spray-in cellulose insulation blocks and fills holes and cracks, effectively reducing air movement through

Boron and the environment.

Boron is a main component in traditionally manufactured fiberglass and cellulose insulation. In fiberglass manufacturing, boron adds flexibility to the glass melt by removing air bubbles from the mixture, ensuring that the glass fibers do not vitrify. In cellulose insulation, boron is used to control fungal (mold) and bacterial growth and to add to fire retardant properties.

Some concern exists concern regarding its long-term availability and sustainable management. According to the US Bureau of Mines, only a 54-year reserve of boron remained in the US given 1995 economics and level of extraction.[1] As supply decreases, the price will rise, bringing home insulation costs even higher than they are now. To counter price increases, some companies are working on developing a boron-free fiberglass insulation material. Even though it will be significantly more difficult to produce, boron-free fiberglass insulation promises the future conservation of this limited resource.

and around the insulation. In addition, the overall production of cellulose requires a relatively low amount of embodied energy — typically 20 to 40 times less than mineral fiber insulation products.

- **Recommendation**: Cellulose insulation can be used in almost all applications. Spread dry cellulose over ceiling joists or spray wet cellulose into wall cavities to increase R-value. Insist on borate-treated cellulose fiber only. Avoid installing during wet months; test the insulation for moisture content before installing drywall.

⚶,♡– Use Bio-based Insulation

- "Bio-based," or soy bean-based, insulation is an open-cell, semi-rigid, water-blown foam that penetrates every cavity and even the tiniest crevices, expanding 100 times its original liquid size for a perfect air-sealed fit. Not only is it an insulated, seamless, windproof barrier that eliminates air infiltration (the number one cause of energy loss in a home), but it emits no VOCs, chlorofluorocarbons (CFCs), or formaldehyde that can harm human health and the environment. It provides excellent acoustical insulation, does not support mold growth, is resistant to pests, is class-1 fire rated, and reduces dust and allergens because there are no loose fibers. Although relatively new to the market, this insulation has been hailed as one the most environmentally friendly insulations.

- **Recommendation**: Use bio-based insulation in place of petroleum-based insulation for walls.

♪ – Use The Blower Door Test to Monitor Leakage

- Your home may be wasting energy and money and you don't even know it! The blower door is a variable speed fan used to pressurize or depressurize a house to measure air leakage. It is mounted in an adjustable frame that fits snugly in a doorway. Using a blower door test and a few other related tools, a technician can determine how much leakage is occuring and where it is. Another test will help you identify duct leakage to maximize duct efficiency, thereby saving money and resources.
- **Recommendation**: A good contractor will have some combination of tools that enables him or her to measure the leakage in your house and ducts to determine if backdrafting is a hazard due to changed house pressures.

♪ – Install Insulation Baffles

- Proper ventilation in attics and roofs is essential to maintaining insulation performance. Insulation becomes much less effective when moisture, trapped under the roof, condenses on the insulation below. Insulation baffles are made of corrugated cardboard, foam, or plastic to ensure that a properly-sized ventilation pathway is maintained between the top of the insulation and the roof decking where the roof meets the outside walls. This prevents insulation degradation that can in turn waste energy.

How much air leakage is OK?

The American Society of Heating, Refrigerating, and Air-Conditioning Engineers (ASHRAE) has set the standard for residential ventilation at a minimum of 0.35 air changes per hour (ACH). This means that all the air in your home is replenished with fresh air from outside a little more than once every 3 hours. If your home rates higher than 0.35 ACH after the blower door test (indoor air is replenished with outdoor air significantly more than once every 3 hours), it is probably leaky.

How Does Air Escape?

During the heating season there is a tendency, called the stack effect, for warm air to rise up through wall cavities and any other cracks or passages into the top floors of a house. This causes low pressure in the lower parts of the house, thus drawing in outside air, and high pressure at the top of the house, forcing air out.

According to the US Department of Energy, 15 percent of air escapes via ducts, 2 percent via electrical outlets, 13 percent via plumbing penetrations, 10 percent via windows, 11 percent via doors, 4 percent via fans and vents, 14 percent via fireplaces, and 31 percent via floors, walls, and ceilings.

- **Recommendation**: Install insulation baffles below the sheathing and above the insulation to maintain proper ventilation in ceilings and attics. They should connect soffit vents to a ridge vent.

Advanced Infiltration Reduction Practices

Air moves in and out of your home through every hole, nook, and cranny. You may have heard that sealing a house traps dangerous pollutants inside. However, simply leaving a house leaky does not guarantee that moisture and pollutants will safely exit the house and allow enough fresh air to enter. More likely, you'll get too much air exchange on days you don't want it; and, the air may carry water vapor as it passes through leaks, causing mold and decay problems in your walls, ceilings, and floors. Without proper sealing and insulation, air can leak into your home through the insulation, causing drafts and loss of heat or cool air. Advanced infiltration reduction practices are essentially techniques that upgrade your insulation to exceed requirements. Note that standard building code requirements are typically "least allowable by law" — any worse would be illegal.

☼ – Caulk, Seal, and Weatherstrip

- Warm air leaking into your home during the summer and out of your home during the winter can waste a substantial portion of your energy dollars. One of the quickest money-saving tasks you can do is caulk, seal, and weatherstrip all seams, cracks, and openings to the outside. The cost of caulking is minimal and you can save ten percent on your energy bill

Table 14.2: How Does Air Escape?

Ducts	15%
Electric Outlets	2%
Plumbing penetrations	13%
Windows	10%
Doors	11%
Fans and Vents	4%
Fireplace	14%
Floors, walls, and ceiling	31%

Source: Office of Energy Efficiency and Renewable Energy. Energy Savers: Insulation [online]. [Cited October 16, 2003].
US Department of Energy, October 6, 2003. <www.eere. energy.gov/consumerinfo/energy-savers/insulation. html>.

by reducing these air leaks in your home.[2] In some cases, you can have payback in as little as a year!

- **Recommendation**: Use water-based caulks and sealants with the least amount of hazardous solvents. If you're building a new living space, or rebuilding old walls, air seal those structures as you go. Use the renovation as an opportunity to do a more extensive sealing job to make the house more comfortable overall. All holes in the outside wall need to be caulked. Look for spots where the outside wall has been breached, where there are dirty spots in the insulation, where different building materials come together, or where seams have pulled apart. You can seal the holes by stapling sheets of plastic over the holes and caulking the edges of the plastic. Be especially vigilant for leaks in doors, windows, plumbing, ducting, electrical wire, and penetrations through exterior walls, floors, ceilings, and soffits over cabinets. For example, it's a good idea to install foam gaskets behind outlets and switch plates on exterior walls. In addition, foam, bronze, or vinyl strips or gaskets attached around the moving parts of doors and windows will reduce air leaks. Note that some caulking won't adhere or flow well in cold weather. In general, caulk is temporary and needs replacement from time to time.

☉ – Increase Insulation Thickness

• If existing insulation is not effective (low R-value), consider improving it. In general, each time you double the R-value of insulation, you cut your conduction heat loss in that area by half. Increased insulation therefore improves comfort, decreases heating and cooling requirements and costs, and makes homes quieter.

• **Recommendation**: Use blown-in insulation for existing drywall applications, or spray-on or batt insulation if drywall is to be removed. Exterior walls can be wrapped with a minimum of one-inch (R-4) rigid foam to increase R-value. Insulate headers to at least R-5 by placing half an inch of rigid foam between the framing material for 2" × 4" walls or two inches in 2" × 6" walls. There are header products on the market that use wood "I" joists pre-insulated to any wall thickness. Wall and ceiling insulation should exceed code by 20 percent.

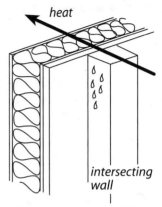

a: Thermal bridge at partition.

☉ – Avoid Thermal Bridging

• When steel wall framing is used, it is important to understand "thermal bridging" and its effect on the overall wall R-value. A thermal bridge is a break in the continuity of insulation in a wall, ceiling, or floor, which allows heat to flow more readily between the inside and outside of your home. Wall studs act as thermal bridges when insulation is installed only between them, because heat or cold can still conduct through the stud itself.

You can minimize thermal bridging by placing a layer of rigid insulation on the exterior of the studs to effectively re-create a continuous layer of insulation. In this way, you improve the energy efficiency and comfort of your home. The R-value can be reduced 40 to 60 percent in metal framed exterior walls without insulated sheathing[3] because metal is an excellent heat conductor.

• **Recommendation**: Minimize the amount of heat the steel framing can conduct through the wall by placing a layer of rigid insulation on the exterior of the studs.

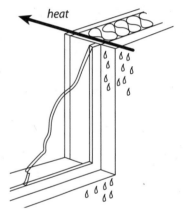

b: Thermal bridge at window.

Fig. 14.2a & b: Thermal bridges. Credit: Jill Haras and Kim Master, redrawn by Jeremy Drought.

☉ – Ensure Adequate Attic Insulation

• Insulating your attic is the easiest and most cost-effective way to insulate your home. You can increase the comfort of your home while reducing your heating and cooling needs by up to 30 percent by investing just a few

Insulation Checklist

- [ ] Avoid CFC- or HCFC-based foam insulation.
- [ ] Use formaldehyde-free, recycled-content fiberglass insulation.
- [ ] Use mineral/rock wool insulation.
- [ ] Use cellulose insulation.
- [ ] Use bio-based insulation.
- [ ] Use the blower door test to monitor leakage.
- [ ] Install radiant barriers.
- [ ] Install insulation baffles.
- [ ] Caulk, seal, and weatherstrip.
- [ ] Increase insulation thickness.
- [ ] Avoid thermal bridging.
- [ ] Ensure adequate attic insulation.

hundred dollars in advanced insulation and weatherization techniques,[4] such as adding attic insulation.

- **Recommendation**: To find out if you have enough attic insulation, measure the thickness of the insulation in various places. If it is less than R-22 (7 inches of fiberglass or rock wool or 6 inches of cellulose), you could benefit by adding more.

CHAPTER 15

Solar Energy

Passive Solar

REMODELING IS THE PERFECT TIME TO IMPROVE passive solar heating in your home. Passive solar is both environmentally friendly and cost-effective. The passive heating techniques involve placing larger windows on south-facing walls and locating thermal mass, such as a concrete slab floor or a heat-absorbing wall, close to the windows. You can cut your heating costs by up to 50 percent compared to the cost of heating the same house without passive solar design.[1] Passive cooling techniques include carefully designed overhangs, windows with reflective coatings, and other natural cooling approaches. Keep in mind that a passive solar house requires careful design and site orientation, which depends on your local climate.

The Role of Windows

Windows play a big role in passive solar design and the energy efficiency of homes. In the summer, windows can allow unwanted heat into the house, and in the winter, windows can account for as much as 25 percent of the heat loss, equaling over $20 billion in electricity costs nationwide.[2] The cost of replacing windows in an existing house is high. However, replacing windows will make a substantial difference in the comfort of your home and may be well worth the cost for you. Also, upgraded windows typically add to the value and sale-ability of your home because of their energy efficiency and resistance to outdoor noise. The labor costs for installing or upgrading windows are equivalent regardless of the quality of window, so buy the best you can afford.

The best windows have the lowest "U-value"

To find the best windows, ask the salesperson to tell you the "U-value" of the windows they offer. U-values is the measure of flow of heat — thermal transmittance — through a material, given a difference in temperature on either side. R-value (used to measure resistance to heat flow in insulation) is the inverse of U-value: U = 1/R. This is sort of like the miles-per-gallon rating for new cars and measured by an independent agency. In the case of U-values, lower is better: the best windows you can buy today have U-values of around U = 0.20. Typical low-e windows are around U = 0.35.

◯ – Install Double-paned Windows

- Windows can be formulated with either single, double, or triple panes of glass. Typically, the more panes of glass, the better your windows will insulate your home. Good double-paned windows not only add to the value of your home, but they also insulate almost twice as well as single-paned windows. Better double-paned window units also tend to have inert gases, such as argon or krypton, installed inside the air space of a double-paned window, to further slow heat transfer. They make the whole house quieter and more comfortable during all seasons, while saving energy and money. Upgrading from single-pane windows to energy-efficient windows can save up to 15 percent of your heating costs. Depending on where you live, this can amount to $50 to $100 per year. Spread over 20 years, this means $1,000 to $2,000.[3] You can get a savings estimate more specific to your home by running the Lawrence Berkeley Laboratory online energy calculator: <www.home energysaver.lbl.gov/>.

- **Recommendation**: Replace single-paned windows with double-paned windows whenever possible. Check with your local utility company for rebate programs.

◯ – Install Low-E (Low-Emissivity) Windows

- Low-E coatings, virtually unnoticeable to the eye, are installed inside the air space of a double-paned window. They help prevent heat from escaping through the glass in winter by reflecting it back in; another type of low-E coating blocks heat from entering the home during summer by reflecting sunlight out of the structure. In some markets, the cost premium of 10 to 15 percent for low-E glass can pay for itself in a few years. Many good window manufacturers nowadays include it as a base standard in their windows.

- **Recommendation**: Use low-E, double-paned windows whenever windows are replaced. Check with your local utility company for rebate programs. Choose one of three types of E-coatings according to your climate:

  - **High transmission low-E**. These windows are best suited for use in cold climates.
  - **Selective transmission low-E**. These products are ideal for homes in mixed climates that have both significant winter heating and summer cooling requirements. The low-E qualities ensure winter performance by allowing daylighting. The selective properties also block most solar infrared energy, thereby keeping the home cooler during summer months.
  - **Tinted low-E**. This glass provides glare control along with a high level of solar heat rejection, thereby helping to control solar gains in hot climates, especially on east and west windows.

💡 – Install Superwindows

- Superwindows are windows with one or two thin plastic films suspended between the panes of glass, effectively making them triple or quadruple glazed windows. (U = .15 to .30) They also reduce ultraviolet rays, which can fade fabric, rugs, and art by up to 95 percent.
- **Recommendation**: Upgrade fixed pane windows to superwindows. The larger the window, the more effective superwindows are in providing comfort.

💡 – Install Glazings Tuned to Orientation

- East-, west-, and south-facing glazing works best if you choose glass with the right glazing characteristics. Glazings tuned to orientation minimize unwanted heat gain and maximize wanted heat gain, which keeps occupants more comfortable and reduces heating and cooling bills. Glass with low solar heat gain coefficient (SHGC) glazing has built-in shading, meaning that it reduces the amount of solar energy that gets in your home.
- **Recommendation**: On the west and east facades, glazings with *low* SHGC should be used to minimize heat gain that will contribute to overheating or higher air conditioning loads. On the south-facing

windows, use glazings with *high* SHGC to increase the amount of solar heat that can warm your home. In most climates, the north-side solar heat gain will be minimal, so SHGC values are not significant. Also, check the visible light transmittance (VT) so that you don't block solar heat at the expense of the light and view you want from the window. Work with your window supplier to get the right combination for your climate.

♀ – Install Low-conductivity Frames

- Most window frames and sashes are made of wood, vinyl, fiberglass, or aluminum. Wood, vinyl, and fiberglass generally insulate better than aluminum or steel frames because they conduct less heat. Wood windows create greater comfort, better energy efficiency, and are an environmentally preferable material.
- **Recommendations**: Choose a low-conductivity framing material, such as wood.

♀, ♡ – Incorporate Daylighting

- See Daylighting in Chapter 13 for more window ideas that maximize daylight while minimizing energy loss.

♀ – Incorporate and Distribute Thermal Mass

- For passive solar heating, thermal mass works in conjunction with south windows to capture and store the sun's heat. Materials such as brick, masonry, poured concrete, or tile soak up most of the heat that hits them, and then gradually release heat when the sun goes down, moderating the temperature of the house. Properly designed passive solar homes that use thermal mass prevent high midday interior temperatures in the summer or on sunny days in the winter, and help keep the home from becoming uncomfortably cool on winter nights.
- **Recommendation**: As a rule of thumb, the thermal mass should be six times the area of direct gain, south-facing glass. For most thermal mass materials, their cost and energy effectiveness increases up to a thickness of about four inches. At a minimum, increase drywall from half an inch to five-eighths of an inch to increase the thermal capacity on walls in sunlit rooms.

☾ – Install Window Coverings and Overhangs

- Carefully designed shading blocks unwanted sun (typically in summer and early fall), but does not block spring, late fall, or winter sun that is desirable for heating. Keep in mind that when sunlight strikes any surface it turns into heat: interior roll-down shades don't keep out much heat because the light is already through the glass. However, if you can't afford to upgrade your windows or are looking for additional ways to conserve energy, shading windows or installing exterior blinds can be cost-effective ways of reducing solar heat gain and air conditioning costs.

- **Recommendation**: In the winter, do not shade south windows between 10 A.M. and 2 P.M. Overhangs are most effective at shading when a two-foot overhang is used 1 to 2 feet above windows within 30 degrees of true south. For east and west windows, it is best to use vertical louvers to prevent overheating. Tight-fitting, insulated window shades help keep heat inside at night. As a minimum in the summer, install white window shades, drapes, or blinds to reflect heat away from the house. Close curtains on south- and west-facing windows during the day.

☾ – Use Landscaping to Shade Windows

- A well-placed tree, shrub, or vine can provide effective shade and reduce overall energy bills. Trees are amazing in the way they can change the local microclimate; not only do they create shade, but they evapotranspirate (or "sweat"), cooling the air in the process. A mature tree with a crown of 30 feet can evapotranspire up to 40 gallons of water in a day — comparable to removing all the heat produced in 4 hours by a small electric space heater.[4] Mature tree canopies reduce the average temperature in suburban areas about 3 degrees Fahrenheit, compared with newer areas with no trees.[5]

- **Recommendation**: Locate deciduous trees to shade the east and west sides of the house, but only areas that are more than 60 degrees east or west of due south from the house. In particular, west and southwest facades should be shaded from low-angle sun that causes overheating. It is not recommended to place deciduous trees on the south side of a home, because even with their bare branches, these trees can block as much as 30 percent of the available winter solar energy. If it is necessary to have trees on the south side of the house, mitigate this effect by locating deciduous trees close to the house and select species with few low branches.

Plant trees, save energy.

Landscaping can save up to 25 percent of a household's energy consumption for heating and cooling.[6] Computer models designed by the US Department of Energy predict that homeowners can save from $100 to $250 annually by the proper placement of only 3 suitable trees.[7] A single tree can provide the same cooling effect as 10 room-sized air conditioners running 24 hours a day.[8] It has also been estimated that a well-designed landscape plan will be paid for, on average, in 8 years.[9] Landscaping is generally more effective for energy-efficiency than installing venetian blinds, plastic window coatings, or heavy reflective coatings on glass, because it blocks heat before it can enter the house.

Active Solar Collectors

Adding a solar heating system to your home is one way to combat increasing energy costs and to raise your home's market value. If the sun shines on your home for most of the day in the winter, it is a potential candidate for a solar heating retrofit. The two major types of active solar retrofits are solar hot water fluid collectors that heat a fluid circulated within them, and air collectors that heat air to be distributed in the house with fans.

Fig. 15.1: This stone wall soaks up the sun during the day and slowly releases heat, warming the home throughout the night. Credit: Kim Master.

The solar collectors are the basic component of both systems. The most common collector for residential water and space heating is a flat-plate collector. It is basically an insulated metal box with a glass or plastic cover — called the glazing — and a dark-colored absorber plate. The glazing allows the light to strike the absorber plate, but reduces the amount of heat that can escape. The sides and bottom of the collector are usually insulated, further minimizing heat loss. Sunlight passes through the glazing and strikes the highly conductive absorber plate, which heats up, changing solar radiation into heat energy. The heat is then transferred to the air or liquid passing through the collector. Table 15.1 opposite lists the common heat distribution systems and the types of solar systems that are compatible with each.

– Install a Solar Space Heater System

- Active solar space heating systems use solar collectors that use the sun to heat liquid or air to then provide space heat. Solar space heating systems are usually designed to provide 30 to 80 percent of heating, depending on geographical location and system type and size.[10] Solar air heaters that directly heat interior air do not require a heat storage component, and can complement most existing heating systems. They are the least expensive and simplest solar technology to install.[11] Active solar space heating systems are most economical in climates that have extended heating

Table 15.1: Heat Distribution System and Compatible Solar Collectors

Heat Distribution System	Compatible Solar Collectors
Forced hot air	All collectors
Hot water radiators	Liquid flat plate, concentrators
Radiant floor (water)	Liquid flat plates

Source: Office of Energy Efficiency and Renewable Energy. Residential Solar Heating Retrofits [online]. [Cited October 8, 2003]. US Department of Energy, November 2002. <www. eere. energy. gov/consumerinfo/refbriefs/ac6.html>.

seasons with many sunny days and/or with high utility rates. They are less cost-effective in areas with cloudy conditions during the winter, such as the coastal Northwest, in areas with short heating seasons, such as Southern California and Florida, or in any area with low prices for electricity and other heating fuels. Solar power will reduce the amount of pollution and greenhouse gases emitted to the atmosphere.

• **Recommendation**: Solar collectors are usually installed on the roof, which means the roof must be in good condition and be capable of support-ing the collectors. Collectors can also be mounted on ground racks, vertically on a south-facing wall, or on an adjacent structure such as a garage. Pipes (for liquid systems) or ducts (for air systems) for transferring heat from the collector(s) to the interior require a roof or wall penetration. You can increase the effectiveness of liquid solar space heating systems by storing the solar heat in a large, well-insulated tank during the day for use at night or on cloudy days.

⌀ – Install a Solar Water Heater

• Solar water heaters have collectors that use the sun's energy to heat water in much the same way as water in a hose left on the lawn gets hot on a sunny day. The heated water is then stored in a tank similar to a conventional gas or electric water tank. Some systems use an electric pump to circulate the fluid through the collectors. Solar water heating systems are also good for the environment. They avoid the harmful greenhouse gases associated with

Calculating solar payback.

More than 1.5 million homes and businesses in the United States have invested in solar water heating systems, and 94 percent of these customers consider the systems a good investment[13] Payback varies widely, but you can expect a simple payback of 4 to 8 years on a well-designed and properly installed solar water heater. (Simple payback is the length of time required to recover your investment through reduced or avoided energy costs.) You can expect shorter paybacks in areas with higher energy costs. After the payback period, you accrue the savings over the life of the system, which ranges from 15 to 40 years, depending on the system and how well it is maintained.[14] You can use the calculator at the following website to explore the energy usage of your water heater, and to estimate whether a solar water heater could save you money: <www.infinite power.org/calc_waterheating.htm>.

electricity production: during a 20-year period, one solar water heater can eliminate over 50 tons of carbon dioxide emissions.[12]

- **Recommendation**: Solar water heaters can operate in any climate. The colder the water, the more efficiently the system operates. In almost all climates, you will need a conventional backup system; in fact, many building codes require you to have a conventional water heater for backup. Always take steps to use less hot water. When you lower the temperature of the hot water you use, you reduce the size and cost of

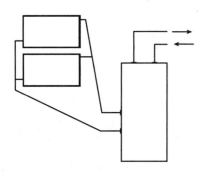

Fig 15.2: Solar water heater system. Credit: Jill Haras & Kim Master, redrawn by Jeremy Drought.

your solar water heater. Look for solar water heater systems certified by the Solar Rating and Certification Corporation (SRCC) or the Florida Solar Energy Center (FSEC). These systems should be installed by a licensed and experienced solar contractor, preferably through a utility program. Vendors and service people who are selling units in cooperation with a utility program are likely to remain in business long enough to repair your system if it ever gives you trouble.

Photovoltaic (PV) Energy

PV panels contain hundreds of small silicon cells that collect the sun's energy and change it into electricity that can be used in the home. This differs from solar space heating and water heating, which is based on solar collectors that change radiation into heat energy using the physics of heat absorption and conduction (see "Active Solar Collectors" in Chapter 5). PV cells are made mostly of silicon, an abundant semi-conductor material in the earth's crust that is also used in computer semi-conductors. One side of the material is electrically positive, the other negative. When light strikes the electrically positive side of the material, the negative electrons are activated in a way that produces an electric current, and voilá! Energy!

Groups of PV cells are electrically configured into modules and arrays, which can be used to charge batteries, operate motors, and to power any number of electrical loads. With the appropriate power conversion equipment (called an "inverter"), PV systems can produce alternating current (AC) compatible with any conventional appliances, and operate in parallel with and interconnected to the utility grid.

Is a solar water heater practical for my family?

Solar water heating is generally sensible for families that use a lot of hot water. Most solar systems are designed to meet half to three quarters of a family's hot water needs, or all of their summer needs. The remainder is supplied by the backup system. Your contractor will need to size the collectors, the storage tank, and the backup system. A rule of thumb is that an efficient collector in good weather will heat between one and two gallons of water per square foot of collector per day. So if your family uses 64 gallons of hot water per day, you would need between 16 and 32 square feet of collector to supply half of the household's needs.

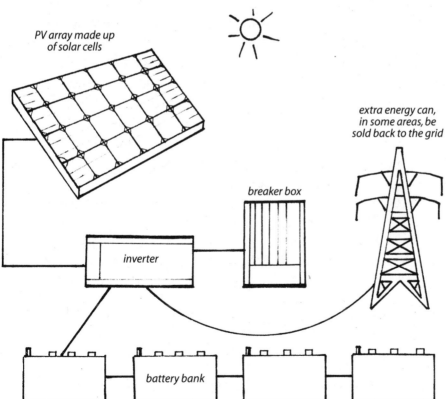

Fig 15.3: A typical PV system consists of a PV array, a control panel, batteries, and an inverter which feeds alternating current into the main service panel or breaker box.
Credit: Jill Haras & Kim Master.

PV array made up of solar cells

extra energy can, in some areas, be sold back to the grid

breaker box

inverter

battery bank

℘ – Install Grid-connected Photovoltaic (PV) Panels

• PV panels can be used as a means to decrease reliance on conventional power plants that contribute to air pollution. PV modules have no moving parts, are virtually maintenance-free, and have a long working life of at least 20 to 30 years.[15] In areas where laws permit net metering, a grid-connected home PV system produces electricity and is allowed to supply the excess to the utility, running the home's meter backward. When the home needs more power than its PVs are producing, it uses utility electricity, running the meter forward again. This means that you don't have to invest in batteries to store your PV power, which represents a large part of the cost of buying and maintaining a PV system. You simply bank your electricity with the utility and withdraw it as needed.

The US Department of Energy estimates the installation of one million solar systems could eliminate carbon dioxide emissions equal to that of 850,000 cars.

Source: Natural Resources Defense Council, "Photovoltaic Cells," <www.nrdc.org>, cited April 10, 2000.

• **Recommendation**: Panels are installed where your home receives the most sunlight — typically on a south-facing roof, but also on the ground or into the home exterior itself. Be sure to find a licensed solar contractor and coordinate your solar installation with a roofing contractor and an electrician. Also research permits, rebates, utility interaction, and regulations.

℘ – Install Off-grid Photovoltaic System

• Off-grid means the system requires batteries. The batteries store energy created by your PV system that you can then use when power from the utility is not available (i.e., if you live too far away from the grid, if it's cloudy, or if the power goes out). Though it costs a bit more than grid-connected PV, off-grid PV makes sense not only for people living in rural areas, but also for people who enjoy a hot cup of coffee when a snow storm, brownout, or thunderstorm knocks the power out. Because I work from home, my off-grid system has saved my business and paid for itself on more than one occasion (see "Solar Saves the Day" on p. 14).

• **Recommendation**: Off-grid systems are typically installed on your roof. If you live more than a quarter-mile from an existing electric line,

Solar Energy Checklist

Passive Solar

❑ Install double-paned windows.

❑ Install low-E windows.

❑ Install superwindows.

❑ Install glazings tuned to orientation.

❑ Install low-conductivity frames.

❑ Incorporate daylighting.

❑ Incorporate and distribute thermal mass.

❑ Install window coverings and overhangs.

❑ Use landscaping to shade windows.

Active Solar Collectors

❑ Install a solar space heater system.

❑ Install a solar water heater.

Photovoltaic (PV) Energy

❑ Install grid-connected PV panels.

❑ Install an off-grid PV system.

off-grid PV is often cost-effective. An off-grid system can also be great for low-level outdoor lighting, electric fence charging, or in applications where small amounts of power are required. Work with a professional to properly size your PV system. You should consider the ability of your system to expand to meet any unexpected demands in the future.

Heating, Ventilation, and Air Conditioning (HVAC)

Heating

O NE OF THE BEST WAYS TO MAKE YOUR HOUSE HEALTHIER and less expensive to operate is to upgrade the heating system. In colder regions, the cost of heating your home is two-thirds of the utility bill. Heating systems in the US are responsible for over a billion tons of CO_2 released into the atmosphere annually, as well as 12 percent of the nation's sulfur dioxide and nitrogen oxides.[1] By combining proper equipment upgrades with appropriate insulation, weatherization, and thermostat settings, you can cut your energy bills and your pollution output in half.[2]

In considering your options for improving or replacing your systems, it helps to know the basic types of systems. The following section will help you determine the system that's best for your home.

Gas and Oil Heating Systems

Gas or oil can be used to generate heat in a furnace or boiler. A furnace, also known as "ducted warm-air" or "forced-air," heats air that is blown through air ducts, which is then delivered to rooms through registers or grills. A boiler heats water or steam that that circulates through pipes to radiators and baseboards. Hot water boiler systems, often called hydronic systems, are more common than older radiators. Radiant floor heating uses hydronic systems to circulate water through copper or plastic tubing in the floor.

In furnaces and boilers, fuel is sprayed into a combustion chamber where it is mixed with air and burned. The combustion products are vented out of your home through a flue pipe. The flames heat a metal box called a heat exchanger. Air is heated in the furnace heat exchanger and water is heated in the boiler heat exchanger.

The efficiency of a gas or oil burner measures how efficiently it converts fuel into useful heat. The most accurate estimation of actual fuel use is the annual fuel utilization efficiency (AFUE).

Electric Heat Systems

There are two types of electric heat: electric resistance heat and electric heat pumps. Electric resistance heat directly converts electricity into heat, like a toaster. Even though it uses all the energy in the electric current, energy losses during electric generation at the power company, in transmission to your home, and in the basic physics of making useful heat from electricity, make it inefficient and expensive. Unlike electric resistance heat, electric heat pumps do not produce heat, but rather move it from one place to another by use of a compressor, like a reversible air conditioner. A heat pump delivers heat more efficiently.

The efficiency of electric systems is measured in a ratio of delivered energy to consumed energy, called the coefficient of performance (COP). A better measure for the Air-Source Heat Pump (see Green Alternative later in this section) is the heating seasonal performance factor (HSPF). This is a ratio of the estimated seasonal heating output (Btus) divided by the seasonal power consumption (watts).

Wood Heat Systems

Wood makes sense in rural areas, and the prices are typically lower than gas, oil or electricity. However, the pollutants have been a problem, causing the EPA to implement pollution emissions regulations. New models are significantly cleaner burning because they use a catalytic converter.

The following recommendations will help you renovate your home with the healthiest, least-polluting, most efficient heating and cooling systems.

℘ – Install 90 Percent or Greater AFUE Forced Air Furnace

• High efficiency furnaces convert gas to heat efficiently. If you insulate your house at the time of furnace replacement, you might be able to buy a smaller capacity furnace and save money. The same holds true for air conditioning and other heating and cooling equipment. Upgrading your furnace or boiler from 65 percent to 90 percent efficiency in an average cold-climate house will save 1.5 tons of CO_2 emissions if you heat with gas, or 2.5 tons of CO_2 emissions if you heat with oil.[3] Dollar savings are

Table16.1: Dollar Savings per $100 of Annual Fuel Cost

AFUE of Existing System	Annual Fuel Utilization Efficiency (AFUE) of New System								
	55%	60%	65%	70%	75%	80%	85%	90%	95%
50%	$9	$16	$23	$28	$33	$37	$41	$44	$47
55%		8	15	21	26	31	35	38	42
60%			7	14	20	25	29	33	37
65%				7	13	18	23	27	32
70%					6	12	17	22	26
75%						6	11	16	2
80%							5	11	16
85%								5	11

To determine savings, find the row corresponding to the old system's AFUE, then choose the number from that row which is in the vertical column corresponding to the new system's AFUE. That number is the projected dollar savings per hundred dollars of existing fuel bills.

Source: Alex Wilson and John Morrill. Consumer Guide to Home Energy Savings.
American Council for an Energy-Efficient Economy, 1991. p. 37.

equally impressive: this same upgrade would save you $27 per $100 of annual fuel costs. If your annual fuel bill is $1,300, then the total yearly savings should be about $27 × 13 = $351. Table 16.1 can help you calculate savings specific to your upgrade.

Forced-air systems also enable you to control humidity, filter air, introduce fresh air from the outside, and quickly change the temperature of an area. However, forced-air systems may cause higher operating costs, noisy operation, distribution of unwanted fumes and particulate matter. However, modification to standard systems may prevent unwanted dehumidification. It is considered far less comfortable than radiant heating.

Furnace tips.

- Bigger is not always better! Too large a system costs more and operates inefficiently.
- Locate furnaces in central, insulated spaces, such as in a combustion closet isolated from conditioned spaces. A central location will reduce duct runs and improve efficiency.
- A programmable thermostat with an "on" switch for the furnace fan will help circulate air and improve indoor air quality.
- Provide balanced or slightly positive indoor pressure using controlled outdoor air ventilation.
- Your home should pass a combustion safety/backdraft test, performed by a qualified technician. This avoids dangerous backdrafting of combustion gases in flue pipes.
- Add a good combination filtration system, which will filter out airborne particulate matter.
- For older furnaces, encapsulate or remove asbestos insulation. Use certified asbestos abatement contractors for removal.

- **Recommendation**: Install a high-efficiency, sealed-combustion furnace instead of a conventional furnace. Gas furnaces generally have the highest ratings, followed by gas boilers and oil furnaces, and finally, oil boilers. Look for EnergyStar® furnaces — as these have high efficiency ratings of 90 percent AFUE or higher.

℘,♡ – Install Sealed-combustion Furnaces

- Sealed-combustion furnaces use fresh air for combustion by directing vents from the outside to the jacket of the appliance. In addition, they do not compete with occupants for oxygen and will not leak carbon monoxide into your home in negative pressure conditions.

- **Recommendation**: Replace your existing furnace with a sealed combustion unit.

℘,♡ – Install Zoned Hydronic Radiant Heating

- Hydronic heating forces hot water through radiators located in different areas or zones throughout the house. It is typically installed as baseboards or in floors, and can be installed in a slab or through open joist cavities. The result is even, comfortable heating that warms the occupants in a room rather than the air in the room. Unlike forced-air, radiant heat will not dry out breathing passages or furniture. There is no unsightly ductwork and radiators provide added space and flexibility in the floor plan. With no fans or blowers, radiant heating is dust-free — virus particles, bacteria, and pet dander fall to the floor instead of circulating in the air. Radiant heating heats objects, therefore the floor stays warm and you stay warm. It costs less than forced air systems because you're not wasting energy heating all the air in the room. Radiant heat can be "zoned" so that it only heats the parts of the house in use at a given time. Typical operating savings range from 30 to 50 percent with a zoned system.[4]

• **Recommendation**: Use hydronic radiant heating instead of forced air heating. The system usually should be designed before construction starts, but retrofitting radiant heat for a remodel is also possible by using flooring that is grooved to fit the tubing. It can be fueled by any heat source.

♀ – Install a Geothermal Heat Pump

• A geothermal heat pump, also known as a "ground-source" or "water source" heat pump, collects heat from the ground or from a large body of water outside your home and concentrates it for inside use. It doesn't actually create heat; the pump literally moves heated air in the winter and cooler air in the summer from one area to another by running it through a compressor. The most common system is called a "closed loop system." High density polyethylene pipe is buried horizontally 4 to 6 feet deep, or vertically 100 to 400 feet deep (as determined by the local water table). The pipes are filled with an environmentally friendly antifreeze or water solution that extracts heat from the earth and carries, or "pumps" it into your home. In the summer, the system can reverse and pump heat from your home into the ground. Although geothermal heat pumps are initially expensive to install, their greater efficiency means the investment can be recouped in 3 to 5 years.[5] They can be installed in many areas of the country; however, it is most effective where there are both heating and cooling loads. Very cold climates are not good candidates for heat pumps.

• **Recommendation**: If you live in a moderate climate, replace your current electric heating system with an EnergyStar® rated geothermal heat pump. It should have a coefficient of performance (COP) rating of at least 2.8, and an energy efficiency ratio (EER) of 13. (See Glossary for definitions of COP and EER.)

Why Use a Geothermal Heat Pump?

- **Free or Reduced-cost Hot Water**. They can provide free hot water when a device called a "desuperheater" transfers excess heat from the heat pump's compressor to the hot water tank. In the summer, hot water is free; in the winter, water heating costs are cut in half.
- **Year-round Comfort**. They are quieter than conventional systems and improve humidity control when they are properly sized. Customer surveys show high levels of user satisfaction, usually well over 90 percent.
- **Design Features**. Geothermal heat pumps are smaller than conventional HVAC systems, meaning you can downsize your equipment rooms and free up space. Geothermal heat pump systems typically utilize existing ductwork for retrofits and provide simultaneous heating and cooling. The lack of rooftop penetration for flues means less potential for leaks, less maintenance, and better roof warranties.

- **Low Environmental Impact**. Geothermal heat pumps use less energy — often created from burning fossil fuels — to operate. According to the EPA, geothermal heat pumps can reduce energy consumption and corresponding emissions up to 44 percent, compared to air source heat pumps (see next listed green alternative) and 72 percent compared to electric resistance heating with standard air conditioning equipment.[6]
- **Low Maintenance**. The average total maintenance cost for buildings with geothermal heat pumps is one-third that of conventional systems, mostly because the piping is underground or underwater.
- **Durability**. The underground piping often has a warranty of 20 to 50 years. There are relatively few moving parts, and those parts are durable, reliable, and sheltered inside in the home.

⌕ – Install an Air-source Heat Pump

- Air-source heat pumps draw heat from outside air during the heating season and "dump" heat outside during the cooling season. There are two types of air-source heat pumps:

  - **Air-to-air heat pump**. This is the most common of the two systems. Air-to-air heat pumps extract heat from the air and transfer it to either inside or outside your home, depending on the season.
  - **Air-to-water heat pump**. This one is used in homes with hydronic heat distribution systems (see "Install Zoned Hydronic Radiant Heat Systems" earlier in this chapter). During the heating season, the heat pump takes heat from the outside air and transfers it to the water in the hydronic distribution system. Although the "cooling" feature is rare, it is possible for the heat pump to extract

heat from the water in the distribution system and "pump" it outside to cool the house.

Air-source heat pumps can be either add-on, all-electric, or bivalent. Add-on heat pumps are designed to be used with another source of supplementary heat, such as an oil, gas, or electric furnace. All electric air-source heat pumps come equipped with their own supplementary heating system in the form of electric-resistance heaters. And bivalent heat pumps use gas- or propane-fired burners to increase the temperature of the air entering the outdoor coil so that the unit can operate at a lower, more efficient temperature.

Although geothermal heat pumps are more efficient, heat pumps in general are the most efficient form of electric heating in moderate climates. Heat pumps provide 3 times more heating than the equivalent amount of energy they consume in electricity and they can reduce the amount of electricity you use for heating by as much as 30 to 40 percent.[7]

• **Recommendation**: If you use electricity to heat your home, consider replacing your current heating system with an energy-efficient Energy-Star® heat pump. If you live in a northern climate, it doesn't make sense to try to meet all your heating needs with an air-source heat pump because as the outdoor temperature drops, so does the efficiency of an air-source heat pump.

$, ☍ – Properly Size Your HVAC System

• A heating system that is too large wastes fuel and money because it keeps cycling on and off. It only runs at peak efficiency for short periods of time and spends most of the time either warming up or cooling down. Many existing systems that were installed in the 1950s and '60s are 2 to 3 times larger than necessary.[8] Properly sized units save you significant amounts of energy and money. Not only can you purchase a smaller, less expensive system, but the smaller system does not have to work as hard to supply your energy, saving you operating costs.

• **Recommendation**: With the exception of steam systems, where the boiler should be sized to the radiator, the best way to determine the proper size of a new heating system is through a Manual "J" heat loss analysis performed by your heating contractor or energy auditor. The system should be sized to no more than 25 percent over the peak hourly demand.

🌲, ♡ – Consider Alternatives to the Fireplace or Wood Stove

- The fireplace is one of the most inefficient heat sources you can use. A fire can exhaust as much as 24,000 cubic feet of air per hour to the outside,[9] which must be replaced by cold air coming into the house from outside. Your heating system must continuously warm this air, which is then exhausted through your chimney. Furthermore, wood smoke from fireplaces and wood burning stoves contains small particles, called particulates, that can be inhaled deep into the respiratory system where they may have a serious impact on health.

  Healthier, more efficient alternatives include pellet stoves, natural gas units, or Environmental Protection Agency (EPA)-certified wood stoves. **Pellet stoves** use small cylinders of compressed sawdust for fuel. They provide heat with significantly less pollution than other types of wood-burning stoves because the stove employs a sophisticated electronic system that provides the right mix of fuel and air. **Natural gas fireplace** equipment uses realistic, natural-looking logs and a controllable gas burner to create a lifelike wood burning fire. With the flip of a switch or the turn of a knob, gas fireplaces provide instant ambiance. Because of their efficient combustion, there is less pollution. Gas logs add no creosote buildup to your chimney and there are no hot ashes or surprising sparks to drop onto your carpeting. Lastly, **EPA-certified wood stoves** are cleaner-burning wood stoves that adhere to established limits on the amount of particulate matter released into the air.

  Purchasing a pellet stove, certified wood stove, or natural gas or propane heating unit increases fuel efficiency and reduces emissions. Upgraded stoves also reduce the health risks associated with older stoves: the amount of particulates emitted by EPA-certified stoves are 70 to 90 percent lower than those from conventional stoves.[10] Some states even offer tax deductions for wood stove upgrades.[11]

- **Recommendation**: Retrofit fireplaces and traditional wood burning stoves with pellet stoves, natural gas units, or EPA certified wood stoves. These units should have direct outside combustion air vented into the insert. Better yet, if you never use your fireplace, plug and seal the chimney flue.

Wood Smoke and Your Health

- According to the US Environmental Protection Agency (EPA), researchers suggest that the lifetime cancer risk from wood stove emissions may be greater than the lifetime cancer risk from exposure to an equal amount of cigarette smoke. Wood stoves are the second highest source of cancer risk from particulate air pollution (after diesel exhaust).[12]
- University of Washington researchers in 1990 found more symptoms of respiratory disease in Seattle preschool kids living in high-wood-smoke residential areas than in children living in areas with lower wood-smoke levels.[13]
- According to a University of Washington study in Seattle, wood smoke can enter homes where stoves are not used. Small particles from wood smoke in homes without wood stoves reach at least 50 to 70 percent of outdoor levels.[14]
- The largest single source of outdoor fine particles entering into our homes in many American cities is our neighbor's fireplace or wood stove.[15]

Turn Down the Temperature

- How you operate the controls of your heating system will have the most significant effect on your heating bills. One family can pay 100 percent more on their utility bill than a family in an identical house. For example, lowering the thermostat to 65 degrees from 70 degrees Fahrenheit saves 10 percent on your utility bill.[16] Likewise, setting your water heater temperature for 120 to 140 degrees Fahrenheit rather than the typical 160 to 180 degrees Fahrenheit, reduces fuel consumption by 5 to 10 percent on your utility bill.[17]
- **Recommendation**: Turn down the thermostat in unused rooms and whenever you don't need heat. Turn down the aquastat setting for hot water boilers to 120 to 140 degrees Fahrenheit. Shut off or turn down the duct dampers (adjustable metal flaps on air ducts that control flow) on the warm air furnace and registers that heat the basement. Turn down dampers and registers to control heat flow to various rooms (but don't cut off too much airflow so that the fan becomes stressed).

Furnace Ducts

Remodeling is a great time to change the way air flows into and out of your home. Air flows through ducts, or a branching network of metal or plastic and fiberglass conduits in the walls, floors, and ceilings from your home's furnace and central air conditioner to each room. Unfortunately these ducts often leak — affecting not only your heating and cooling bills, but also your indoor air quality, the safety of combustion equipment, and the potential for mold and decay. You should not focus exclusively on the part of the house you are remodeling; since the house operates as a system, sealing leaks in a room in another part of the house may make your remodeled room more comfortable.

Nationwide, over 20 percent of the hot or cold air that is supposed to go to rooms in your house never gets there — even in brand new homes.[18] Each year, US residential duct leakage cost consumers $5 billion.[19] This energy loss is equivalent to the potential annual oil production from the Arctic National Wildlife Refuge or the annual energy consumption of 13 million cars.[20] The carbon dioxide uptake of seven billion trees is needed to offset the global warming impacts of this energy waste.[21]

♀ – Use Duct Mastic or Aeroseal™ Instead of Duct Tape

- Duct tape loses its effectiveness and starts leaking in three to five years. Alternatively, duct mastic maintains the seal for decades. Also consider spraying a sort of non-toxic aerosolized chewing gum called Aeroseal™ into ducts that automatically lodges in cracks and seals them up. Well-sealed ductwork improves the performance of heating and cooling systems, making you more comfortable by cooling or heating the house more quickly, delivering more hot or cold air, and distributing heating and cooling more uniformly throughout your house. Sealing also reduces the entry of dust, excess humidity, automotive exhaust, radon gas, fumes from stored paints, solvents, pesticides, and combustion gases. Duct sealing can reduce heating and cooling energy use by up to 30 percent with annual utility bill savings of up to $300. In the US annually, this saves more than $1 billion and is the equivalent to the displacement of ten giant power plants.[22]
- **Recommendation**: Use duct mastic at every duct joint or Aeroseal™ for the whole system. It is particularly important for return ducts to be sealed.

♀ – Install Ductwork Within the Conditioned Space

- Duct runs in exterior wall cavities or in uninsulated spaces, like attics, garages, crawl spaces, and sometimes basements, lose significant amounts of warm or cool air because of insufficient insulation in these spaces.

Locating the ducts in conditioned space reduces energy loss and improves occupant comfort.

- **Recommendation**: All ductwork for heating or cooling should be run through conditioned space inside the insulated envelope. You can either move the ductwork to an insulated space, or insulate the space with the ductwork. Locate supply and return registers on walls or ceilings to help prevent debris and dust from falling into the ducts. Do not locate a wall return close to the floor because it acts like a vacuum cleaner, covering the heating and cooling coils with dust, dirt, and pet hair. Ensure that a well-sealed vapor barrier exists on the outside of the insulation on cooling ducts to prevent moisture buildup. Finally, protect and clean the ducts as you remodel; debris and dust from construction can cause allergic reactions in occupants.

◯ – Eliminate Panned Joist Space or Ceiling Cavities Used as Air Returns

- Using building panned joist space or cavities as an air handler return is common practice. These spaces are invariably leaky, resulting in significant energy losses. Installing a tight, fully-ducted system significantly reduces duct leakage and unintended infiltration on the return system. This in turn improves indoor air quality — especially if the air handler is located in the garage where car exhaust and gas/paint vapors can seep into your home.
- **Recommendation**: Install a tight, fully-ducted return system. If possible, abandon air handler returns in panned joist space or ceiling cavities.

◯, ♡ – Provide a Jumper/Transfer Duct

- Significant airflow is delivered to the master bedroom where the return airflow path typically passes. If the door is closed, this depressurizes the rest of the house and can lead to indoor air quality problems related to toxic combustion gases. A jumper transfer duct will reduce indoor air quality risks associated with house pressurization.
- **Recommendation**: Install a jumper or transfer duct between the hall and master bedroom to allow the air to return to the air handling unit.

♡ – Clean All Ducts Before Occupancy

- Debris and dust from construction can cause allergic reactions in occupants. Children are especially sensitive to micro particulates, like drywall dust.
- **Recommendation**: Seal all ducts in construction areas to reduce dirt entering the system, then clean all ducts before occupancy.

Ventilation

While you're considering upgrading your heating system, think about improving the air quality in your house. To a greater extent than with any other building system or component, heating and cooling methods can be a major cause of sick building syndrome — a sickness characterized by nasal congestion, lethargy, fatigue, irritability, difficulty concentrating, dizziness, headaches, lightheadedness, nausea, burning mucous membranes, and irritation of the eyes, nose, and throat. According to the World Health Organization, up to 30 percent of new or remodeled buildings have unusually high occupant health complaints.[23]

Ventilation systems can be a logical add-on since you can connect ventilation equipment to the furnace's forced-air system while you're having the new heating equipment installed and tuned. It is especially important if you're going to be air sealing your house as part of your renovation. For example, sealing windows or upgrading your insulation can reduce fresh air ventilation, increasing the potential for moisture or indoor pollution problems since you're filling cracks that used to allow air flow in and out. Minimizing air leakage makes a home more efficient; however, controlled ventilation also eliminates the stale and polluted air issue associated with "tight," air-sealed homes, while preserving their energy-efficiency.

There are four types of ventilation systems to consider:

- **Natural Ventilation**: Houses built before the 1980s have no controlled ventilation; they rely on windows and natural air flow through cracks. With natural ventilation, there is no way to ensure that you get fresh air when it is needed and to turn it off when it's not.

- **Exhaust-only Ventilation**: These are typically bathroom or kitchen exhaust fans. If your house is leaky, this is sufficient ventilation. However, if your house is "tight" it will need a more planned supply of fresh air. It is a good idea to look for exhaust fans with timer controls and humidity sensors. Timer controls allow you to turn on the fan when moisture level is high and have it turn itself off later. Humidity sensors turn the fan off when the humidity in the room reaches a certain level set on the humidistat.

- **Balanced Ventilation**: These systems have both ventilation fans, to remove moist and polluted air, as well as fresh air intakes to replace the air being vented. For example, balanced ventilation might include kitchen and bathroom exhaust fans in combination with a fresh air intake joined to your furnace's air circulation system.

- **Central Ventilation**: This mechanical fan exhausts stale air while pulling in outdoor air at the same rate. Additionally, many of these units recover heat or cold from indoor air before exhausting it to the outside and then transfer that heat or cold to the incoming fresh air stream.

The recommendations that follow will help you install the most efficient and healthy ventilation system for your home.

♡ – Incorporate Variability in Temperature and Ventilation Control
- The typical mechanical engineer strives to eliminate variability in human-made environments with thermostats, humidstats, and photosensors in order to maximize the conditions under which a significant portion of diverse people will feel comfortable according to a standard equation. Some newer systems found internationally can vary temperature and ventilation to achieve higher human comfort; this may be the future direction for controlling indoor air. People tend to be happier, healthier, and more alert under subtly dynamic conditions.
- **Recommendation**: Variable systems involve microchip control that delivers air in seemingly random gusts.

♡ – Install Operable Windows for Natural Ventilation
- Excess moisture can cause serious mold and decay problems in kitchens, bathrooms, and laundry rooms. Open windows can help keep smoke, cooking odors, and moisture from building up and causing problems in your home.

- **Recommendation**: Install operable windows in kitchens, bathrooms and laundry rooms; these are often required by code.

♡ – Vent the Kitchen Range Hood to the Outside

- Gases, smoke, and other combustion by-products (such as unburned hydrocarbons) often result from cooking. Not only do these particles smell, but they are unhealthy to breathe. Stovetop range hoods expel gases, steam, smoke, and other combustion by-products to the outside, preventing indoor air pollution, overheating, and excess moisture in your kitchen. Given the combustion gas issues associated with gas appliances, they are especially important for gas stovetops.
- **Recommendation**: Range hoods are positioned directly above the cooking surface and carry odors and pollutants out through the wall or roof via an air duct. For stoves or cook tops on an island, canopy hoods can be hung down from the ceiling. Try to avoid less effective vents that draw the exhaust downward into a duct. To minimize ventilation noise, look for models with "sone" ratings of 3.5 or below.

♡ – Properly Vent Laundry Dryer

- Poor ventilation in the laundry room can lead to the buildup of moisture, cleaning chemicals, synthetic material toxins, and carbon monoxide poisoning. An EnergyStar® exhaust fan will sufficiently exhaust unwanted moisture, mold, and potentially toxic detergent/fabric softener particulates from your living space.
- **Recommendation**: Vent laundry room exhaust fan to the outside, whether or not the room has an operable window. Make sure the exhaust opening on the exterior of your house isn't blocked. If airflow from the dryer is blocked either in the duct or at the opening, the dryer can overheat and cause a fire. If you have a choice of where to install your dryer, position it so that the ductwork is short and free of twists and turns; that will minimize the risk of lint buildup inside the ductwork. Use metal dryer ductwork. Rigid ducts are best, but flexible metal ductwork can be used where needed for turns. Never use foil or plastic ducts, which tend to kink and sag, creating pockets where lint or condensation can accumulate.

♡ – Install a Bathroom Exhaust Fan

- Bathrooms create a lot of moisture which can lead to unhealthy mold growth or building decay if it gets into the home's structure. Lingering odors are also a telltale sign of improper ventilation. Bathrooms are generally small, so a simple exhaust fan is capable of keeping them fresh and getting rid of the moisture generated by bathing and showering.

- **Recommendation**: To be effective, the bathroom fan must be connected directly to an outside vent or to your heat recovery ventilation with a well-installed air duct. Do not vent fans into the attic. (Refer to Install Heat Recovery Ventilator later in this chapter.) Fans tend to make a lot of noise, so look for a bathroom fan with a "sone" rating of 2 or lower.

♡ – Install Separate Garage Exhaust Fan

- According to the US Environmental Protection Agency, an attached garage is a significant contributor to poor indoor air quality.[24] Car exhaust contains many known carcinogens that can migrate into living spaces when doors are opened to the garage, or simply by leaking through cracks in the building envelope. An exhaust fan creates a healthier indoor environment by reducing the potential hazard of car exhaust from entering the house.

- **Recommendation**: Install an exhaust fan on the opposite wall from the door to the house. It can be wired to an electric garage door or put on a timer to run for 15 minutes after door has been opened or closed. In addition, make sure you seal common leaks between the garage and the living space, such as the bottom plate of wall between spaces and any penetrations through the wall.

♡ – Install a Carbon Monoxide (CO) Alarm

- Any fuel fired appliance, vehicle, tool, or other device has the potential to produce dangerous levels of carbon monoxide: high concentrations of CO can kill in less than five minutes. Continued exposure can cause irreversible damage to the nervous system, personality deterioration, and severe memory loss.[25] CO alarms protect you and your family from the serious dangers associated with carbon monoxide poisoning.

- **Recommendation**: CO alarms should meet Underwriters Laboratories, Inc. standards. They should have a long-term warranty, be easily self tested, and should reset to ensure proper operation. Battery-powered devices should have the batteries changed yearly. Buy the best brands you can find because less expensive brands tend to be ineffective.

𝒫 – Install Attic Ventilation Systems

- Attic ventilation lowers the temperature between the roofing and the ceiling insulation. In homes without air conditioning, attic ventilation systems are capable of venting out air that can be as hot as 160 degrees Fahrenheit. This improves occupants' comfort, saves electricity used for air conditioning, and increases the life of the roof assembly.
- **Recommendations**: Any unvented attic or under-ventilated attic is a candidate for ventilation retrofit. Install a balanced system of ridge and soffit natural ventilation. Solar powered attic fans are also effective since they run for free when the sun is creating the most heat.

𝒫 – Install a Whole House Fan

- Whole house fans work by continuously replacing warm indoor air that goes out through the fan with cooler outdoor air brought in through open windows. Moving large volumes of hot air out and cooler air in can achieve indoor comfort at higher temperatures without the need for air conditioning. In fact, an average whole house fan uses one-tenth the electricity of an air conditioning unit.[26]
- **Recommendation**: The fan must be mounted in a hallway ceiling on the top floor of a house. An airtight seal is required to prevent air infiltration or exfiltration when the fan is off. The fan should also have an insulated housing that prevents heat loss in winter. Fans should be sized to produce between 4 to 5 AHC within the home. Note that whole house fans are not well suited in extremely humid climates because they draw unwanted outdoor moisture inside the home. They should only be operated with windows open to prevent carbon monoxide backdrafting. (Refer to "Backdrafting" in the glossary.)

𝒫,♡ – Install a Heat Recovery Ventilator (HRV)

- An HRV, also known as an air-to-heat exchanger, is a boxy-looking device, usually installed near the furnace. In this system, heat or cold is captured from the exhausted air stream and transferred to the incoming air. HRVs can recover up to 85 percent of the heat or cold in the outgoing air stream,[27] and typically contain filters that keep particulates such as pollen and dust from entering your home.

- **Recommendation**: The HRV unit can be installed to operate independently or designed into the heating, ventilation, and air conditioning systems to capture exhausted air from bathrooms, kitchens, and the returns of forced air furnaces. HRVs are particularly appropriate for well-insulated homes. In climates with excessive outdoor humidity, an energy-recovery ventilator (ERV) is more suitable because they dehumidify the incoming air to eliminate unwanted moisture.

♡ – Install a High Efficiency Particulate Arresting (HEPA) Filter

- Forced-air heating and ventilation systems often transport a tremendous load of dust, mold spores, and other irritants that are carried in the air circulating around your house. The EPA has identified these micro-particulates as a leading cause of respiratory discomfort. Most filters on forced-air heating systems only remove the larger particles from the air in your home, therefore they protect your furnace, but not your health. However, a high-efficiency, particulate-arresting filter uses a complex air flow system to remove a higher percentage of particles. HEPA filters can remove 99.9 percent of the microparticulates from your air.[28] Other effective filter types include electrostatic (which remove 95 percent of airborne particles),[29] electronic (remove 95 percent of airborne particles),[30] and pleated media (remove 40 to 95 percent of airborne particles).[31] These filters help protect your furnace and your health.

- **Recommendation**: HEPA filters are installed in the return air stream at the air handler, which needs to be designed to handle the reduced air pressure caused by the filter. Some units have an air conditioning setting for the fan that will handle the retrofit filter. In other cases, high efficiency air filters slow down the air flow enough to cause your furnace blower motor to overheat. To counter this, the furnace's air flow might have to be increased.

♡ – Purchase a Portable Room Air Cleaner

- Portable room air cleaners use a fan to draw air through the machine, where particles are either trapped in a HEPA filter, or, for electronic precipitator models, in electrically-charged metal collector plates. A cleaner's success is

Source control tips.

Before shelling out money for an air purifier, consider source control. Source control prevents small particles and fumes from becoming a problem in the first place. Here are a few source control ideas:

- Damp-wipe surfaces.
- Vacuum regularly.
- Wash bedding and curtains in hot water to kill dust mites.
- Remove carpets (traps dust).
- Replace or seal cabinetry or furniture made of pressed wood (contains formaldehyde).
- Service combustion appliances regularly to decrease combustion gases.
- Use low-emission paints and varnishes.
- Maintain low humidity.
- Prevent moisture buildup indoors.
- Leave shoes at the door.

based on its "clean air delivery rate," or how quickly it filters out particles such as dust and tobacco smoke. Some models also contain carbon filters to eliminate odors. According to John Bower, the owner of the Healthy House Institute, air purifiers are merely a "band-aid approach which can reduce concentrations of pollutants but do not eliminate them." Although the cost of the best models average only $60.34,[32] annual energy bills can reach $200 for certain models.

- **Recommendation**: For homes without central air, consider portable room air cleaners.

♡ – Install a Humidification/Dehumidification System

- Humidity above 50 percent creates excess moisture and unhealthy mold growth. This makes your house look and smell bad. Certain micro-organisms cause building materials, not to mention your health, to deteriorate (refer to Moisture and Mold in Chapter 5). High humidity makes you feel hotter and you respond by setting the thermostat on your air conditioner lower than if the air were drier, which uses more energy than needed. On the other hand, humidity below 30 percent leaves your skin, hair, and mucous membranes dried out and you become more vulnerable to colds. This typically occurs in the winter when the furnace lowers the humidity. A dehumidification system will help keep humidity below 50 percent. To keep humidity above 30 percent during the winter or in dry environments, some experts recommend a steam or trickle type humidifier.
- **Recommendation**: When providing central heating and cooling, install whole house humidification and dehumidification. Check your humidifier for leaks monthly.

Air Conditioning (AC)

Air conditioning is more complicated than heating because instead of using energy to create heat, ACs use energy to take heat away. Like a refrigerator, there is a compressor on the outside filled with a special fluid called a refrigerant. This fluid changes from a liquid to a gas (and vice versa), all the while absorbing or releasing heat. In this way, it carries heat from a cooler place and dumps it into a hotter place. This process is not so easy: it requires a *lot* of energy. In fact, the Rocky Mountain Institute estimates that in peak summer months, about 50 percent of electricity used in the United States is devoted to powering ACs. This equates to about 100 million tons of carbon dioxide emissions (primary culprit of global climate change) from power plants every year.[33]

Typically, there are three types of ACs: room air conditioners, central air conditioners, and electric heat pumps. Room air conditioners are mounted in windows or walls to cool just one room. The whole house can be cooled by a central air conditioning system that uses the same duct system as the forced air heating systems. Alternatively, you can cool your home with a heat pump that works like an air conditioner, but also can be used for heating during the winter.

Surprising air conditioning fact.

You might assume that people turn on their air conditioner (AC) when they feel hot. You might also assume they set the AC at a temperature that feels most comfortable. Don't assume anything! In reality, only 25 to 35 percent of people actually associate "cooling" with feeling good in hot weather. Most people run their AC occasionally, and often in ways unrelated to comfort. Their usage depends on six other factors: household schedules; folk theories about how air conditioners work (many people think the thermostat is a valve that makes the cold come out faster); general strategies for dealing with machines; complex belief systems about health and physiology; noise aversion; and (conversely) wanting white noise to mask outside sounds that might wake the baby.

Source: Paul Hawken, Amory B. Lovins, and L. Hunter Lovins, Natural Capitalism: Creating the Next Industrial Revolution, *Little Brown, 1999, p.265.*

What is your air conditioner costing you?

If you use central air conditioning, the air conditioner will probably be the biggest energy user in your home, even though it is used only a few months out the year.[34] You can get a very rough idea of what your air conditioner is costing you by subtracting the electric portion of your bill in a spring month when you aren't using your air conditioner from the electric portion of the bill in the summer when you do use it. This gives you the monthly cost. Multiply this by the number of months you use your air conditioner to arrive at your approximate annual cost. In hot climates, your annual air conditioner cost can easily exceed a thousand dollars.

However, keep in mind the purpose of an air conditioner should be to keep home occupants cool — and there are many ways to stay cool beyond conventional, refrigeration-based air conditioning. Remodeling is the perfect time to consider incorporating shading, natural ventilation, thermal mass, and fans that reduce the need for costly, resource-consuming air conditioners. The following recommendations will help you reduce cooling loads in your home that reduce energy use by 20–50 percent.[35]

💡 – Eliminate Sources of Unwanted Heat

- The three sources of unwanted heat in your home are: heat that conducts through your walls and ceiling from outside air; heat given off by lights and appliances inside your home; and sunlight that shines through your windows. When you eliminate the sources of heat in your home, you minimize the need to install costly, energy-consuming cooling appliances.
- **Recommendation**: To reduce heat that conducts through your walls and ceilings to the interior of your home, you can insulate, tighten, and ventilate your house. To reduce heat used for lights, refrigerators, stoves, washers, dryers, dishwashers, and other household appliances, use energy-efficient appliances and lights that produce less waste heat. To reduce solar gain through windows, shade east and west windows that add the largest amount of unwanted summertime heat.

💡, 🤍 – Install Operable Windows and Fans

- Natural ventilation can help you get rid of unwanted heat: simply open screened windows at night. Window fans are another low-cost option that should be installed on the downwind side of the house facing out. Doors should be open and a window should be open in each room to allow air to flow. They will not work as well in homes with long, narrow halls or small rooms with multiple partitions. In addition, ceiling paddle fans improve interior comfort by circulating cold and warm air. They can be adjusted either to draw warm air upward during summer months or to push it downward during the winter. Ceiling fans must be supported adequately between ceiling joists. Ceiling fans run on 98 percent less electricity than most air conditioners,[36] thereby substantially reducing your home energy costs. In northern US states and mountainous areas, ventilation can eliminate the need for air conditioning all together, especially in spring and fall seasons. Moreover, operable windows have been associated with enhanced learning in children: in a California study commissioned by Pacific Gas and Electric, it was shown that classrooms with operable windows were associated with seven to eight percent greater improvements in test scores than students in classrooms with fixed windows.[37]

• **Recommendation**: In order for ventilation to cool the house most effectively, the temperature of incoming air should be at least 5 degrees Fahrenheit lower than indoor air; therefore, fans are most effective at night and on cooler days.

𝒫 – Install Evaporative Cooling Units

• Sometimes called "swamp coolers," evaporative coolers work by blowing house air over a damp pad or by spraying a mist of water into the house air. The dry air evaporates moisture and cools off. This is the same process as a breeze that makes you feel cooler when you get out of a swimming pool. Even though evaporative cooling units use a significant amount of water, they are often preferable to energy-intensive AC units. They cost about half as much to install as central air conditioners and use about one quarter as much energy; sophisticated swamp coolers employ wood fiber or paper as the water medium and come complete with a blower for under $100.[38] However, they require more frequent maintenance than refrigerated air conditioners and they're suitable only for areas with low humidity.

• **Recommendation**: Install an evaporative cooling unit in place of an air conditioner. They are most practical in dry, hot climates. Replace evaporative media twice per year.

𝒫 – Install Ductless ("split systems") Air Conditioning Units

• In a ductless system, copper tubing and electrical wiring to indoor units connects to a separately installed outdoor unit. Refrigerant is pumped from the outdoor condenser coil and compressor through the tubing to the indoor unit or units. A fan then quietly distributes cool air across the unit's evaporator coil. Unlike an air conditioner, the amount of cold entering each individual room in the house can be controlled by a thermostat or it can be regulated by remote control; this is called "zone control." They offer quiet air conditioning as well as a heating option; are

Fig 16.1: Evapourative cooling unit. As dry air absorbs moisture, the air temperature is lowered inside the building.
Credit: David Johnston.

Strategies to Avoid Using Your Air Conditioner

- Use natural or forced ventilation at night.
- Keep the house closed up tight during hot days. Don't let in unwanted heat and humidity.
- Don't cool unoccupied rooms.
- If possible, relocate your freezer to the basement or garage where it won't contribute its waste heat to your living space.
- Try to plan your meals carefully to minimize the use of the oven on the hottest days.

- Increase comfort at warmer levels by lowering the humidity: use a bathroom exhaust fan when you shower; vent your range hood to the outside; use a laundry room exhaust fan; locate your clothes washer in the garage or basement; don't dry firewood in your basement; don't vent your clothes dryer inside; and put those house plants outside during the summer.
- When possible, delay heat- and moisture-generating activities like dishwashing and clotheswashing until the evening on hot days.

simple to control, small, and efficient; and come in a variety of shapes and sizes to fit any design scheme.

- **Recommendation**: Install a ductless system in place of standard central AC. A contractor needs to set up the outdoor condenser coil and compressor, like a standard AC unit, but instead of installing an indoor evaporator coil unit and ductwork and registers, he or she will drill holes for the incoming and outgoing piping, hang the indoor fan coil units and temperature controllers in the room, and then connect the piping and power lines. Be sure to sure to seal the penetrations through the wall with foam or caulk. Ductless systems are also available as heat pumps to provide economical heating.

𝒟 – Use Your AC Wisely!

- Not only do air conditioners use a lot of energy while in operation, but simply installing the unit can cost anywhere from a few hundred dollars to $10,000. A switch to high-efficiency air conditioners and implementation of measures to reduce cooling loads in homes can reduce cooling energy use by 20 to 50 percent.[39]

  Central air conditioners are rated according to their seasonal energy efficiency ratio (SEER), which is the seasonal cooling output in Btu divided by the seasonal energy input in watt-hours for an average US climate. The

Heating, Ventilation, and Air Conditioning Checklist

Heating

- [ ] Install a 90 percent or greater AFUE forced air furnace.
- [ ] Install sealed combustion furnaces.
- [ ] Install zoned, hydronic, radiant heating.
- [ ] Install a geothermal heat pump.
- [ ] Install an air-source heat pump.
- [ ] Properly size your HVAC system.
- [ ] Consider alternatives to fireplaces and wood stoves.
- [ ] Turn down the temperature.

Furnace Ducts

- [ ] Use duct mastic or Aeroseal™ instead of duct tape.
- [ ] Install ductwork within the conditioned space.
- [ ] Eliminate panned joist space or ceiling cavities used as air returns.
- [ ] Provide a jumper/transfer duct.
- [ ] Clean all ducts before occupancy.

Ventilation/Filtration

- [ ] Incorporate variability in temperature and ventilation control.
- [ ] Install operable windows for natural ventilation.
- [ ] Vent the kitchen range hood to the outside.
- [ ] Properly vent laundry dryer.
- [ ] Install a bathroom exhaust fan.
- [ ] Install separate garage exhaust fan.
- [ ] Install a carbon monoxide (CO) alarm.
- [ ] Install attic ventilation systems.
- [ ] Install a whole house fan.
- [ ] Install a heat recovery ventilator (HRV).
- [ ] Install a HEPA (high efficiency particulate arresting) filter.
- [ ] Purchase a portable room air cleaner.
- [ ] Install a humidification/dehumidification system.

Air Conditioning

- [ ] Eliminate sources of unwanted heat.
- [ ] Install operable windows and fans.
- [ ] Install evaporative cooling units.
- [ ] Install ductless air conditioning units.
- [ ] Use your AC wisely!

efficiency of room air conditioners and heat pumps are measured by the energy efficiency ratio (EER), which is the ratio of the cooling output (in British thermal units or Btus) divided by the power consumption (in watt-hours). The ratings are posted on an Energy Guide Label that must be conspicuously attached to all new air conditioners. Many air conditioners are also participants in the voluntary EnergyStar® labeling program that only labels products with high EER and SEER ratings.

- **Recommendation**: Look for a quiet EnergyStar® AC unit with minimum SEER 12/EER 9, variable speeds, a fan-only option, and a filter when you replace your AC. These options help energy-efficiency, humidity, and noise. Also, correctly size your unit — a ton of cooling should equate to a minimum of 600 square feet or more of living space if your home is already somewhat designed for energy efficiency. Strategically place the AC registers in a shaded spot and only in rooms where it's needed. Install multiple-zone, programmable thermostats set at 78 degrees Fahrenheit or higher. You will save a minimum of 3 to 5 percent on AC costs for each degree you raise the thermostat.[40]

CHAPTER 17

Water Heating

Replacing Your Water Heater

A FTER HEATING AND COOLING, water heating is typically the largest energy user in the home.[1] Therefore, it pays to choose a replacement for your worn-out existing water heater carefully. You should not simply buy the cheapest model available; rather, take into account its life cycle cost, or how much it will cost after 13 years (see chart at end of section). The key is to plan ahead!

If your water heater is about ten years old, have your plumber assess how much useful life it has left. You may need to replace it now before it starts leaking or the burner stops working. Often it even makes economic sense to replace an old inefficient model with a high-efficiency one before the old model fails. The energy savings alone could pay for the new water heater after just a few years. You'll dump fewer pollutants into the air and dollars down the drain.

In general, you want your water heater to satisfy your specific hot water needs and to use as little energy as possible. Energy performance between different products can be compared using energy factors (EF) — products with higher EF ratings are more efficient.

Energy Factor (EF) is the rate of the amount of the fuel's energy (under average conditions) that comes out of the tap in hot water. In other words, if a gas burning heater could use all the energy in the gas for heating water, and could then get that water to the tap without losing any heat, the tank would have an EF of .10. The EF takes into account heat lost up the flue, heat lost from the water sitting in the tank, and how much of the fuel's energy gets captured in the first place. The most efficient gas-fired water heaters will have EFs of .62 or higher, while some units heated by electric heating elements, which produce no combustion waste to escape up a flue, have EFs above .96.

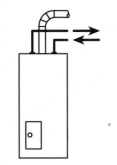

17.1a: Storage water heater.

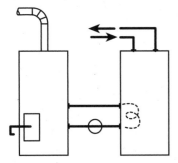

17.1b: Indirect water heater.

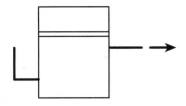

17.1c: Heat pump water heater.

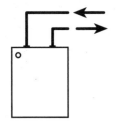

17.1d: Tankless water heater.

Fig 17a, b, c, e & f: Types of water heaters.

The following options will help you identify the most energy-efficient water heaters for your home.

🖉,♡ – Install a High-efficiency Sealed-combustion Fossil-fuel Water Heater

• Water heaters that burn propane, natural gas, or oil are the most common water heaters. They typically use house air for combustion and exhaust the combustion byproducts by natural draft. However, there is a risk of backdrafting, in which combustion gases such as carbon monoxide are drawn back into the home as a result of negative pressure in the house. Backdrafting is fairly common: according to the Colorado Department of Health, an average of more than six people in Colorado are killed and 50 people accidentally poisoned by CO produced from furnaces, gas appliances, and kerosene heaters every year. In tests by home energy raters in Colorado, as many as 30 percent of new homes had backdrafting problems. These numbers represent only the cases that are properly diagnosed and reported to the state.[2] Sealed-combustion units eliminate the risks associated with backdrafting. They also save energy because they do not waste heated or cooled indoor air. High EF models further save energy because they have a greater heat transfer surface, or a more submerged combustion chamber so that heat is transferred to the water and does not escape into the air.

• **Recommendation**: Look for a sealed-combustion water heater with a high EF.

🖉,♡ – Install a High-efficiency Electric Resistance Storage Water Heater

• An electric-resistance water heater has electric heating elements submerged in a storage tank. Like all storage water tanks, electric resistance models work by pulling water from the top of the water heater, when you turn the hot water tap, while cold water flows into the bottom to replace it. The hot water is always there, ready for use, but heat is lost through the walls of the storage tank even when no hot water is being used. Although electric resistance is typically the most expensive way to heat your home, these heaters are easy to install and last longer than fossil-fueled tanks. They require no special venting, supply air, or gas lines. Some electric-resistance heaters even come with plastic-lined or cement-lined tanks that eliminate corrosion. From an indoor air quality standpoint, electric water heaters do not pose risk of combustion gas backdrafting. In general, electric water heaters are a good option if you don't use much hot water or if electric rates are low.

- **Recommendation**: Look for new EF, electric storage water heaters that contain higher levels of insulation around the tank to substantially reduce standby heat loss.

Standby losses.

The heat conducted and radiated from the walls of the tank — and in gas-fired water heaters — through the flue pipe. These standby losses represent 10 to 20 percent of a household's annual water heating costs.

Source: US Department of Energy, "Demand (Tankless or Instantaneous) Water Heaters," Home Energy Magazine, <www.eere.energy.gov/consumerinfo>, cited October 13, 2003.

🔦 – Consider a Tankless Water Heater

- Most houses have storage water heaters, which allow many users to draw hot water at once. However, these hot water heaters lose significant energy through standby losses. You can avoid standing tank losses by installing a tankless water heater (also called a "flash," "demand," or "instantaneous" water heater). Tankless water heaters heat water only as it is used rather than having a tank in which hot water is stored. A tankless unit has a heating device that is activated by the flow of water when a hot water valve is opened by turning the faucet.

Tankless water heaters are quick and reliable. Whereas storage tank water heaters last ten to 15 years, most tankless models last more than 20 years; moreover, their easily replaceable parts may extend their life by many more years. Tankless units save energy by not heating water until necessary, thereby reducing energy consumption 20 to 30 percent.[3] Their capacity to provide hot water is virtually unlimited and you never have to wait for hot water to arrive.

Tankless water heaters are available in propane, natural gas, or electric models. They come in a variety of sizes for different applications, such as a whole-house water heater, a hot water source for a remote bathroom or hot tub, or as a boiler to provide hot water for a home heating system.

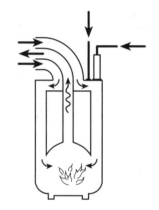

17.1e: Sealed combustion gas water heater.

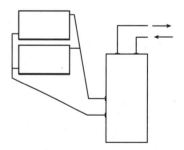

17.1f: Solar water heater.

Credit: Jill Haras & Kim Master, redrawn by Jeremy Drought.

Table 17.1: Power Supply Requirements

	Gas	Electric
Storage H20 Heater	40,000 Btu/h	6,000 watts
Tankless Heater	170,000 Btu/h	28,000 watts

Source: Home Energy Magazine. *No Regrets Remodeling. Energy Auditor and Retrofitter, 1999. p.120.*

- **Recommendation**: In general, a tankless water heater is a good option if you don't have many people in your house, if you're adding on faucets far from the current water heater, if you are remodeling a guest cottage or summer home, or as a back up for a solar water heater. It can also be useful to install one as a booster heater for a dishwasher because you can keep your storage water heater set at a low temperature (saving energy) and the booster's small size lets you fit it in small spaces.

◌ – Consider a Heat Pump Water Heater

- Heat pumps move heat from the surrounding air into the water. The heat pump is backed up by electric heating elements in the water tank for times when demand outruns supply. Heat pumps are similar to air conditioners, except that they can reverse their cycle to pump heat into a home in addition to removing heat. Heat pumps are expensive, but they are a good alternative if electricity is your only available source of energy. A heat pump water heater uses one-third to one-half as much electricity as a conventional electric-resistance water heater,[4] thereby saving 25 to 45 percent of the cost of heating water.[5]
- **Recommendation**: Heat pump water heaters may be purchased as integral units with their own storage tanks (called one-piece systems), or they may be added on to electric-resistance water heaters. Heat pump systems should be designed and installed by experts indoors because they can freeze up if the temperature drops below 45 degrees Fahrenheit. They should also be in an unconfined space, since they need lots of surrounding air from which to extract heat.

Table 17.2: Evaluating Water Heaters

Energy Source	Efficiency	Energy Cost	Cost/yr	Life (yrs)	Cost/13 yrs
Conventional Gas Storage Water Heater	55%	$425	$190	13	$2,895
High Efficiency Gas Storage Water Heater	60%	$500	$174	13	$2,762
Conventional Electric Storage	90%	$425	$454	13	$6,327
High Efficiency Electric Storage	95%	$500	$430	13	$6,090
Tankless (gas)	70%	$650	$160	20	$2,503
Tankless (electric)	100%	$600	$404	20	$5,642
Heat Pump	200%	$1,200	$204	13	$3,852
Indirect Water Heater	75%	$700	$148	30	$2,227
Solar Water	N/A	$3,000	$144	20	$3,822

Source: Alex Wilson and John Morrill. Consumer Guide to Home Energy Savings.
American Council for an Energy-Efficient Economy, 199 1. p. 147.

$, ⏾ – Consider Combined (Indirect) Hot Water and Heating Systems

• If you have a boiler or a heat pump for home heating, you can use it to provide hot water in what is called a combined, or indirect, system. For boilers, hot water is circulated through a heat exchanger in a separate insulated tank. Since hot water is stored in an insulated storage tank, the boiler does not have to turn on and off as frequently, improving its fuel economy. For heat pumps, a heat reclaimer can move waste heat from an air conditioner, air source heat pump, or ground source heat pump into the water tank. When used in combination with new, high-efficiency boilers or heat pumps, indirect water heaters are usually the least expensive way to provide hot water.

• **Recommendation**: An indirect water heater can be used in combination with a heat pump or added to an existing boiler. If you are building a system from components, make sure you are working with contractor and manufacturer's representative to size it correctly. For example, you can usually use a smaller storage tank for tap water because a space-heating boiler heats up water very quickly and needs fewer reserves.

𝒫 – **Install a Solar Water Heater**

Refer to "Install a Solar Water Heater" in Chapter 15.

Installing a Water Heater

Energy factor.
The portion of the energy going into the water heater that gets turned into usable hot water under average conditions. Takes into account heat lost through the walls of the tank, up the flue, and in combustion.

It is probably a good idea to hire a plumber to replace your water system because working on a hot water system is dangerous. Sources of hazard include hot water, gas, and electricity. Ask for bids from several contractors and evaluate the warranties, service, reputation, and price. Make sure your contractor has experience with the type of system you want to put in. The following recommendations will help you (or your contractor) properly install your water heater:

𝒫 – **Use the Smallest Water Heater Possible**

• An oversized tank is less efficient because you are heating more water than you need and its increased surface area loses more heat. A smaller, more efficient water heater will save you energy and money.

• **Recommendation**: Get help from a professional to figure out the correct size of water heater for your home.

𝒫 – **Locate Water Heater Close to Bathroom, Kitchen and Clothes Dryer**

• Locating water heaters close to the areas where you need hot water will allow for shorter distribution runs that minimize heat loss. This way, no one in your household has to waste time and water waiting for hot water to arrive.

• **Recommendation**: Install your heater close to where you need hot water.

𝒫 – **Set the Water Temperature to 120 Degrees Fahrenheit**

• Typically, water heaters have three temperature settings: high, medium, and low. These settings correspond roughly, depending on the age and condition of the water heater, to about 160 degrees Fahrenheit for high, 140 degrees

Fahrenheit for medium, and 120 degrees Fahrenheit for low. Most people have the temperature set to medium, or around 140 degrees Fahrenheit. If it is not already set there, you should consider lowering the thermostat to 120 degrees Fahrenheit.

Hazards of hot tap water.
Hot tap water is a scalding hazard, especially to children and seniors. Scalding occurs in two seconds at 150 degrees Fahrenheit while it takes ten minutes to be scalded by 120 degrees Fahrenheit water.

Higher temperatures lead to scalding, heat loss, and corrosion. Each drop of ten degrees Fahrenheit in temperature will generally save 3 to 5 percent on your water heating costs.[6]

- **Recommendation**: When you go away for several days, turn the thermostat down to the lowest possible setting, or better yet, turn the water heater off. When you are home, turn the water heat down to the lowest setting. Water heater thermostats are notoriously inaccurate, so just turn yours down little by little, each time giving the tank a few days to adjust until the water is hot enough for your needs. Electric water heaters often have two thermostats — one each for the upper and lower heating elements. These should be adjusted to the same level to prevent one element from overloading and wearing out prematurely. Note that not all households can implement this measure, due to an automatic dishwasher without a booster heater or the need for a very large amount of hot water.

Upgrading Your Water Heater

Even if you aren't going to purchase a new water heater, there are many ways to save a lot of energy and money with your existing system. Here a few simple recommendations.

⌀ – Install Hot Water Jacket Insulation

- Water heating is usually one of the largest energy users in the house, accounting for 15 to 40 percent of your utility bill.[7] To save energy, install water heater jacket insulation. Jacket insulation is an insulated wrapper that goes around the hot water tank; once it is secured in place it will reduce standby losses by 25 to 45 percent.[8] Insulation blankets are easy to install, and their cost (a mere $10 to $25), will be recouped within a year. In other words, a hot water jacket could save you hundreds of dollars over the lifetime of your heater.

- **Recommendation**: Wrap the tank with a fiberglass water heater blanket to reduce standby losses (the energy lost through the walls and up the flue of a hot water tank). Make sure you buy an insulation blanket designed for your type of water heating system. For most older water heaters, it will be cost-effective to wrap it with an R-12 blanket, but these can be hard to find. R-6 (try two of them) will do if you can't get an R-12. With gas heaters, take special precautions not to block the air intake opening and to keep the insulation from touching the flue; this is essential for the heater to function properly and to avoid a fire hazard. Note that some superinsulated water heater models recommend not adding a blanket, so double-check the manufacturer's recommendations as the blanket may void the warranty in these cases. Insulation blankets can be found in hardware and home stores.

𝒫 – Install Heat Traps

- Hot water is buoyant, so it tends to rise up from the tank into the pipes where it loses heat. When it cools, the water sinks back down into the tank. This tendency, known as "thermosyphoning," should be eliminated whenever possible to minimize heat loss and save energy. Heat traps are simple, one-way valves that can resolve thermosyphoning by preventing the cooling water in the pipes from settling down again into the tank where it is replaced by hot water. A heat trap will reduce standby losses by about 25 to 45 percent,[9] and while it is easiest to install when replacing a heater, the savings are large enough to justify adding one to your existing system at any time.

- **Recommendation**: Insert a heat trap where the pipes enter the tank. They cost about $3 and are best installed when your water heater is being replaced.

𝒫 – Install a Timer on Your Water Heater

- Consider a timer for your water heater that turns it off when not in use, typically between 10 or 11 P.M. to 6 A.M. Even though some energy is required to heat up the water, regardless of whether it is heating a little bit at a time (throughout the night), or all at once (in the morning), the heat

Water Heating Checklist

Replacing Your Water Heater

❏ Install a high-efficiency sealed-combustion fossil-fuel water heater.

❏ Install a high-efficiency electric resistance storage water heater.

❏ Consider a tankless water heater.

❏ Consider a heat pump water heater.

❏ Consider combined (indirect) hot water and heating system.

❏ Install a solar water heater.

Installing a Water Heater

❏ Use the smallest water heater possible.

❏ Locate water heater close to bathroom, kitchen, and clothes dryer.

❏ Set the water temperature to 120 degrees F.

Upgrading Your Water Heater

❏ Install hot water jacket insulation.

❏ Install heat traps.

❏ Install a timer on your water heater.

losses through the tank walls and pipes wastes energy overnight and adds to the cost. A simple timer costs about $30 plus installation and may take about a year to pay for itself. Timers for gas water heaters are not nearly as useful or cost-effective because the gas pilot is always running, limiting the savings.

• **Recommendation**: For electric water heaters, install a timer that can automatically turn the hot water off at night and on in the morning, perhaps an hour or so before you get up in the morning, so your hot shower is ready to go when you are.

CHAPTER 18

Appliances

APPLIANCES ACCOUNT FOR ABOUT 20 percent of household energy usage.[1] Therefore, keeping appliances in excellent condition, even if that means replacing them, can greatly reduce your energy consumption. The Energy Guide label on appliances provides a fast and easy way to compare the energy efficiency of most home appliances. The label helps you find certified energy-efficient appliances that meet strict energy-efficiency guidelines set by the Environmental Protection Agency and US Department of Energy. These appliances can save families about a third on their appliance energy bill with similar savings of greenhouse gas emissions, without sacrificing features, style, or comfort.

♀, ⚒ – Install Horizontal-axis (H-axis) Washing Machines

• The standard American vertical-axis design washing machine, accounting for 23 percent of indoor water use, agitates clothes in a big tubful of water at 400 to 500 revolutions per minute (rpm). In contrast, horizontal washing machines load from the front and rotate on a horizontal axis, spinning clothes in and out of the water at 1,000 rpm or faster to tumble them clean.[2] New generation vertical axis machines can be as efficient as horizontal axis models.

Although current prices of horizontal axis machines are $200 to 500 higher than vertical axis machines, the benefits of the horizontal design far outweigh this initial cost. Compared to a

Beware of phantom loads!

Many modern appliances remain partially on when they appear to be turned off. Anything with a clock — VCRs, coffee makers, microwave ovens — all use a small amount of power all the time. And anything that uses a "power cube" in the AC socket, such as answering machines and electric toothbrushes, use very tiny amounts of power, maybe only a watt or two, but they stay turned on and running 24 hours per day. Unplug your appliances or the surge arrester whenever possible. We also recommend that power cubes be kept on plug strips or switchable outlets that can be switched off when not in use.

ten-year-old model, an EnergyStar® qualified washer can save up to $120 per year on your utility bills. Through superior design and system features, EnergyStar® qualified clothes washers clean clothes using 50 percent less energy and 15 to 22 fewer gallons of water per load than standard washers.[3] The front-loading configuration of most H-axis washers allows easy stacking of washer and dryer to fit into tight spaces. Quiet operation, in at least some of the machines, allows placement in occupied rooms. Clothes get cleaner using less detergent than standard vertical-axis machines, and there is less wear-and-tear on clothing.

- **Recommendation**: Install an EnergyStar® H-axis washing machine whenever washing machines are replaced. Be wary that vertical axis machines may have high EnergyStar® ratings because they are rated in comparison to other conventional vertical axis machines, not the significantly more efficient H-axis washing machines.

⏚ – Use an Energy-efficient, Sealed-combustion Clothes Dryer (or No Dryer!)

- Like most appliances, clothes dryers can be gas or electric. Where natural gas is available, gas dryers have been historically less expensive and more efficient to operate, but have a higher initial cost. Energy-efficient sealed-combustion dryers save energy and money over time. Check with your local utility and state energy office for rebates and other incentives that may be awarded for installing energy-efficient appliances. Clothes lines provide free clothes-drying that wastes no energy.

- **Recommendation**: Provide an outdoor clothesline to dry your clothes on; this eliminates the need to replace or buy a clothes dryer in the first place. If you do need to buy a dryer, look for an energy-efficient sealed-combustion unit. Also, consider clothes dryers with moisture and temperature sensors that

EnergyStar®

The EnergyStar® label is the government's seal of approval. It was created by the US Department of Energy and the US EPA; they set the criteria in order to help shoppers for large and small home appliances identify the most energy-efficient products on the market. EnergyStar®-labeled appliances exceed existing federal efficiency standards, typically, by 10 to 30 percent, and as much as 110 percent for some appliances. Customers can be assured that the appliance being purchased is a high-performance product which will reduce the operating cost of that appliance or product every month during the course of its lifetime.

shut off automatically, or those which offer energy-saving settings. Locate your dryer in a heated space to improve efficiency. Ensure the outside dryer exhaust vent is clean and that the flapper on the outside hood opens and closes freely so that cold or humid air will not blow into your house. Your dryer vent should be made of metal, have a short run, and be free of bends or kinks for smooth and fast venting.

⌀, ♨ – Replace a Standard Dishwasher with a Low Water Use Model

- Water heating comprises 80 percent of the energy consumed by a dishwasher. In general, models that use less water use less energy. Energy-Star® dishwasher models are quiet, easy to use, and reliable — and they use at least 25 percent less energy than is called for by government standards.[4] In effect, replacing a ten-year-old dishwasher with an Energy-Star® qualified dishwasher can save you more than $30 a year in energy costs.[5] Based on electric and gas savings alone, energy efficient models are cost efficient when they are purchased for a premium of less than $80 to $240 (electric water heating) and $50 to $130 (gas water heating) more than standard models. Water savings provide additional savings.

- **Recommendation**: Install low water use models whenever dishwashing machines are replaced. Select EnergyStar® qualified dishwashers with light, medium, and heavy wash settings; an energy saving "no heat dry" or "air-dry" switch; and a booster or supplemental water heater that allows the water heater to be set to 120 degrees Fahrenheit. Also consider an ultrasonic dishwasher that adjusts your water use to match the dirtiness of the load.

⌀ – Look for the EnergyStar® Label on Gas Appliances

- Consider the life cycle costs when calculating the benefit of buying a gas appliance. In general, although gas appliances are more expensive to purchase, they are less expensive to operate and better for the environment. The EPA is working with hundreds of voluntary manufacturer partners to provide more efficient appliances with special EnergyStar® labels. EnergyStar® appliances can save the typical US household about 30 percent of its appliance energy bills with a 30 percent internal rate of return. Over the next 15 years, full adoption of EnergyStar® appliances could save American households as much as $100 billion.[6]

Discarding old refrigerators and freezers.

Old refrigerators and freezers contain CFCs, a hazardous chemical. Only discard refrigerators where trained personnel are available to recover the gas for reuse or proper disposal.

• **Recommendation**: Look for EnergyStar® labels on every new gas appliance you purchase. However, in light of unwanted gas stove combustion particulates that affect indoor air quality, consider electric stove tops (preferably with ceramic glass covers). Better yet, use the most efficient option — induction cook tops. Induction elements use electromagnetic energy to heat the pan without any heating element that can cause skin burns.

– Buy the Best Available Technology Refrigerators

• Most in-service refrigerators are poorly insulated boxes with their inefficient compressor mounted at the bottom, so its heat rises up into the food compartment. They typically have an undersized and dust-clogged condenser on the back. Internal heaters prevent "sweating" caused by thin insulation, while inefficient lights, fans, and defroster coils inside generate still more heat. A typical refrigerator uses so much electricity that the coal burned to generate it would about fill the whole inside of the refrigerator every year.[7] An automatic ice-maker is a popular refrigerator feature that increases energy use by 20 percent.[8]

EnergyStar® qualified models use at least 10 percent less energy than required, and 40 percent less energy than the conventional models sold in 2001. Replacing a 10-year-old refrigerator bought in 1990 with a new EnergyStar® qualified model would save enough energy to light the average household for over three months and eliminate over 300 pounds of pollution each year.[9] As a result of more insulation, tighter door seals, larger coil surface area, and improved compressors and motors, food stays fresher for longer, the units make less noise, and they are more reliable.

• **Recommendation**: Replace your old refrigerator (more than 10 years old) with the best available technology refrigerators. EnergyStar® top-freezer models, are the most energy-efficient and repair-free of the configurations offered.[10] Don't buy a unit bigger than you need. You need 12 cubic feet for the first two people in the house; for 3 or 4 people, you many need 14 to 17 cubic feet. For each extra person, add another 2 cubic feet.

Appliance Checklist

❑ Install horizontal-axis (H-axis) washing machines.

❑ Use an energy-efficient sealed-combustion clothes dryer (or no dryer!).

❑ Replace your standard dishwasher with a low water use model.

❑ Look for the EnergyStar® label on gas appliances.

❑ Buy the best available technology refrigerators.

Placement of the refrigerator is very important. Direct sunlight and close contact with hot appliances will make the compressor work harder. More importantly, heat from the compressor and condensing coil must be able to escape freely, or it will cause the same problem. Don't suffocate the refrigerator by enclosing it tightly in cabinets or against the wall. The proper breathing space will vary depending on the location of the coils and compressor on each model — something important to know before the cabinets are redesigned.

CHAPTER 19

Interior Materials/Finishes

ALTERING INTERIOR CABINETS, SHELVING, COUNTERTOPS, flooring, painting, and wallpapering are popular ways people use to fix up the interior of their homes. While they do not entail structural changes, interior modifications can significantly affect the look and feel of your home since it may be the largest visual component. More importantly, the finishes you use can have a serious impact on not only the health of the structural materials but also the health of your family.

Nearly all the toxins covered by US environmental regulations are found in highest concentrations indoors. In the February 1998 issue of *Scientific American*, authors Wayne Ott and John Roberts revealed that indoor air contains at least 5 (but typically 10 or more) times higher concentrations of pesticides than outside air. The authors cite another study that alarmingly reports, "In more than half the households ... surveyed, the concentrations of 7 toxic organic chemicals [which are thought to induce cancer in humans] were above the levels that would trigger a formal risk assessment for residential soil at a Superfund site."[1] Regrettably, environmental laws for indoor air quality are negligible compared to laws for Superfund sites; therefore, your health depends on *you* to make healthy choices for your home.

Here are some strategies to help you plan healthy, cost-effective finishing touches on your home.

♡ – Use Formaldehyde-free Materials

- Particleboard, interior grade plywood, and medium density fiberboard (MDF) are typically used for cabinets, counter tops, stair treads, and shelving. Unfortunately, they are made from wood fibers and an adhesive that contains urea formaldehyde, a suspected human carcinogen. The formaldehyde offgases, or releases into the air in your home, for years after installation. Exterior grade plywood uses phenolic resins that offgas much less formaldehyde than interior plywood. Some boards made from

Fig. 19.1: Reclaimed wood countertop adds an attractive, durable finish to this remodeled kitchen. Credit: Ethan Kaplan, Leger Wanaselja Architecture, San Francisco, CA.

Surviving Home Remodeling with Allergies

Follow these tips from the National Association of the Remodeling Industry (NARI) to reduce allergens during your home remodeling project:

- Inform your remodeling contractor during your first meeting about any allergies.
- Seal off the area to be remodeled with plastic sheets; leave plastic sheeting up until cleanup from the job is complete. When sheeting is removed and discarded, avoid carrying it through the home.
- Close all vents in the room being remodeled so that dust won't travel through the air ducts.
- Open windows in the room being remodeled at least a crack and set up a fan to exhaust air toward the outside.
- Keep family and pets out of the work area. They can track dust and mites all over the house.

- Vacuum and sweep daily to prevent dust from spreading.
- Change your furnace filters often since it will pick up more particles than usual.
- Plan for removing debris. Using a chute out the window is ideal since it will eliminate the possibility of tracking materials through the house.
- If you are having a new floor installed, consider using a hard surface floor such as wood, ceramic, and slate instead of carpeting. Carpeting is an ideal home for dust mites, which is a leading cause of allergies in the home.
- Before making a decision, talk to your remodeling contractor for recommendations. Be sure if you choose hardwood floors, to discuss the effects of fumes that may be emitted when finishing the floor.
- Consider staying out of the house for a couple days to let it properly air out after installing a floor.

What is "offgasing?"

Offgasing is the release of gases or vapors into the air. Whenever possible, avoid all forms of offgasing in your home. The term "outgasing" is similar, but refers to the release of gases or vapors in a controlled vacuum-like environment, such as a research lab.

agricultural waste are superior to wood-based particleboard in moisture resistance and structural properties, and enable you to reuse a former waste product. In general, by eliminating particleboard in your home, you reduce your family's exposure to harmful formaldehyde offgasing.

- **Recommendation**: Whenever possible, eliminate new particleboard inside your house by using solid wood for stair treads, exterior grade plywood for shelving, and formaldehyde-free MDF for cabinets and countertops substrate. Other alternatives include FSC-certified plywood and boards made from agricultural waste, such as wheatboard (a straw-based particleboard) manufactured with a non-formaldehyde and emission-free binder.

♡ – Seal All Exposed Particleboard or MDF

- Use non-toxic, low permeability paint to seal exposed particleboard or MDF and encapsulate the formaldehyde. This will slow the release of harmful gases; however, when the sealant wears down or chips, formaldehyde might still be released into your home. Sealants simply lower the formaldehyde emissions rate for a longer period of time.
- **Recommendation**: Whenever MDF or particleboard is used, seal all exposed edges of cabinets, undersides of counter tops, stairs, shelving, etc., with at least two coats of non-toxic, low-permeability latex paint prior to installation.

Fig 19.2: Formaldehyde-free cabinets will make your home healthier. Credit: Environmental Home Center.

Flooring

– Select FSC-certified Wood Flooring

- FCS-certified wood flooring comes from forests that are managed in accordance with sustainable forest practices that will assure the long-term availability of these precious woods while protecting ancient, old-growth forests. Certified wood flooring products are available in a wide variety of domestic and exotic species.
- **Recommendation**: Use FSC-certified wood in place of conventional hardwood flooring.

– Use Rapidly Renewable Flooring Materials

- Bamboo and cork flooring are alternatives to hardwood flooring. Bamboo is a fast growing grass that can grow up to 60 feet in several months, though it should be left to mature internally for a total of 4 to 6 years before it is harvested. Bamboo is as durable as wood. Cork is made from bark peeled from cork oak trees and uses the waste left over from making wine bottle corks. The cork regenerates quickly, and since cork oaks are never actually cut down to supply the cork, this process is fairly sustainable. Cork naturally resists fire and moisture, radiates warmth, and absorbs sound.

- **Recommendation**: Use rapidly renewable flooring materials in place of conventional hardwood. In dry climates, let the flooring acclimate for 30 days before installation.

🌲 – Use Recycled-content Tile

- Tile is an inert and healthy floor material. Recycled-content tile is available in ceramic and glass. Like most tile, ceramic and glass tile are dense materials, thereby significantly reducing the amount of moisture and stains that are absorbed into the tile. Ceramic and glass tile are uniquely "green" because of their high recycled-content and durability that minimizes harmful impacts on natural resources. Recycled ceramic tiles often use mining waste (called feldspar), while recycled glass tiles use everything from windshield glass to recycled bottles. Both ceramic and glass are as durable as their traditional "virgin" counterparts.
- **Recommendation**: Install recycled-content tiles wherever conventional tiles are specified. Grouts (those areas between tiles) are porous and can harbor mold and mildew, and should therefore be sealed with a low-toxic sealer where exposed to water.

🌲, ♡ – Replace Vinyl Flooring with Natural Linoleum

- Sheet vinyl is a popular flooring for kitchen and utility areas because it is easy to clean, inexpensive, and provides a soft walking surface. In hot or humid climates requiring air conditioning, the vinyl will trap moisture that can promote delamination and mold growth or rot. Moreover, vinyl chloride fumes emitted from the flooring are carcinogenic. Natural linoleum is a healthier alternative manufactured from only natural, renewable materials such as linseed oil, pine resins, and cork. Linoleum is low-toxic, anti-microbial, easy to repair, durable, and stain resistant. It can last up to 40 years whereas vinyl lasts seven to ten years and hardens with age.
- **Recommendation**: Use natural linoleum in place of vinyl flooring.

🌲 – Install Recycled-content Carpet and Underlayment

- Carpeting accounts for 70 percent of all floor coverings in the US and has been associated with a growing number of health and environmental problems. Alternatively, recycled-content carpet is made from recycled

Fig. 19.3: No, that's not a rug! This woman is seated comfortably on natural linoleum. Credit: Forbo Flooring Marmoleum Linoleum.

What's Wrong With My Carpet?

In a typical carpet, toxic chemicals may be found in the fiber bonding materials, dyes, backing glues, fire retardant, latex binder, fungicide, and antistatic and stain resistant treatments.[2] During a congressional hearing in 1992 on the potential risk of carpets, the US EPA stated that a typical carpet sample contains at least 120 chemicals, many of which are known to be neurotoxic. Offgassing from new carpeting can persist at significantly high levels for up to three years after installation.[3] The most common carpet backing, synthetic latex, contains approximately 200 different chemicals which contribute to the unpleasant and harmful "new carpet smell."[4] Most underpads are made of foamed plastic or synthetic rubber and contain petroleum products.[5] This enhances our dependence on oil production and the numerous hazards associated with oil, such as spilling and leaks. Once discarded, traditional carpeting and backings are neither renewable nor biodegradable.[6] According to the EPA, over 2.5 million tons of carpeting and rugs were dumped in 2000.[7]

PET plastic bottles, recycled wool, nylon, or recycled cotton. Approximately 40 2-liter soda bottles are recycled per square yard of carpeting. Recycled carpet is often more resilient and colorfast than carpet made from virgin fibers; it is available in broadloom or carpet tiles; it does not differ in appearance or performance; and the price is comparable to conventional carpet. Recycled-content underlayment and padding are also available. Tack-down installations are easier to remove than glue-down; they do not destroy the floor surface, and the carpeting can be partially recycled, saving resources and diverting waste from landfills. Taking precautions (see

Minimize carpet-related health issues.

- Choose your carpeting as early as possible and lay it out somewhere so it will have the most time to air out.
- Avoid carpeting containing permanent stain resistance treatment.
- To remove old carpeting, vacuum the old carpet thoroughly to reduce the dust level.
- Open doors and windows to provide fresh air during the process.
- Use non-toxic and odor free shampoos and maintain carpets regularly to prevent mold, bacteria, dust, and pesticide buildup.
- If carpet or pad gets wet, dry as quickly as possible to prevent microbial growth.
- Choose stretching and tacking over adhesives.
- Imported wool carpets are often treated with highly toxic mothproofing pesticides. Therefore, an expensive 100% wool carpet does not necessarily mean a safer carpet.
- Never use wall-to-wall carpet in bathrooms, kitchen, laundry rooms, or mechanical rooms because carpeting in these areas inevitably becomes damp, inviting mold and bacterial infestation.

"Minimize Carpet-Related Health Issues") will help keep you and your family healthy during and after carpet installation.

- **Recommendation**: Use recycled-content carpet, underlayment, and padding in all applications where conventional carpet is installed. Some major carpet manufacturers have instituted take-back programs, which can help keep carpets and their chemicals out of the general waste stream. Most of these manufacturers even recycle carpet and backing into new materials.

$, ⛰, ♡ – Use Exposed Concrete as the Finish Floor

- For slab-on-grade additions, the concrete can be smooth, finished, scored in various patterns, and stained with pigments to make an attractive finish floor. This approach is especially appropriate for radiant, in-floor heating systems, and for use as a "thermal mass" for passive solar heating and cooling. (See Passive Solar in Chapter 15.) When slab is used as a floor finish, it eliminates the need to use additional resources and potentially toxic flooring materials. Concrete is also durable and easy to clean.
- **Recommendation**: Use this approach for finished basements or additions on slab construction. Finish must be designed and constructed when slab is being poured.

Paints and Wood Finishes

$, ⛰ – Calculate Paint Needs Beforehand

- Get a rough idea of how much paint you will need for your project so that you do not waste money or paint materials. This also allows you to mix a more uniform-colored paint. If you know the quantities you will need, you can make sure your contractor gives you a fair estimate if you are out-sourcing the job.
- **Recommendation**: Each gallon of paint covers about 400 square feet; still, double-check the label to be sure. Base your calculations (and contractors' bids) on two coats, not one. Allow for some leeway. Figure on buying 10 percent extra for spills and other waste, and for touchups down the road. Buy everything at once. Make sure all the paint comes from the same manufacturing lot; check the numbers on the cans. Also make sure

How to Avoid Hazardous Effects of Paints and Finishes

- To prevent skin irritations, wear rubber gloves.
- If you are sensitive to fumes, wear a respirator mask that is especially designed to filter them out.
- Do not use any solvent-containing products around children or if you are pregnant.
- Avoid applying finishes inside the home. Work in an open garage or outdoor workspace, and transport finished products inside after the finish has had several days to dry and lose its odor.
- When applying finishes indoors, open all doors and windows and use ceiling and window fans to circulate fresh air into the room. Don't occupy the room for two or more days afterward, until the finish has dried and fumes have dissipated.

the paint is color-mixed at the same time; otherwise, the color may be off slightly from one batch to another. Better still, for uniformity, mix all the paint together in five-gallon buckets. A handy paint calculator is available online at: <www.truevalue.com/paint/>.

♡ – Practice Safe, Paint/Finish Application

- Keep in mind that even the least hazardous finishes may still cause skin or respiratory irritation in some individuals.
- **Recommendation**: To avoid paint/finish hazards altogether, consider designing surfaces that don't require painting, such as integrally pigmented plaster walls and natural wood trim. Follow manufacturers' application instructions to ensure proper coverage, long-term performance, and safe use.

♡ – Understand the Finishes' Label

- Labels do not always list all the ingredients, making it difficult to figure out what you're purchasing. However, reading it carefully can still help you avoid particularly hazardous chemicals, such as formaldehyde, petroleum distillates, and mineral spirits.
- **Recommendation**: Look for clues, including words like "Danger," "Warning," or "Caution." Labels with "Danger" or "Poison" typically indicate that the product is the most dangerous to use, while "Warning" or "Caution" on the labels means the product has a slight or moderate risk associated with it. The phrase next to these key words describes the nature of the hazard, such as "combustible" or "harmful if swallowed." Most

Volatile Organic Compounds (VOCs).

VOCs are chemicals that evaporate into the air and react with sunlight to form ground-level ozone. Formaldehyde and solvents are VOCs and are some of the most dangerous pollutants in household building products. VOCs include carcinogens, endocrine disrupters, central nervous system disrupters, and sensitizers. For more information, refer to "Volatile Organic Compounds" in Chapter 5.

finishes are combustible, and a few do contain nerve-damaging neurotoxins, so it's best to avoid "May affect the brain or nervous system." Don't assume "organic" ingredients indicate the product is safe. Although in grocery stores it refers to foods grown without synthetic pest-icides, in chemistry it simply means the chemicals are carbon-based, including VOCs, some of which can cause brain damage or cancer.

The VOC line is also important to decipher. If no "low-VOC" products are available where you're shopping, you can compare VOC numbers on the labels to choose the lowest offgasing brand.

♡ – Use Water-based or Latex Paint

- Alkyd or solvent-based paints offgas numerous VOCs and contain over 300 toxic, including 150 carcinogenic, ingredients. Water-based or latex paints use water as a thinner or base, therefore there are no harmful solvents to evaporate.

- **Recommendation**: Use water-based or latex paints instead of alkyd or solvent-based paints. If you are going to use alkyd paints in situations that need its tough, water-repellent finish, look for 100 percent acrylic low-emission paints as a healthier choice, or confine alkyd use to exterior jobs or small-scale applications. It needs to be used with great care due to the strong terpenes (VOCs derived from plants).

♡ – Use Low- or No-VOC and Formaldehyde-free Paint (less than 160 grams per liter)

- Most paint releases VOCs into the home. All paint will continue to volatize in the form of dry paint "microflakes" for years after the home is completed. Low- and no-VOC products reduce the danger of the flakes since they are manufactured without mercury or mercury compounds, or pigments of lead, cadmium, chromium, or their oxides. Low- or no-VOC paint reduces the emissions of VOCs into the home, improving indoor air quality and reducing the formation of urban smog. Low toxicity paints also reduce health and environmental hazards.

- **Recommendation:** Paints with low- or no-VOC content are available from most major manufacturers and are applied like traditional paint products with a brush, roller, or spray gun. Every finish and color is available in low-VOC paints. Choose lower gloss, lighter-colored paints because higher gloss and darker-colored paints contain more VOCs. Consider oil-based paint only for use over bare wood, or over old surfaces with layers of old, chalking, oil-based paint. Even on these surfaces, investigate latex opt-ions first — VOCs make up more than 50 percent of oil-based paint.

Painting secrets.

- With latex paint, prime the brush or roller by dipping it in water and shaking out the excess. The tools will pick up paint more easily.
- Work from painted sections toward unpainted sections — from wet to dry — and back over the same stroke. But do not over brush.
- Low-luster, semigloss, and gloss paints tend to highlight surface irregularities. Use flat paint to help hide imperfections in your home's wood siding.
- If you're painting over stucco or cement, first mist the surface with a garden hose.

 – **Use Natural Paints**

- Natural paints are derived from substances such as citrus and balsam, as well as minerals, and are petroleum-free. Unfortunately, they often contain terpenes, which are VOCs derived from plants. However, an important health reason to use natural paints is to avoid toxic pesticides such as biocides and fungicides. They also ensure the use of natural dyes and pigments that are free of toxic heavy metals.
- **Recommendation:** Choose to apply natural paints wherever you would use interior paint.

Painting tips for the chemically sensitive.

When choosing paint for a chemically sensitive person, no product can be considered safe until that person has tested it. Suppliers of paint to the chemically sensitive are aware of this need and can provide their products in small sample packages for such tests. Even so-called non-toxic and zero-VOC paints release trace amounts of chemicals to the air that some people may find irritating. Increased ventilation is always a good idea for areas being painted, and paint fumes should never be allowed to circulate through a building's HVAC system. With certain paints, respirators with charcoal filters for organic compounds should be used.

♡ – Use Low-VOC, Water-based Wood Finishes

- Conventional solvent-based wood finishes can offgas dangerous VOCs for months. Low-VOC finishes, such as waterborne urethane and acrylic, are lower in toxic compounds compared to conventional solvent-based finishes while providing similar durability. Also consider finishes made with beeswax or carnauba wax.
- **Recommendation**: Low-VOC wood finishes can be used in most applications where solvent-based finishes are typically used. Avoid epoxies and oil-based formulas with dryers. Low-VOC, water-based finishes may still contain solvents, biocides, and harmful chemicals (though far fewer than oil-based formulas), therefore you should ventilate by opening windows and turning on fans during application.

⚘ – Use Darker Stains and Sealants for Exterior Wood Finish

- Besides darkening wood, stains also protect wood from the sun's ultraviolet (UV) rays. The more pigment, the more protection from UV light. Although clearer sealants contain fewer VOCs, a longer-lasting darker stain actually minimizes exposure to VOCs because it does not require as much re-staining to protect the wood as clearer varieties do.
- **Recommendation**: Apply darker stains wherever you choose to finish exterior wood without paint. Keep in mind that paint is preferable to stain due to the higher levels of pesticides in stain.

⚘, ♡ – Make Your Own Paints and Stains

- You can make your own paints and stains with natural ingredients and pigments, thereby avoiding biocides and minimizing your exposure to VOCs.
- **Recommendation**: Natural dyes made of plants will make light to dark pastel colors when mixed with paint. Minerals can make darker dyes. Start with one cup pigment to a gallon of paint; increase the amount of pigment as desired. For stains, make a strong dye bath and use as many coats as necessary to reach the desired color.[8]

$, ⚘ – Deal Properly with Leftover Finish

- Unfortunately, all too often leftover paint is dumped illegally into sewers or into the ground. However, there are ways to dispose of paint and its toxic elements so they do not leach into our ground water and drinking supply. Oil paints that must be discarded should be treated as hazardous waste.

Recycled paint reduces environmental impacts because the main raw material is leftover paint from other users. It's usually a bargain, too, with prices around $10 per gallon. Although recycled paints may not be the best choice if you are sensitive to small quantities of mildewcides or legal amounts of VOCs collected from exterior paints, buying recycled paint is a great way to eliminate waste, minimize groundwater contamination, and save money. Saving and donating paint also reduces hazardous paint waste.

• **Recommendation**: Buy paints from companies that actively encourage or sponsor paint collection and recycling programs. Small amounts of leftover paint should be carefully labeled and stored for use in touching-up damaged areas in the future. Donate leftover paint to a local low-cost housing group, community assistance organization, or theater organization that can make use of the paint. Or contact local solid waste or hazardous waste management authorities to find out about opportunities to recycle paint.

🌲, ♡ – Use Solvent-free Adhesives

• Solvent-free adhesives reduce toxic gases such as aromatic hydrocarbons or solvents that contribute to air pollution, whereas solvent-based adhesives offgas toxic compounds for months. Most major manufacturers have new "environmentally friendly" formulations. Solvent-free adhesives are often stronger, emit fewer pollutants, and reduce the potential harmful impacts on the health of the occupants and installers.

• **Recommendation**: Use solvent-free products in place of standard adhesives for all interior applications such as installation of flooring, countertops, wall coverings, paneling, and tub/shower enclosures.

Interior Materials/Finishes Checklist

❏ Use formaldehyde-free materials.

❏ Seal all exposed particleboard or MDF flooring.

❏ Select Forest Stewardship Council (FSC)-certified wood flooring.

❏ Use rapidly renewable flooring materials.

❏ Use recycled-content tile.

❏ Replace vinyl flooring with natural linoleum.

❏ Install recycled-content carpet and underlayment.

❏ Use exposed concrete as the finish floor.

Paints and Wood Finishes

❏ Calculate paint needs beforehand.

❏ Practice safe, paint/finish installation.

❏ Understand the finishes' label.

❏ Use water-based or latex paint.

❏ Use low- or no-VOC and formaldehyde-free paint.

❏ Use natural paints.

❏ Use low-VOC, water-based wood finishes.

❏ Use darker stains and sealants for exterior wood finish.

❏ Make your own paints and stains.

❏ Deal properly with leftover finish.

❏ Use solvent-free adhesives.

Epilogue: 2025

EVERY STORY NEEDS A HAPPY ENDING; ours ends like this. The year is 2025. In the early years of the 21st century a wake-up call spread like wildfire throughout the world. People realized that they could make a difference in the rapid decline of all natural systems. One by one, family by family, community by community, people started making choices. They saw the important role they played in the web of life and realized that every product they bought either preserved opportunities for their children's futures — or diminished them. So they refused to buy wood products from old and endangered forests. They no longer bought products that were developed at the expense of other animals: as a result, food such as dolphin-free tuna and products that were not tested on animals became the norm.

Recycling has become so common that businesses have sprung up all over the country to take the feedstock of the previous waste stream and turn it into products that can be used, reused, and recycled back into something useful. People have found new and innovative ways to harness the energy of the sun to create goods and services that provide a high quality of life for all, using petrochemicals to produce medicines and innovative, long-lived plastics that allow us to keep people healthier and living longer.

In 2025, we use water sparingly; all the bounty from the sky is captured as rain water and used for most of our daily uses. It is hard to believe that at one time it was squandered and flushed away! Our battery technology is so advanced that we simply leave our electric cars and bicycles in the sun to recharge them. Windmills cover the plains and lie just over the horizon on the coasts, providing electricity to the grid, making it much more stable, and splitting water into oxygen and hydrogen to power large industries and the high-speed trains that connect us like a web.

Perhaps most exciting is what has happened in our communities. The population pressure on our cities made us realize that what we really wanted was the opportunity to connect with each other. Streets closed to traffic are transformed into gardens. The old gas-powered cars and trucks have been relegated to the outskirts so that people can walk or ride bikes in town. The public transit system has grown at an astounding rate. Outdoor plazas have become full of tent-covered markets and free music. People enjoy each other's company rather than competing for parking places. There is a lightness of being in the cities and crime is down because the new economy has given rise to millions of new jobs manufacturing sustainable technologies.

Our homes are more than our castles. We produce more energy than we use so each of us bolsters the self-sufficiency of our local municipal utility. We grow food year round in our passive solar greenhouses. Our local currency allows us to trade goods and services with each other so we can use our standard money for those things we can't produce locally. Local businesses address the basic needs that keep our communities prosperous and balanced. We take care of our own now — our elderly, our community's children, the unfortunate, and the immigrants that give our community so much diversity.

It all happened so quickly that it is hard to know exactly how it came to be. A realization swept across the country that seemed so logical! No longer is there a constant stream of our precious capital out of the community and into the pockets of multinational corporations. Once each of us made our priorities clear through our purchasing decisions, the conversation changed from sustainability to flourishability. We didn't want only to survive the global changes, we wanted to regenerate the natural system that supports us all. When it became clear that we needed the health of our natural systems to transform carbon dioxide back into oxygen, to maintain the diversity of species that keeps the forests and plains healthy, and to purify the air so that all might breathe, it all seemed like common sense.

I'm happy to report that we are optimistic about the future: in 2025, we can see the forests and not just the trees.

Further Resources

THE FOLLOWING LIST OF INTERNET SITES provides valuable information on a variety of topics pertinent to the green building process. You might start with the following website which is one of the most comprehensive summaries of green building websites and easily searched sources of information: <**www.afford ablegreenbuilding.org**>

Government Resources

- **Ask an Energy Expert: Energy-efficiency and Renewable Energy Clearinghouse (EREC)**
  <**www.eere.energy.gov/informationcenter**>
  Provides free general and technical information to the public on a wide spectrum of topics and technologies pertaining to energy-efficiency and renewable energy.

- **Canada Mortgage and Housing Corporation (CMHC)**
  <**www.cmhc-schl.gc.ca**>
  Promotes improvements in designs, techniques, technologies, products and materials that will result in the construction and renovation of housing and communities that are healthier for occupants; more energy-efficient; more resource-efficient; healthier for the global environment in the long-term; and more affordable to create, operate and maintain.

- **Energy Efficiency and Renewable Energy Network**
  <**www.eren.doe.gov**>
  Information on a comprehensive range of energy-related technologies, featured sites, and specialized resources.

- **Energy Information Administration**
<www.eia.doe.gov>
Information on all forms of energy, including petroleum, natural gas, coal, nuclear, renewables, and alternative fuels. Also provides environmental, international, and forecast information about electricity.

- **Energy Star Program**
Participate in the Energy Star program, search for builders in the program and learn about Energy Star appliances.

- **Environmental Protection Agency (EPA) Indoor Air Quality**
<www.epa.gov/iaq/>
Covers general indoor air quality topics, including radon, asthma, and mold. Also highlights an extensive list of EPA publications.

- **Lawrence Berkeley National Laboratory**
<www.eande.lbl.gov/btp/btp.html>
Includes research on all aspects of building energy-efficiency.

- **Office of Building Technology, State and Community Programs**
<www.eren.doe.gov/buildings/builders.html>
Building energy codes, partnership opportunities (including the Building America program), emerging green building technologies, building materials, building systems, and whole-building design.

- **Partnership for Advancing Technology in Housing (PATH)**
<www.pathnet.org>
PATH aims to dramatically improve the quality, cost effectiveness, durability, safety, and disaster resistance of housing in the US.

- **Smart Communities Network**
The "Green Buildings" section includes descriptions of building principles, rating systems, and municipal codes and ordinances that promote sustainable building. Articles, publications and educational materials are available, as well as a collection of inspirational "success stories."

- **US Department of Energy Building Technologies Program**
<www.eere.energy.gov/buildings/>
Brief introduction to many building-related topics. Includes links to Department of Energy programs as well as initiatives on energy-efficiency and sustainability in buildings.

- **Whole Building Design Guide**
<www.wbdg.org>
An effort by the US government to consolidate design guidelines into one easily maintained and current reference. Web resources managed by the Sustainable Buildings Industry Council (SBIC) and the National Institute of Building Science (NIBS).

Other Resources

- **American Solar Energy Society**
  <www.ases.org>
  Dedicated to advancing the use of solar energy for the benefit of US citizens and the global environment. Publishes *Solar Today Magazine*.

- **American Wind Energy Association**
  <www.awea.org>
  A national trade association that promotes wind energy as a clean source of electricity for consumers around the world.

- **Architects, Designers and Planners for Social Responsibility (ADPSR)**
  <www.adpsr.org>
  A national non-profit organization of architects, designers, planners and related professionals. ADPSR's efforts are directed towards arms reduction, protection of the natural and built environment, and socially responsible development.

- **Bay Area Build It Green**
  <www.build-green.org/>
  A program designed to provide Bay Area homeowners, home buyers, remodelers, and builders a trusted resource for information on Green Building and its various applications. A local effort with far reaching effects, this organization is dedicated to creating healthier, more durable and energy-efficient homes. An excellent reference site.

- **Boulder Green Building Guild**
  <www.bgbg.org>
  An association of building professionals dedicated to promoting healthier, resource-efficient homes and workplaces. Their primary goal is to make green building common practice.

- **Builders Without Borders**
  <www.builderswithoutborders.org/>
  An international network of ecological builders who form partnerships with communities and organizations around the world to create affordable housing out of local materials. Its site includes free, downloadable educational materials about natural building, as well as descriptions of the organization's current projects and upcoming activities.

- **Building Green**
  <www.buildinggreen.com>
  An independent company committed to providing accurate, unbiased, and timely information designed to help building-industry professionals and policy makers improve the environmental performance, and reduce the adverse impacts, of buildings. Publishers of *Environmental Building News* and *GreenSpec*.

- **BuildingEnvelopes.org**
  <www.buildingenvelopes.org>
  A joint venture of MIT and Harvard, this site includes the latest research on sustainability and energy-related building practices.

- **Building Science Corporation (BSC)**
<www.buildingscience.com>
BSC specializes in building technology consulting. Their focus is preventing and resolving problems related to building design, construction and operation.

- **Center for Resourceful Building Technology (CRBT)**
A project of the National Center for Appropriate Technology, CRBT promotes resource-efficiency in building design, materials selection and construction practices. Includes a database of resource-efficient building products, demonstration sites, publications, and educational materials.

- **Forest Certification Resource Center**
Offers clear, well-organized background materials on forest product certification. Its primary feature is a series of databases tied directly to the Forest Stewardship Council's official listing of certified organizations. Also features a toolkit that helps designers specify certified wood.

- **Co-op America's Solar Catalyst Group**
<www.solarcatalyst.com>
Focused on building demand for solar and lowering the price of solar technologies.

- **Database of State Incentives for Renewable Energy**
<www.dsireusa.org>
A comprehensive source of information on state, local, utility, and selected federal incentives that promote renewable energy.

- **Ecological Design Institute**
Education in the form of public presentations, lectures, and workshops promoting ecological design.

- **Energy and Environmental Building Association (EEBA)**
<www.eeba.org>
EEBA promotes the awareness, education and development of energy-efficient, environmentally responsible buildings and communities. Diverse membership base includes architects, builders, developers, manufacturers, engineers, utilities, code officials, researchers, educators, and environmentalists.

- **Florida Solar Energy Center**
<www.fsec.ucf.edu/index.htm>
View the work of some of the US's leading researchers on energy conservation and comfort in hot, humid climates.

- **Forest Stewardship Council (FSC)**
An independent, non-profit organization that promotes responsible forest management by evaluating and accrediting certifiers, encouraging the development of national and

regional forest management standards, and providing public education about independent, third-party certification.

- **Forest World**
  <www.forestworld.com/>
  Provides contact and certification information for manufacturers of forest products. Categorizes over 6,000 industry-related sites by description, company name, or category. Also includes information about more than 900 species of wood, including uses, distribution, physical properties, and environmental profiles.

- **Global Energy Marketplace (GEM)**
  <www.solstice.crest.org/gem.html>
  A searchable renewable energy and energy-efficiency web encyclopedia. Includes information on greenhouse gas mitigation.

- **Green Building**
  <www.greenbuilding.com>
  A detailed online resource for green building websites and publications.

- **GreenClips**
  <www.greenclips.com/>
  A bimonthly e-mail newsletter with news about sustainable building design.

- **The Green Guide**
  <www.thegreenguide.com>
  Consumer source for practical everyday actions benefiting environmental and personal health. The information is presented in a way that is brief and focused, with all facts rigorously researched and verified.

- **Greenguard**
  <www.greenguard.org>
  A registry for products that have been tested and found to meet applicable government and non-government organization standards for emissions to ensure healthy indoor air quality.

- **Health House**
  <www.healthhouse.org>
  A national education program created by the American Lung Association of Minnesota to raise the standards for better indoor environments. Includes indoor air quality home checklists and training programs.

- **Housing Zone**
  <www.housingzone.com>
  Residential construction website that caters to builders, remodelers, architects, consumers, and manufacturers. Includes an online buyer's guide, store, classes, research, events, and many other helpful features.

- **Institute for Energy Information**
  <www.ieionline.com>;
  <www.energysuperstore.com>
  A non-profit corporation providing energy management information, energy analysis programs, and solution applications to the public. Their online store, <www.Energysuperstore.com>, sells a wide range of energy-efficient products hand-picked by energy experts as the best in their class.

- **Lighting Research Center**
  <www.lrc.rpi.edu>
  Technical information relating to the design and specification of energy-efficient lighting.

- **National Association of Home Builders (NAHB)**
  <www.nahb.org>
  Industry-based organization catering to large home builders. Massive collection of documents on case studies, indoor air quality, land development, and more. Experts can answer green building questions through the ToolBase Service.

- **Natural Resources Defense Council (NRDC)**
  <www.nrdc.org/cities/building/>
  The Cities and Green Living section of NRDC's site has information and resources for green building, including descriptions and photographs of the organization's environmentally-friendly offices in New York, Los Angeles, San Francisco and Washington, DC.

- **New Buildings Institute**
  <www.newbuildings.org>
  Promotes energy-efficiency through policy development, research, guidelines, and codes. Includes links to guidelines, reports, and tools for energy-efficient lighting, architecture, and mechanical systems.

- **NSF International**
  <www.nsf.org/>
  A non-profit public health and safety company, NSF develops national standards, provides learning opportunities through its Center for Public Health Education, and provides third-party conformity assessment services for water filters, bottled water, and other consumer goods.

- **Oikos**
  <www.oikos.com/>
  A site devoted to serving professionals whose work promotes sustainable design and construction. Includes an extensive green products directory, recent news stories, a bookstore, green project articles, and marketing information.

- **Rocky Mountain Institute**
  <www.rmi.org>
  Data and resources on green energy, green development, case studies, and consulting services. Also includes a bookstore.

- **Scientific Certification Systems (SCS)**
  <www.scs1.com>
  Provides independent, third-party certification of environmental and food safety achievements.

- **Smart Growth Network**
  <www.smartgrowth.org>
  A coalition of developers, planners, government officials, lending institutions, and others involved in the development process of US cities

and states. They encourage development that serves the economy, community, and environment.

- **SmartWood**
  <www.smartwood.com>
  SmartWood, a program of the Rainforest Alliance, works to reduce the negative impact of commercial forestry by rewarding its seal of approval to responsible forest managers.

- **Society of Building Science Educators**
  <www.sbse.org>
  An association of university educators in architecture who support excellence in the teaching of environmental science and building technologies.

- **Solar Energy Industries Association (SEIA)**
  <www.seia.org>
  SEIA is the national trade association of solar energy manufacturers, dealers, distributors, contractors, installers, architects, consultants, and marketers, concerned with expanding the use of solar technologies in the global marketplace.

- **Solar Living Institute**
  <www.solarliving.org>
  A non-profit organization that offers hands-on workshops on renewable energy, ecological design, sustainable living practices, and alternative construction techniques.

- **Sustainable Buildings Industry Council (SBIC)**
  <www.sbicouncil.org>
  A non-profit organization whose mission is to advance the design, affordability, energy performance, and environmental soundness of residential, institutional, and commercial buildings in the US Includes links to green building guidelines and software.

- **Sustainable Sources**
  <www.greenbuilder.com/>
  A one-stop online resource center for green building, sustainable agriculture, and responsible planning. Includes a thorough "Sustainable Building Sourcebook" full of detailed recommendations to foster the implementation of environmentally responsible practices in home-building.

- **Union of Concerned Scientists**
  <www.ucsusa.org>
  A non-profit alliance of concerned citizens and scientists across the country, connecting scientific insights with environmental and social advocacy.

- **United States Green Building Council (USGBC)**
  <www.usgbc.org>
  Promotes understanding, development, and implementation of green building policies, technologies, and design practices. Extensive information about the LEED Green Building Rating System. Members gain access to product database case studies and directories.

- **WaterWiser**
  <www.awwa.org/waterwiser/>
  The American Water Works Association's water-efficiency clearinghouse website. Offers reliable information on water conservation strategies.

- **What's Working**
  <www.whatsworking.com>
  The author's website! A design and consulting firm specializing in energy conservation, environmental construction technology, and sustainable community development. Services include policy development, program design and development, design consultation, green building materials spacifications, marketing, media relations and training.

Tools

- **ATHENA Environmental Impact Estimator**
  <www.athenaSMI.ca>
  A modeling tool designed to make life-cycle analysis more accessible to building designers.

- **Building for Environmental and Economic Sustainability (BEES)**
  <www.bfrl.nist.gov/oae/software/bees.html>
  Free software that gives environmental and economic performance data for 65 building products.

- **Building Energy Tools Directory**
  <www.eere.energy.gov/buildings/tools_directory>
  A collection of 210 software tools with pro/con descriptions. Searchable by function or computer platform.

- **DOE Building Energy Software Tools**
  <www.eren.doe.gov/buildings/energy_tools/>
  EnergyPlus, a building-energy simulation program, and other DOE-sponsored software.

- **EcoAdvisor**
  <www.ecoadvisor.com>
  Graphical and animated tools that illustrate the workings of various energy systems in buildings.

- **Ecological Footprint**
  <www.redefiningprogress.org/programs/sustainabilityindicators/ef/>
  Provides the most complete comparison of natural resources demand and supply available. Download for more detailed consumption categories, including recycling, energy use, and even how much coffee you drink.

- **Energy-10 Software from the Sustainable Buildings Industry Council**
  <www.sbicouncil.org/enTen/index.html>
  Simulates the design of your project and calculates whole-building performance, including thermal, HVAC, and daylighting components.

- **EPA Global Warming**
  <www.epa.gov/globalwarming>
  Under "Tools," you can estimate greenhouse gas emissions for various activities. Additionally, you can measure and analyze energy consumption and emission reduction opportunities.

• **Green Building Advisor**
<www.buildinggreen.com/orders/gba_info.html>
Identifies green design strategies in building projects.

• **GreenSpec**
<www.buildinggreen.com/bg/gsmenu/index.jsp>
Contains detailed listings for more than 1,700 green building products with environmental data, manufacturer information, and links to additional resources.

• **Safe Climate**
<www.safeclimate.net>
Information and resources to stop climate change, including an online tool to calculate your carbon footprint.

• **Scorecard**
<www.scorecard.org>
Scorecard's searchable database includes information about pollution in your neighborhood, various chemicals' health effects, current toxic chemical regulatory controls, and more.

• **Target Finder**
<www.energystar.gov/index.cfm?c=target_finder. bus_target_finder>
An extension of the EPA's Portfolio Manager (provides rough calculations of energy performance for existing buildings), this simple online calculator helps architects estimate energy targets for new buildings.

• **The Power Scorecard**
<www.powerscorecard.org>
A rating mechanism that assesses the environmental impact of different types of electricity generation. Gives users comprehensive environmental information on green power providers.

• **WasteSpec**
<www.tjcog.dst.nc.us/cdwaste.htm>
Model specifications for construction, waste reduction, reuse, and recycling. Available free online.

Indoor Air Pollutants — Where To Find Them and What To Do

Particulates

Asbestos

Sources:
- Vinyl floor tiles and vinyl sheet flooring.
- Patching compounds and textured paints.
- Ceilings.
- Pipe insulation and furnace ducts.
- Wall and ceiling insulation.

Remedies:
- Take precautions when you tear up, remove, cut, scrape or sand any materials you suspect may contain asbestos.
- If the asbestos material is isolated or doesn't pose a threat, leave it alone; seal it with paint or (in small areas) duct tape; cover it with a new surface such as wallboard.
- If it is deteriorating or must be removed, have an experienced contractor do it using precautions to avoid exposure to fibers or from spreading them into the house.

Dust

Sources:

- Carpets
- Upholstered furniture.
- Pets.
- Fireplaces.
- Smoking.
- Heating ducts.
- Doors, windows, and air leaks that allow particles to enter from outdoors.
- Exposed, worn, or damaged surfaces on building materials such as particleboard or vinyl-asbestos flooring.

Remedies:

- Install a central vacuum system vented to the outside.
- Install a ventilation system with a filter for incoming air.
- Add an effective air filter to your forced-air heating system.
- Remove carpets or other sources of dust and allergens.
- Reduce moisture levels causing mold and mildew problems.
- Caulk and weatherstrip thoroughly to prevent outdoor irritants from getting in.
- Do without pets or build a facility outside for them.

Lead

Sources:

- Old house paint (especially that made before 1950) on interior and exterior of home, notably, on window frames, window and door trim, railings, baseboards, radiators, walls.
- Plumbing pipes and the service connection into homes (before 1950s).
- Solder used to join water pipes in modern homes.
- Miscellaneous household items (ceramic pottery, lead crystal, hobby materials.

Remedies:

- Have old paint lab-tested for high lead content, especially before major renovations.
- Leave paint undisturbed if it is not accessible to children and if it is in good condition.
- Do not create any dust from leaded paint when young children or pregnant women are living in the house.
- Replace doors, windows, or trim covered with lead paint, or strip and repaint them away from the house.
- If you want paint removed, have it done by a trained contractor. This work will be dusty. Move the family out of the house during the work and clean up thoroughly before moving back in.
- Have your water tested for lead content. Recommended levels are below 10 parts per billion. If results show high lead levels replace old plumbing; install a point-of-use water purification device in your kitchen for drinking and cooking; secure a better source of lead-free water for drinking and cooking.

Combustion Gases

Types of Gas

- Carbon Monoxide
- Formaldehyde
- Nitrogen Dioxide
- Sulfur Dioxide
- Carbon Dioxide
- Hydrogen Cyanide
- Nitric Oxide
- Benzo(a)pyrene

Sources of Gas

- Gas or oil furnaces.
- Boilers.
- Hot water heaters.
- Gas fire places.
- Wood burning fireplaces.
- Coal-burning stoves.
- Kitchen ranges.
- Clothes dryers.
- Space heaters.
- Wall heaters.
- Central heating systems.
- Kerosene space heaters.

Remedies:

- Use all-electric appliances (except electric self-cleaning ovens).
- Use other fuel-burning alternatives such as solar or steam heat.
- Install a balanced ventilation system in the house; beware of negative pressure and downdrafts.
- Ensure that fuel-burning appliances are induced-draft, sealed-combustion units. Small enclosed rooms should have direct air supply from exterior.
- Seal ductwork with mastic, including cracks or leaks in stovepipe.
- Have a qualified service person check for chimney obstructions like creosote buildup.
- Have your furnace maintained regularly by a competent mechanical contractor — look out for cracked heat exchanger.
- Eliminate wood or gas fireplaces. Alternatively, install tight-fitting doors on fireplaces or install an efficient fireplace insert. Replace older wood stoves with new cleaner-burning devices.
- Install a quiet range hood, exhausted to the exterior, that is close to the cooking surface. This is especially important for gas stoves.
- Install a certified carbon monoxide detector or smoke alarm in rooms with a fuel burning appliance that uses a chimney.
- Make sure stoves are installed and fitted properly.

Volatile Organic Compounds (VOCs)

Formaldehyde

Sources:

- Composite wood products, including particle board, fiberboard, and plywood.

- Furniture, cupboards, cabinets made from composite wood products.
- Upholstery.
- Carpeting.
- Insulation.
- Paints and finishes.
- Glues, cleaners, waxes, and other household products.

Remedies:
- Choose building materials that don't contain urea formaldehyde; for example, solid wood instead of particleboard or medium density fiberboard.
- If solid wood is too expensive, use construction-grade (softwood) plywood, which uses more stable phenol formaldehyde glues, rather than finishing-grade plywood (hardwood) made with urea formaldehyde glues.
- Seal all exposed surfaces and edges on materials that are manufactured with high formaldehyde content. Use plastic laminates or multiple coats of low toxicity, water-based, acrylic sealers designed to reduce emissions from wood, gypsum board and grout, and to reduce the moisture permeability of surfaces. Aluminum foil inside cabinets is also an effective barrier.
- Improve the ventilation in your house.
- Use a heavy duty filter designed for formaldehyde removal.
- Buy spider plants to put in a room with formaldehyde.

Pesticides

Sources:
- Herbicide (weed killer, lawns and garden, turf, algae control).
- Insecticide (mosquito, flea, ant, roach, lice, mite, termite control; lawn treatment; pet products).
- Fungicide (paints, plastics, wood preservatives, grout, lawn treatment, carpet treatment).

Remedies:
- Do not treat soil under the building.
- Eliminate standard building products that contain biocides.
- If a pest must be eliminated, first see if its current access to nourishment and habitat can be limited. For example, if you have ants, you might clean up crumbs from the floor and counters and caulk the cracks.
- If a pest must be trapped or killed, consider the most benign methods first. Least toxic chemicals are employed as a last resort.

A well-renovated home will also be pest resistant because it incorporates the following features:

- Weather tightness.
- Appropriate grading and drainage.
- Provisions made for the prevention of excess moisture buildup from within, including extraction fans and windows that allow cross-ventilation.
- Dry wood without rot or infestation used in construction.

- Exterior wood appropriately treated for prevailing climatic conditions.
- All openings screened.
- Ground cover, leaves, chip and wood piles, and other potential insect habitats kept at a distance from the building.
- Avoid carpeting where pesticides can become trapped for years, protected from degradation caused by sunlight and bacteria.

Vinyl Chloride

Sources:
- Municipal drinking waters.
- PVC pipes.
- Vinyl flooring.
- Adhesives.
- Swimming pools.
- Upholstery.
- Wall coverings.
- Countertops.

Remedies:
- Use natural linoleum instead of vinyl flooring.
- Avoid PVC piping.
- Use safer plastic alternatives, like melamine formaldehyde use in countertops.

Radioactive Contaminants

Radon

Sources:
- Floor drains and sumps.

- Joints where basement walls and floor come together.
- Cracks in basement walls and floors.
- Holes in the foundation wall for pipes and wiring.
- Exposed earth or rock surfaces in the basement
- Well water.

Remedies:
- The US EPA and the Surgeon General recommend that all homes be tested for their radon levels below the third floor.
- Install a sub-slab ventilation system.
- Cover exposed earth with a polyethylene air barrier.
- Seal all cracks and joints in the foundation wall and floor slab with caulking or foam.
- Install a self priming drain or gas trap in floor drains leading to a sump or to drainage tiles.
- Remove radon from well water using activated charcoal filters or aeration units.

Environmental Tobacco Smoke

Cigarette, Pipe, and Cigar Smoke

Sources:
- Exhaled from lungs of smokers while smoking in the house.

Remedies:
- Stop smoking in the house — the only really healthy choice.

- Create a separate smoking room, with its own ventilation system and air seals to keep the smoke from spreading throughout the house.
- Install an effective ventilation system with a supply of outside air and a particulate filter.

Molds

Sources:

- Basement or crawl space.
- Kitchen.
- Bathrooms.
- In, and under, carpet and rugs on cold floors.
- On window frames or below windows.
- In, and on, furniture against outside walls.
- Inside wall cavities where there is dampness or condensation.
- In damp or unventilated storage areas.
- Closets, especially ones adjacent to exterior walls.
- Around plumbing leaks.
- Near roof or wall leaks.

Remedies:

- Fix basement plumbing and leaks.
- Do not store porous, absorbent materials such as cardboard, newspaper, or books in the basement. Keep the floors and walls clear.
- In winter, do not turn on the humidifier unless relative humidity falls below 30 percent.
- Provide better general ventilation, and spot ventilation in damp areas.
- Insulate fresh air ducts and cold water pipes to prevent condensation.

- Use air-vapor barriers to keep wall cavities dry.
- Remove carpets and rugs from cold floors, such as basements.
- Remove obstacles obstructing air flow in damp areas.
- Eliminate piles of newspapers, clothing and other materials in damp areas that can give molds a place to grow.

Moisture

Sources:

- Kitchen.
- Bathroom.
- Laundry room.
- Basements and crawl spaces.

Remedies:

- Repair leaks and cracks in basement and eliminate all other sources of moisture.
- Use a hygrometer to monitor humidity levels.
- Provide direct ventilation with an outside exhaust for moisture sources such as stoves, showers, and clothes dryers.
- Do not hang clothes to dry in the basement unless the area is well ventilated, with stale air being exhausted to the outdoors.
- Install a balanced, whole-house ventilation system that controls moisture by bringing in drier outside air, where possible.
- Provide an outside area for drying firewood.

Metric Conversion Tables

If You Know:	Multiply By:	To Find:
Inches	25.4	millimeters
Inches	2.54	centimeters
Feet	0.3048	meters
Yards	0.9144	meters
Miles	1.609	kilometers
Fluid ounces	28.4	milliliters
Gallons	4.546	liters
Ounces	28.350	grams
Pounds	0.4536	kilograms
Short tons	0.9072	metric tons
Acres	0.4057	hectares
Square inches	6.452	square centimeters
Square yards	0.836	square meters
Cubic yards	0.765	cubic meters
Square miles	2.590	square kilometers
Millimeters	0.039	inches
Centimeters	0.39	inches
Meters	3.281	feet
Meters	1.094	yards
Kilometers	0.622	miles
Milliliters	0.035	fluid ounces
Liters	0.220	gallons
Grams	0.035	ounces
Kilograms	2.205	pounds
Metric tons	1.102	short tons
Hectares	2.471	acres

If You Know:	Multiply By:	To Find:
Square centimeters	0.155	square inches
Square meters	1.196	square yards
Cubic meters	1.307	cubic yards
Square kilometers	0.386	square miles

Temperature

- To convert degrees Fahrenheit to degrees Celsius: (F° subtract 32) divided by 1.8.
- To convert degrees Celsius to degrees Fahrenheit: (°C multiplied by 1.8) add 32.

Endnotes

Chapter 5 Changing the World ...

1. US Department of Energy, Office of Building Technology State and Community Programs. *BTS Core Databook*. US Department of Energy, 2000. Charts 1.1.2 and 1.1.3.

2. Energy Information Administration. *Renewable Energy Annual 2001*. Department of Energy, November 2002.

3. Tracy F. Rysavy. "Renewables: Safety, Security, Sustainability." *Co-Op America Quarterly*, Summer 2002. pp. 9 to 11.

4. Intergovernmental Panel on Climate Change. *Climate Change 1995: The Science of Climate Change*. Cambridge University Press, 1996.

5. Environmental Protection Agency. *Inventory of U.S. Greenhouse Gas Emissions and Sinks: 1990 to 2000*. U.S. Environmental Protection Agency, Office of Atmospheric Programs, April 2002.

6. Environmental Protection Agency. *Global Warming: Climate* [online]. [Cited October 2, 2003]. Environmental Protection Agency, January 7, 2000. <http://yosemite.epa.gov/oar/globalwarming.nsf/content/climate.html>.

7. Tracy F. Rysavy. "Renewables: Safety, Security, Sustainability." *Co-Op America Quarterly*, Summer 2002, p. 10.

8. Dick Tompson. "Melt Away Future." *Time Magazine*, November 1997, pp. 38 to 40.

9. Tyler G. Miller. *Living in the Environment: Principles, Connections, and Solutions*. Wadsworth, 1998, p. 398.

10. William P. Cunningham and Barbara Woodworth Saigo. *Environmental Science: A Global Concern*. Wm. C. Brown Publishers, 1997, pp. 4945.

11. William D. Browning and Joseph J. Romm. *Greening the Building and the Bottom Line*. Rocky Mountain Institute, 1998.

12. Paul Hawken. *The Ecology of Commerce*. HarperCollins Publishers, 1993, pp. 178 to 79.

13. Adam Serchuk. "The Environmental Imperative for Renewable Energy." *Renewable Energy Policy Project*, April 2000.

14. Adam Serchuk. "The Environmental Imperative for Renewable Energy." *Renewable Energy Policy Project*, April 2000.

15. Adam Serchuk. "The Environmental Imperative for Renewable Energy." *Renewable Energy Policy Project*, April 2000.

16. Lester R. Brown. *Eco-Economy*. W.W. Norton, 2001, p. 37.

17. Energy Information Administration. *Renewable Energy Annual 2001*. Department of Energy, November, 2002.

18. US Department of Energy. *Wind Economics* [online]. [Cited September 26, 2003]. Energy Efficiency and Renewable Energy, February 2, 2003. <www.eere.energy.gov/RE/wind_economics.html>.

19. American Wind Energy Association. *Wind Energy Basics* [online]. [Cited September 26, 2003]. Wind Energy Fact Sheets, 2003. <www.awea.org/pubs/factsheets/WindEnergyBasics-2003-0324.pdf>.

20. American Wind Energy Association. *U.S. Wind Industry on Track to Grow More Than 25 percent in 2003* [online]. [Cited May 8, 2003]. News Release, May 8, 2003. <www.awea.org/news/news030508gro.html>.

21. Danish Wind Industry Association. *Windpower Note: The Energy Balance of Modern Wind Turbines* [online]. [Cited September 26, 2003]. Background information note, December 5, 1997. <www.windpower.org/en/publ/enbal.pdf >.

22. Lester R. Brown. *Eco-Economy*. W.W. Norton, 2001, p. 103.

23. Bjorn Lomborg. *The Skeptical Environmentalist: Measuring the Real State of the World*. Cambridge University Press, 2001, p. 134.

24. Dawn Levy. "Harnessing the Wind." *Stanford Report*, May 21 2003, p. 1.

25. Dawn Levy. "Harnessing the Wind." *Stanford Report*, May 21 2003, p. 1.

26. American Wind Energy Association. *Wind Energy and Economic Development: Building Sustainable Jobs and Communities* [online]. [Cited September 26, 2003]. Wind Energy Fact Sheet: Benefits of Wind Energy, 2002. <www.awea.org/pubs/factsheets/EconDev.PDF>.

27. American Wind Energy Association. *Wind Energy Fast Facts* [online]. [Cited September 26, 2003]. Wind Energy Publications, 2003. <www.awea.org/pubs/factsheets/FastFacts2003.pdf>.

28. National Wind Coordinating Committee. *Wind Performance Characteristics* [online]. [Cited October 2, 2003]. The Wind Issue Briefs, No. 10, January 1997. <www.nationalwind.org/pubs/wes/ibrief10.htm>.

29. American Wind Energy Association. *Wind Energy Fast Facts* [online]. [Cited October 2, 2003]. Wind Energy Publications, 2003. <www.awea.org/pubs/factsheets/FastFacts2003.pdf>.

30. Electric Power Research Institute. *Wind Forecasting System Development and Testing: Expanding Wind Energy Capacity by Eliminating a Barrier to Competitive Operation* [online]. [Cited October 2, 2003]. Electric Power Research Institute, 2001. <www.epri.com/corporate/products_services/project_opps/gen/1001460.pdf>.

31. Bjorn Lomborg. *The Skeptical Environmentalist: Measuring the Real State of the World*. Cambridge University Press, 2001, p. 133.

32. Energy Information Administration. *Renewable Energy Annual 1999 with Data for 1998.* US Department of Energy, March 2000.

33. Energy Information Administration. *Renewable Resources in the U.S. Electricity Supply* [online]. [Cited September 25, 2003]. US Department of Energy, June 5, 2003. <www.eia.doe.gov/cneaf/electricity/pub_summaries/renew_es.html>.

34. Lester R. Brown. *Eco-Economy.* W.W. Norton, 2001, p. 109.

35. Solarbuzz. *Fast Solar Energy Facts* [online]. [Cited September 26, 2003]. Solarbuzz, 2003. <www.solarbuzz.com>.

36. Bjorn Lomborg. *The Skeptical Environmentalist: Measuring the Real State of the World.* Cambridge University Press, 2001, p. 133.

37. Solar Energy Society of Canada, Inc. *Passive Solar Energy* [online]. [Cited September 26, 2003]. Solar Energy Society of Canada, 1999. <www.solarenergysociety.ca/passive.htm>.

38. Canadian Solar Industries Association. *Product Guide on Solar Water Heating* [online]. [Cited September 26, 2003]. Canadian Solar Industries Association, August 27, 2003. <http://cansia.ca/solarheat.html>.

39. Energy Efficiency and Renewable Energy Clearinghouse. *Solar Water Heating* [online]. [Cited September 26, 2003]. A Reference Brief, ToolBase Services, March 1996, revised February 2000. <www.toolbase.org/tertiaryT.asp?TrackID=&DocumentID=3216&CategoryID=949>.

40. Stuart Christie and Justin Sherrard. *Assessing the Sustainability of Biomass Resources* [online]. [Cited September 25, 2003]. Henley Publishing, Edition 2, Sustainable Development International. <www.sustdev.org/energy/articles/energy/edition2/sdi2_6_1.pdf>.

41. Stuart Christie and Justin Sherrard. *Assessing the Sustainability of Biomass Resources* [online]. [Cited September 25, 2003]. Henley Publishing, Edition 2, Sustainable Development International. <www.sustdev.org/energy/articles/energy/edition2/sdi2_6_1.pdf>.

42. Lester R. Brown. *Eco-Economy.* W.W. Norton, 2001, p. 110.

43. Michael T. Eckhart. *Renewable Energy in the U.S.* [online]. [Cited September 25, 2003]. World Council for Renewable Energy, June 13–15, 2002. <www.world-council-for-renewable-energy.org/downloads/WCRE-Eckhart.pdf>.

44. Tracy F. Rysavy. "Renewables: Safety, Security, Sustainability." *Co-Op America Quarterly*, Summer 2002, p. 9.

45. Energy Information Administration. *Renewable Energy Annual 2001.* US Department of Energy, November, 2002.

46. Tracy F. Rysavy. "Renewables: Safety, Security, Sustainability." *Co-Op America Quarterly*, Summer 2002, p. 11.

47. Lester R. Brown. *Eco-Economy.* W.W. Norton, 2001, p. 106.

48. *Economist.* "When Virtue Pays a Premium." Vol. 346 no. 8064 (April 18, 1998), pp. 57 to 58.

49. Lester R. Brown. *Eco-Economy.* W.W. Norton, 2001, p. 141.

50. Adam Serchuk. "The Environmental Imperative for Renewable Energy." *Renewable Energy Policy Project*, April 2000.

51. Sandra F. Mendler and William Odell. *The HOK Guidebook to Sustainable Design.* John Wiley and Sons, 2000.

52. Alex Wilson. "Dealing with Construction Waste: Innovative Solutions for a Tough Problem." *Environmental Building News*, November/December 1992, p. 1.

53. Environmental Protection Agency Municipal and Industrial Solid Waste Division. *Characterization of Building-Related Construction and Demolition Debris in the United States.* Franklin Associates, June 1998. See page ES-3, <www.epa.gov/epaoswer/hazwaste/sqg/c&d-rpt.pdf>

54. The Peaks to Prairies Pollution Prevention Information Center. *Construction and Demolition Waste Management Guide* [online]. [Cited September 25, 2003]. Montana State University Extension Service, US Environmental Protection Agency Region 8, Terrachord, 2002. <http://peakstoprairies.org/p2bande/Construction/C&DWaste/index.cfm>.

55. US Environmental Protection Agency. *Municipal Solid Waste: Basic Facts* [online]. [Cited September 25, 2003]. February 11, 2003. < www.epa.gov/epaoswer/non-hw/muncpl/facts.htm>.

56. Paul Hawkin, Amory B. Lovins, and L. Hunter Lovins. *Natural Capitalism: Creating the Next Industrial Revolution.* Little, Brown and Company, 1999.

57. Smart Growth Network. *Residential Construction Waste: From Disposal to Management* [online]. [Cited September 25, 2003]. Sustainable Communities Network, January 6, 2000. <www.smartgrowth.org/library/resident_const_waste.html>.

58. Ann Edminster. *Using Less Wood in Buildings* [online]. [Cited October 2, 2003]. Resource Conservation Alliance, Woodconsumption.org. <www.woodconsumption.org/products/annedminster.html>.

59. American Forest and Paper Association. *U.S. Forest Facts and Figures, 2001* [online]. [Cited September 25, 2003]. American Forest and Paper Association, Active Matter, 2002. <www.afandpa.org/Content/NavigationMenu/Forestry/Forestry_Facts_and_Figures/forest_health.pdf>.

60. Bob Villa. *Using Wood for Responsible, Renewable Building and Remodeling* [online]. [Cited September 29, 2003]. BVWebTiles, 2003; Bob Villa, 2001. <www.bobvila.com/ArticleLibrary/Subject/Framing/Engineered_Wood/ResponsibleWood.html>.

61. Paul Hawken. *The Ecology of Commerce.* HarperCollins, 1993, p. 29.

62. Darin Lowder and William Biddle. "How Much Lumber in a House." *Housing Economics*, April 1997.

63. Paul Hawken. *The Ecology of Commerce.* HarperCollins, 1993.

64. Paul Hawken. *The Ecology of Commerce.* HarperCollins, 1993, p. 29.

65. Helen Walsh, et al. *Not All Forest Certification Programs Measure Up* [online]. [Cited October 2, 2003]. Environmental Building News Archives, Vol. 12, no. 6 (June 1999); Building Green, 2003. <www.buildinggreen.com/features/12-4/letter.cfm>.

66. Rainforest Relief. *Tropical Woods* [online]. [Cited October 2, 2003]. Rainforest Relief, 2002. <www.rainforestrelief.org/What_to_Avoid_and_Alternatives/Rainforest_Wood/What_to_Avoid_What_to_Choose/By_Tree_Species/Tropical_Woods.html>.

67. *Green Building Technical Assistance, Green Resource Center, Fact Sheet.* Alameda County Waste Management Authority, 2003.

68. Dwight Holmes, Larry Strain, Alex Wilson, and Sandra Leibowitz. *GreenSpec*. E Build, 1999, p. 96.

69. Alex Wilson. "Disposal: The Achilles' Heel of CCA-Treated Wood." *Environmental Building News*, March 1997, p. 1.

70. Dwight Holmes, Larry Strain, Alex Wilson, and Sandra Leibowitz. *GreenSpec*. E Build, 1999, p. 98.

71. Alameda County Waste Management Authority. *Green Building Technical Assistance Fact Sheet*, 2003.

72. Dwight Holmes, Larry Strain, Alex Wilson, and Sandra Leibowitz. *GreenSpec*. E Build, 1999, p. 109.

73. Nadav Malin. "Recycled Plastic Lumber." *Environmental Building News*, July/August 1993, p. 1.

74. Nadav Malin. "Recycled Plastic Lumber." *Environmental Building News*, July/August 1993, p. 16.

75. Dwight Holmes, Larry Strain, Alex Wilson, and Sandra Leibowitz. *GreenSpec*. E Build, 1999, p. 109.

76. Dwight Holmes, Larry Strain, Alex Wilson, and Sandra Leibowitz. *GreenSpec*. E Build, 1999, pp. 111 to 112.

77. Sandra F. Mendler and William Odell. *The HOK Guidebook to Sustainable Design.* John Wiley and Sons, 2000.

78. Canadian Wood Council. *Water Pollution Index* [online]. [Cited September 29, 2003]. Canadian Wood Council, 1995 to 2002. <www.cwc.ca/environmental/bulletins/bulletin_4/water.html>.

79. Bjorn Lomborg. *The Skeptical Environmentalist: Measuring the Real State of the World.* Cambridge University Press, 2001, p. 149.

80. Rocky Mountain Institute. *Water Efficiency for your Home* [online]. [Cited October 2, 2003]. Rocky Mountain Institute, 1995, 2003; Intrcomm Technology. <www.rmi.org/images/other/W-WaterEff4Home.pdf>.

81. Bjorn Lomborg. *The Skeptical Environmentalist: Measuring the Real State of the World.* Cambridge University Press, 2001, p. 149.

82. *Natural Home*, May/June 1999, p. 63.

83. Bjorn Lomborg. *The Skeptical Environmentalist: Measuring the Real State of the World.* Cambridge University Press, 2001, p. 150.

84. Office of the Federal Environmental Executive. *The Need for Green Building* [online]. [Cited October 1, 2003]. Office of the Federal Environmental Executive: Promoting sustainable environmental stewardship throughout the federal government, 2003. <www.ofee.gov/sb/state_fgb_2.pdf>.

85. Scott Chaplin. *Water Efficiency: The Next Generation.* Rocky Mountain Institute, 1998.

86. *Natural Home*, May/June 1999, p. 63.

87. *Natural Home*, May/June 1999, p. 63.

88. Peter Casey, Clement Solomon, Colleen Mackne, and Andrew Lake. *Water Efficiency: A Project Funded by the U.S. Environmental Protection Agency.* National Small Flows Clearinghouse, 1998.

89. Frits Van der Leeden, Fred L. Troise, and David Keith Todd. *The Water Encyclopedia.* Lewis Publishers, 2nd ed., 1990.

90. *Natural Home.* May/June, 1999, p. 63.

91. City of Nanaimo. *Water Saving Ideas* [online]. [Cited September 29, 2003]. Public Works, 2003. <www.city.nanaimo.bc.ca/a_public/water_intro.asp#ideas>.

92. City of Nanaimo. *Water Saving Ideas* [online]. [Cited September 29, 2003]. Public Works, 2003. <www.city.nanaimo.bc.ca/a_public/water_intro.asp#ideas>.

93. City of Nanaimo. *Water Saving Ideas* [online]. [Cited September 29, 2003]. Public Works, 2003. <www.city.nanaimo.bc.ca/a_public/water_intro.asp#ideas>.

94. City of Nanaimo. *Water Saving Ideas* [online]. [Cited September 29, 2003]. Public Works, 2003. <www.city.nanaimo.bc.ca/a_public/water_intro.asp#ideas>.

95. City of Nanaimo. *Water Saving Ideas* [online]. [Cited September 29, 2003]. Public Works, 2003. <www.city.nanaimo.bc.ca/a_public/water_intro.asp#ideas>.

96. *Natural Home.* May/June 1999, p. 62.

97. *Natural Home.* May/June 1999, p. 63.

98. Guy Gugliotta and Eric Pianin. "Environmental Protection Agency: Few Fined for Polluting Water." *Washington Post.* June 6, 2003, p. A01.

99. Ed Olson. "Bottled Water: Pure Drink or Pure Hype?" *National Resources Defense Council,* 1999.

100. *Natural Home.* May/June 1999, p. 58.

101. Miguel Bustillo. "Water Bottles are Creating a Flood of Waste: The state launches a campaign urging Californians to recycle the plastic containers." *Los Angeles Times,* May 28, 2003, p. B1

102. David Suzuki. *Feeding our senses is important to health* [online]. [Cited September 29, 2003]. David Suzuki Foundation, 2002; Environmental News Network, 2003. <www.enn.com/news/enn-stories/2002/06/0625 2002/s_47531.asp>.

103. The American College of Allergy, Asthma, and Immunology. *New Survey Reveals Allergies Nearly Twice as Common as Believed — Afflicting More Than One-Third of Americans* [online]. [Cited September 29, 2003]. American College of Allergy, Asthma, and Immunology, 1999. <http://allergy.mcg.edu/news/survey.html>.

104. Doris Rapp. *Is This Your Child?* William Morrow and Company, 1991, p. 35.

105. American Lung Association Health House. *Asthma: Tips on eliminating common "triggers" in the home* [online]. [Cited October 1, 2003]. American Lung Association of Minnesota, 2003; Clientek. <www.healthhouse.org/tipsheets/asthma.asp>.

106. *American Journal of Respiratory and Critical Care Medicine,* 1998, 158: 320–334.

107. US Centers for Disease Control. *Forecasted State-Specific Estimates of Self-Reported Asthma Prevalence — United States, 1998* [online]. [Cited September 29, 2003]. Morbidity and Mortality Weekly Report, December 4, 1998; (47) 47: 1022 to 1025; updated May 2, 2001. <www.cdc.gov/mmwr/preview/mmwrhtml/00055803.htm>.

108. Barbara Lippiatt and Gregory Norris. *Selecting Environmentally and Economically Balanced Building Materials.* National Institute of Standards and Technology Special Publication 888, Second International Green Building Conference and Exposition, 1995, p. 37.

109. World Health Organization. *Health and Environment in Sustainable Development: Five Years after the Earth Summit: Executive Summary* [online]. [Cited September 30, 2003]. World Heath Organization, 1997. <www.who.int/environmental_information/Information_resources/htmdocs/execsum.htm>.

110. Chad B. Dorgan, Charles E. Dorgan, Marty S. Kanarek, and Alexander J. Willman. "Health and Productivity Benefits of Improved Indoor Air Quality." *ASHRAE Transactions*, Vol. 104, Part 1; 1998.

111. Alan Hedge, William R. Sims, and Franklin D. Becker. *Cornell University: Lighting the Computerized Office* [online]. [Cited September 30, 2003]. Department of Design and Environmental Analysis, New York State College of Human Ecology, Cornell University, October 1989, September 1990. <http://ergo.human.cornell.edu/lighting/lilstudy/lilstudy.htm>

112. H.H.S. Yu and R.R. Raber. *Ventilation and Filtration Requirements Based on Fanger's "Perceived Good Air Quality" and ASHRAE Standard 62-1989.* American Society of Heating, Air Conditioning, and Refrigeration Engineers, September 4 to 8, 1991, pp. 358 to 62.

113. Canada Mortgage and Housing Corporation. *Healthy Housing Renovation Planner.* Canada Mortgage and Housing Corporation, 1999, p. 14.

114. Wayne R. Ott and John W. Roberts. "Everyday Exposure to Toxic Pollutants." *Scientific American*, February 1998, p. 90.

115. American Lung Association Health House. *Is Your Home Making You Sick?* [online]. [Cited September 30, 2003]. American Lung Association of Minnesota, 2003; Clientek. <?>.

116. John Harte, Cheryl Holdren, Richard Schneider, and Christine Shirley. *Toxics A to Z: A Guide to Everyday Pollution Hazards.* University of California Press, 1991, p. 134.

117. Wisconsin Department of Health and Family Services Division of Public Health. *Information on Toxic Chemicals: Lead* [online]. [Cited September 30, 2003]. Agency for Toxic Substances and Disease Registry, Public Health Service, US Department of Health and Human Services; December 2000. <www.dhfs.state.wi.us/eh/ChemFS/pdf/lead2.pdf>.

118. John Harte, Cheryl Holdren, Richard Schneider, and Christine Shirley. *Toxics A to Z: A Guide to Everyday Pollution Hazards.* University of California Press, 1991, p. 334.

119. American Lung Association Health House. *Lead: Tips for identifying and eliminating lead in the home* [online]. [Cited October 1, 2003]. American Lung Association of Minnesota, 2003; Clientek. <www.healthhouse.org/tipsheets/lead.asp>.

120. Debra L. Dadd. *The Nontoxic Home and Office.* Jeremy P. Tarcher, 1992, p. 164.

121. Steve Andrews. "A House Under Pressure." *Home Builder.* January 1997, pp. 13 to 14.

122. Steve Andrews. "A House Under Pressure." *Home Builder.* January 1997, pp. 13 to 14.

123. American Lung Association Health House. *Carbon Monoxide: Tips for detecting the gas and preventing tragedy* [online]. [Cited October 1, 2003]. American Lung Association of Minnesota, 2003; Clientek. <www.healthhouse.org/tipsheets/carbonmonoxide.asp>.

124. Debra L. Dadd. *The Nontoxic Home and Office.* Jeremy P. Tarcher, 1992.

125. John Harte, Cheryl Holdren, Richard Schneider, and Christine Shirley. *Toxics A to Z: A Guide to Everyday Pollution Hazards.* University of California Press, 1991, p. 319.

126. John Harte, Cheryl Holdren, Richard Schneider, and Christine Shirley. *Toxics A to Z: A Guide to Everyday Pollution Hazards*. University of California Press, 1991, p. 319.

127. John Harte, Cheryl Holdren, Richard Schneider, and Christine Shirley. *Toxics A to Z: A Guide to Everyday Pollution Hazards*. University of California Press, 1991, p. 319.

128. US General Accounting Office. "Lawn Care Pesticides: Risks Remain Uncertain While Prohibited Safety Claims Continue." *U.S. Government Printing Office*. March 1990, pp. 4 to 5.

129. Wayne R. Ott and John W. Roberts. "Everyday Exposure to Toxic Pollutants." *Scientific American*, February 1998.

130. Wayne R. Ott and John W. Roberts. "Everyday Exposure to Toxic Pollutants." *Scientific American*, February 1998.

131. Wayne R. Ott and John W. Roberts. "Everyday Exposure to Toxic Pollutants." *Scientific American*, February 1998.

132. Jack Leiss and David Savitz. "Home Pesticide Use and Childhood Cancer: A Case-Control Study." *American Journal of Public Health*, February 1995, pp. 249 to 252.

133. Ellen Gold, et al. "Risk Factors for Brain Tumors in Children." *American Journal of Epidemiology*, Vol. 109 (1979), pp. 309 to 319.

134. American Cancer Society. "Drug Free Lawn." *American Cancer Society*, 1993.

135. Sierra Club of Canada. *Pesticide Fact Sheet: The truth about pesticides* [online]. [Cited October 1, 2003]. Sierra Club of Canada, 2003. <www.sierraclub.ca/national/pest/pesticide-truth.html>

136. David J. Hunter and Karl T. Kelsey. "Pesticide Residues and Breast Cancer: The Harvest of a Silent Spring?" *Journal of the National Cancer Institute*, April 28, 1993, pp. 598 to 599.

137. Debra L. Dadd. *The Nontoxic Home and Office*. Jeremy P. Tarcher, 1992, p. 152.

138. American Lung Association Health House. *Secondhand Smoke: Tips for reducing your exposure to secondhand smoke* [online]. [Cited October 1, 2003]. American Lung Association of Minnesota, 2003; Clientek. <www.healthhouse.org/tipsheets/secondhandsmoke.asp>.

139. Canada Mortgage and Housing Corporation. *Healthy Housing Renovation Planner*. Canada Mortgage and Housing Corporation, 1999, p. 13.

Chapter 6 ... One Room at a Time

1. Alex Wilson. "Daylighting: Energy and Productivity Benefits."*Environmental Building News*, Vol. 8, no. 9 (September 1999), pp. 11 to 13.

2. "No Regrets Remodeling." *Home Energy Magazine*. Energy Auditor and Retrofitter, 1997, p. 163.

3. David Philipps. "PCs are not PC, Computer Waste Is Piling Up" [online]. [Cited October 25, 2003]. *Columbia News Service*, April 3, 2002. <www.jrn.columbia.edu/studentwork/cns/2002-04-03/272.asp>.

4. David Philipps. "PCs are not PC, Computer Waste Is Piling Up" [online]. [Cited October 25, 2003]. *Columbia News Service*, April 3, 2002. <www.jrn.columbia.edu/studentwork/cns/2002-04-03/272.asp>.

5. David Philipps. "PCs are not PC, Computer Waste Is Piling Up" [online]. [Cited October 25, 2003]. *Columbia News Service*, April 3, 2002. < www.jrn.columbia.edu/studentwork/cns/2002-04-03/272.asp>.

Chapter 7 Job Site and Landscaping

1. Wayne R. Ott and John W. Roberts. "Everyday Exposure to Toxic Pollutants." *Scientific American*, February 1998, p. 90.

2. Wayne R. Ott and John W. Roberts. "Everyday Exposure to Toxic Pollutants." *Scientific American*, February 1998, p.90.

3. Doris Rapp. *Is This Your Child?* William Morrow and Company, 1991, p. 25.

4. Doris Rapp. *Is This Your Child?* William Morrow and Company, 1991, p. 353.

5. John F. Straube. *Moisture Fundamentals and Mould.* OBEC Seminar, November 17, 1999.

6. US Green Building Council. *Industry Statistics* [online]. [Cited October 2, 2003]. BuildingGreen, Inc., 2001. <www.usgbc.org/Resources/industry_statistics.asp>.

7. Noel Grove. "Recycling." *National Geographic*, July 1994.

8. Sandra F. Mendler and William Odell. *The HOK Guidebook to Sustainable Design.* John Wiley and Sons, 2000.

9. John E. Young and Aaron Sachs. "Creating a Sustainable Materials Economy." State of the World, *Worldwatch Institute*, 1995.

10. Paul Hawkin, Amory B. Lovins, and L. Hunter Lovins. *Natural Capitalism: Creating the Next Industrial Revolution.* Little, Brown, 1999, p. 219.

11. Department of Environmental Protection. *Landscaping to Save Energy in Winter* [online]. [Cited October 2, 2003]. Montgomery County Department of Environmental Protection, October 16, 2002. <www.montgomerycountymd.gov/mc/services/dep/greenman/windy.htm>.

12. Diane Relf. *Conserving Energy with Landscaping* [online]. [Cited October 2, 2003]. Virginia Cooperative Extension, 2001. <www.ext.vt.edu/pubs/envirohort/426-712/426-712.html#L2>.

13. Diane Relf. *Conserving Energy with Landscaping* [online]. [Cited October 2, 2003]. Virginia Cooperative Extension, 2001. <www.ext.vt.edu/pubs/envirohort/426-712/426-712.html#L2>.

Chapter 8 Foundation

1. Roy A. Grancher. "US Cement: Acquisition and Concentration." *Cement Americas.* July 1, 2000.

2. Vanessa A.T. Hoy. "Green Building Foundations." *Permanent Buildings and Foundations.* December 1, 2001, p. 12.

3. National Association of Home Builders. *Design Guide for Frost-Protected Shallow Foundations.* US Department of Housing and Urban Development, June 1994.

4. National Association of Home Builders. *Design Guide for Frost-Protected Shallow Foundations.* US Department of Housing and Urban Development, June 1994.

5. Nadav Malin. "The Fly Ash Revolution: Making Better Concrete with Less Cement." *Environmental Building News*, Vol. 8 no. 6 (June 1999), p. 10.

6. Dwight Holmes, Larry Strain, Alex Wilson, and Sandra Leibowitz. *GreenSpec*. E Build, 1999, p. 52.

7. Bjorn Lomborg. *The Skeptical Environmentalist: Measuring the Real State of the World*. Cambridge University Press, 2001, p. 183.

Chapter 9 Structural Framing

1. Will Warburg. MDF: *The Rest of the Tree* [online]. [Cited October 6, 2003]. This Old House Ventures, Inc., 2003; Time Media Company. <www.thisoldhouse.com/toh/knowhow/interiors/article/0,16417,195108,00.html>

2. Paul Hawkin, Amory B. Lovins, and L. Hunter Lovins. *Natural Capitalism: Creating the Next Industrial Revolution*. Little, Brown, 1999, pp. 183 to 184.

3. Paul Hawkin, Amory B. Lovins, and L. Hunter Lovins. *Natural Capitalism: Creating the Next Industrial Revolution*. Little, Brown and Company, 1999, p. 183.

Chapter 11 Roofing

1. Paula Baker, Erica Elliott, and John Banta. *Prescriptions for a Healthy House: A Practical Guide for Architects, Builders and Homeowners*. InWord Press, 1998, p. 100.

2. Dwight Holmes, Larry Strain, Alex Wilson, and Sandra Leibowitz. *GreenSpec*. E Build, 1999, p. 141.

3. Alex Wilson. "Low-Slope Roofing: Prospects Looking Up." *Environmental Building News*, Vol. 7, no. 10 (November 1998), p. 1.

4. Canada Mortgage and Housing Corporation. *Building Materials for the Environmentally Hypersensitive*. Canada Mortgage and Housing Corporation, 1999, p. 6.

5. Alex Wilson. "Low-Slope Roofing: Prospects Looking Up." *Environmental Building News*, Vol. 7, no. 10 (November 1998), p. 13.

6. National Renewable Energy Laboratory. *Cooling Your Home Naturally* [online]. [Cited October 7, 2003]. US Department of Energy, October 1994; August 7, 2003. <www.eere.energy.gov/consumerinfo/factsheets/coolhome.html>.

7. Paul Hawkin, Amory B. Lovins, and L. Hunter Lovins. *Natural Capitalism: Creating the Next Industrial Revolution*. Little, Brown, 1999, p. 226.

8. Paul Hawkin, Amory B. Lovins, and L. Hunter Lovins. *Natural Capitalism: Creating the Next Industrial Revolution*. Little, Brown, 1999, p. 108.

9. Katrin Scholz-Barth. *Green Roofs: Stormwater Management from the Top Down* [online]. [Cited October 7, 2003]. Environmental Design and Construction; Business News Publishing, 2003. <www.edcmag.com/edc/cda/articleinformation/features/bnp__features__item/0,4120,18769,00.html>.

Chapter 12 Plumbing

1. Paul Hawkin, Amory B. Lovins, and L. Hunter Lovins. *Natural Capitalism: Creating the Next Industrial Revolution*. Little, Brown , 1999, p. 223.

2. Paul Hawkin, Amory B. Lovins, and L. Hunter Lovins. *Natural Capitalism: Creating the Next Industrial Revolution*. Little, Brown , 1999, p. 224.

3. Paul Hawkin, Amory B. Lovins, and L. Hunter Lovins. *Natural Capitalism: Creating the Next Industrial Revolution*. Little, Brown, 1999, p. 220.

4. Dwight Holmes, Larry Strain, Alex Wilson, and Sandra Leibowitz. *GreenSpec*. E Build, 1999, p. 246.

5. Paul Hawkin, Amory B. Lovins, and L. Hunter Lovins. *Natural Capitalism: Creating the Next Industrial Revolution*. Little, Brown, 1999, p. 220.

6. Paul Hawkin, Amory B. Lovins, and L. Hunter Lovins. *Natural Capitalism: Creating the Next Industrial Revolution*. Little, Brown, 1999, p. 227.

7. Dwight Holmes, Larry Strain, Alex Wilson, and Sandra Leibowitz. *GreenSpec*. E Build, 1999, p. 245.

8. Marq Devilliers. *Water: The Fate of Our Most Precious Resource*. Houghton Mifflin , 2000, p. 157.

9. Nebraska Energy Quarterly. *Hot Water Use Can Affect Utility Bills* [online]. [Cited October 7, 2003]. Nebraska Energy Office, Winter 2000. <www.state.ne.us/home/NEO/winter2000/winter002.html>.

10. Nebraska Energy Quarterly. *Hot Water Use Can Affect Utility Bills* [online]. [Cited October 7, 2003]. Nebraska Energy Office, Winter 2000. <www.state.ne.us/home/NEO/winter2000/winter002.html>.

11. Canada Mortgage and Housing Corporation. *Healthy Housing Renovation Planner*. Canada Mortgage and Housing Corporation, 1999, p. 270.

12. Paul Hawkin, Amory B. Lovins, and L. Hunter Lovins. *Natural Capitalism: Creating the Next Industrial Revolution*. Little, Brown, 1999, p. 222.

13. Global Green USA. *Top 20 Cost-effective Ways to Green an Affordable Housing Project* [online]. [Cited October 7, 2003]. Tree Media Group; Global Green USA, 2003. <www.globalgreen.org/programs/20ways.html>.

14. Nadav Malin. "Pedal Controls Save Water, Time." *Environmental Building News*. Vol. 8, no. 6 (June 1999), p. 6.

15. Paula Baker, Erica Elliott, and John Banta. *Prescriptions for a Healthy House*. Inword Press, 1998, p. 154.

Chapter 13 Electrical

1. Penn State. *The Mueller Report: Moving Beyond Sustainability Indicators to Sustainability Action at Penn State* [online]. [Cited October 23, 2003]. Penn State Green Destiny Council, n.d. <www.bio.psu.edu/greendestiny/publications/gdc-mueller_report.pdf>.

2. Southern Minnesota Municipal Power Agency. *Interesting Facts: Lighting* [online]. [Cited October 23, 2003]. Southern Minnesota Municipal Power Agency. 2002. <www.smmpa.org/athome/enst_home_fact.asp>.

3. Sofia Perez. *Light Bulbs* [online]. [Cited October 7, 2003]. Green Guide Institute, 2003. <www.thegreenguide.com/reports/product.mhtml?id=8>.

4. Sofia Perez. *Light Bulbs* [online]. [Cited October 7, 2003]. Green Guide Institute, 2003. <www.thegreenguide.com/reports/product.mhtml?id=8>.

5. E Source. *E Source Lighting Atlas* [online]. [Cited October 23, 2003]. E Source, December 1997. <www.esource.com/public/products/atlas_light.asp>.

6. Annie Berthold-Bond. Special *Energy Saving Tips Issue* [online]. [Cited October 23, 2003]. Care2.com, October 15, 2002. <www.care2.com/newsletters/care2lifestyle/2002/10-15.html>.

7. Olympus. *The Physics of Light and Color: Incandescent Light Sources* [online]. [October 23, 2003]. Olympus Corporation, 2003. <www.mic-d.com/curriculum/lightandcolor/lightsources.html>.

8. Sofia Perez. *Light Bulbs* [online]. [Cited October 7, 2003]. Green Guide Institute, 2003. <www.thegreenguide.com/reports/product.mhtml?id=8>.

9. Sofia Perez. *Light Bulbs* [online]. [Cited October 7, 2003]. Green Guide Institute, 2003. <www.thegreenguide.com/reports/product.mhtml?id=8>.

10. Lester R. Brown. *Eco-Economy*. W.W. Norton , 2001, p. 101.

11. Connecticut Light and Power. *Home Lighting Tips* [online]. [Cited October 23, 2003]. The Connecticut Light and Power Company; Northeast Utilities Service Company, 1998 to 2003. <www.cl-p.com/clmres/energy/lightingtips.asp>.

12. Christine Geltz. *Building an Energy Efficient Home Office* [online]. [Cited October 7, 2003]. Home Energy Magazine Online, May/June 1993. <http://hem.dis.anl.gov/eehem/93/930512.html>.

13. Evan Mills. *Energy Myths* [online]. [Cited October 7, 2003]. Home Energy Magazine, 2002. <homeenergy.org/hewebsite/consumerinfo/myths/webmystemp.html>

14. Goddard Space Flight Center. *Science Question of the Week* [online]. [Cited October 23, 2003]. NASA, October 5, 2001. <www.gsfc.nasa.gov/scienceques2001/20011005.html>.

Chapter 14 Insulation

1. Alex Wilson. "Insulation Materials: Environmental Comparisons" [online]. [Cited October 23, 2003]. *Environmental Building News*, Vol. 4, no. 1 (January/February 1995). <www.buildinggreen.com/features/ins/insulation.cfm#RTFToC1>.

2. Questar Gas. *Energy Conservation Tips to Make Your Home (And Your Wallet) More Comfortable* [online]. [Cited October 23, 2003]. American Gas Association, June 12, 2003. <www.questargas.com/AboutNaturalGas/EnergyTips/energysavingtips.htm>.

3. Canadian Wood Council. *Laboratory Research Results* [online]. [Cited October 23, 2003]. Canadian Wood Council, 1995 to 2002. <www.cwc.ca/publications/tech_bulletins/tech_bull_3/lab.html>.

4. Lafayette Utilities System. *Energy Efficiency Tips* [online]. [Cited October 23, 2003]. Lafayette Consolidated Government; Firefly Digital, n.d. <www.lus.org/site.php?pageID=49>.

Chapter 15 Solar Energy

1. Office of Energy Efficiency and Renewable Energy. *Energy Savers: Solar Heating and Cooling* [online]. [Cited October 8, 2003]. US Department of Energy, October 6, 2003. <www.eere.energy.gov/consumerinfo/energy_savers/heatcool.html>.

2. Pennsylvania Department of Environmental Protection. *Flood Recovery: Adding Energy Efficiency* [online]. [Cited October 8, 2003]. Commonwealth of Pennsylvania, Department of Environmental Protection, February 5, 1997. <www.dep.state.pa.us/dep/deputate/watermgt/GENERAL/FLOODS/fs1959.htm>.

3. Time Energy, Inc. *FAQ: Windows* [online]. [Cited October 23, 2003]. Time Energy, 2002. <www.timeenergy.com/html/faqwindows.html>.

4. Brian Pon. *Cooling Comes Naturally: What is Evapotranspiration?* [online]. [Cited October 8, 2003]. Heat Island Group, April 27, 2000. <http://eetd.lbl.gov/HeatIsland/LEARN/Evapo>.

5. Mudit Saxena. *Microclimate Modification: Calculating the Effect of Trees on Air Temperature* [online]. [Cited October 8, 2003]. Masters Thesis; Arizona State University, 2001. <www.sbse.org/awards/docs/Saxena.pdf>.

6. Pioneer Thinking. *Lawn Care/Landscaping* [online]. [Cited October 8, 2003]. Pioneerthinking.com, 1999 to 2003. <www.pioneerthinking.com/landscape.html>.

7. Pioneer Thinking. *Lawn Care/Landscaping* [online]. [Cited October 8, 2003]. Pioneerthinking.com, 1999 to 2003. <www.pioneerthinking.com/landscape.html>.

8. John Tillman Lyle. *Regenerative Design for Sustainable Development.* John Wiley and Sons, 1994, p. 102.

9. Pioneer Thinking. *Lawn Care/Landscaping* [online]. [Cited October 8, 2003]. Pioneerthinking.com, 1999 to 2003. <www.pioneerthinking.com/landscape.html>.

10. Office of Energy Efficiency and Renewable Energy. *Residential Solar Heating Retrofits* [online]. [Cited October 8, 2003]. US Department of Energy, November 2002. <www.eere.energy.gov/consumerinfo/refbriefs/ac6.html>.

11. Office of Energy Efficiency and Renewable Energy. *Residential Solar Heating Retrofits* [online]. [Cited October 8, 2003]. US Department of Energy, November 2002. <www.eere.energy.gov/consumerinfo/refbriefs/ac6.html>

12. Pacific Northwest National Laboratory. *Solar Energy Can Heat Your Water with Less Energy and Money* [online]. [Cited October 8, 2003]. Battelle; US Department of Energy, July 2003. <www.pnl.gov/conserve-energy/tips/may03.stm>.

13. Office of Energy Efficiency and Renewable Energy. *Energy Savers: Solar Water Heating* [online]. [Cited October 8, 2003]. US Department of Energy, October 6, 2003. <www.eere.energy.gov/consumerinfo/energy_savers/waterheat.html>.

14. Office of Energy Efficiency and Renewable Energy. *Information Resources: Solar Water Heating* [online]. [Cited October 8, 2003]. US Department of Energy, October 6, 2003. <www.eere.energy.gov/consumerinfo/factsheets/solrwatr.html>.

15. Natural Resources Defense Council. *Photovoltaic Cells: Alternative energy technologies hold the key to curbing air pollution and global warming* [online]. [Cited October 8, 2003]. Natural Resources Defense Council, April 20, 2000. <www.nrdc.org/air/energy/fphoto.asp>.

Chapter 16 Heating, Ventilation, and Air Conditioning

1. Alex Wilson and John Morrill. *Consumer Guide to Home Energy Savings*. American Council for an Energy-efficient Economy, 1991, p. 37.

2. Office of Energy Efficiency and Renewable Energy. *Heating and Cooling* [online]. [Cited October 9, 2003]. US Department of Energy, October 6, 2003. <www.eere.energy.gov/consumerinfo/energy_savers/heatcool.html>.

3. Alex Wilson and John Morrill. *Consumer Guide to Home Energy Savings*. American Council for an Energy-efficient Economy, 1991, p. 37.

4. Robert Starr. *Radiant Heat* [online]. Environmental Design and Construction, May 1, 2003; Business News Publishing, 2003. <www.edcmag.com/CDA/ArticleInformation/features/BNP__Features__Item/0,4120,97615,00.html>.

5. Washington State University Energy Program. *Ground Source Heat Pumps — Residential* [online]. [Cited October 9, 2003]. Toolbase Services, National Association of Home Builders, 2001 to 2003. <www.toolbase.org/tertiaryT.asp?DocumentID=3215&CategoryID=963>.

6. US Department of Energy. *Economics of Geothermal Heat Pumps* [online]. [Cited October 24, 2003]. Energy Efficiency and Renewable Energy; U.S. Department of Energy, August 13, 2003. <www.eere.energy.gov/consumerinfo/heatcool/hc_space_geothermal_economics.html>.

7. Santee Electric Cooperative. *Energy Saving Tips* [online]. [Cited October 9, 2003]. Santee Electric Cooperative, 2003. <www.santee.org/save/energytips.html>.

8. Alex Wilson and John Morrill. *Consumer Guide to Home Energy Savings*. American Council for an Energy-efficient Economy, 1991, p. 50.

9. Office of Energy Efficiency and Renewable Energy. *Heating and Cooling* [online]. [Cited October 9, 2003]. US Department of Energy, October 6, 2003. <www.eere.energy.gov/consumerinfo/energy_savers/heatcool.html>.

10. State of Idaho Department of Environmental Quality. *Wood Stove Replacements and Tax Deduction Information* [online]. [Cited October 9, 2003]. Idaho Department of Environmental Quality, 2000; May 12, 2003. <www.deq.state.id.us/air/WoodStove_Replace.htm>.

11. State of Idaho Department of Environmental Quality. *Wood Stove Replacements and Tax Deduction Information* [online]. [Cited October 9, 2003]. Idaho Department of Environmental Quality, 2000; May 12, 2003. <www.deq.state.id.us/air/WoodStove_Replace.htm>.

12. Puget Sound Clean Air Agency. *Facts About Burning Wood* [online]. [Cited October 9, 2003]. Puget Sound Clean Air Agency, November 2002. <www.pscleanair.org/burning/30-17-fact_sheet-burning.pdf>.

13. Puget Sound Clean Air Agency. *Facts About Burning Wood* [online]. [Cited October 9, 2003]. Puget Sound Clean Air Agency, November 2002. <www.pscleanair.org/burning/30-17-fact_sheet-burning.pdf>.

14. Puget Sound Clean Air Agency. *Facts About Burning Wood* [online]. [Cited October 9, 2003]. Puget Sound Clean Air Agency, November 2002. <www.pscleanair.org/burning/30-17-fact_sheet-burning.pdf>.

15. Wayne R. Ott and John W. Roberts. "Everyday Exposure to Toxic Pollutants." *Scientific American*, February 1998.

16. Alex Wilson and John Morrill. *Consumer Guide to Home Energy Savings*. American Council for an Energy-efficient Economy, 1991, p. 90.

17. Alex Wilson and John Morrill. *Consumer Guide to Home Energy Savings*. American Council for an Energy-efficient Economy, 1991, p. 90.

18. Home Energy Magazine. *No Regrets Remodeling*. Energy Auditor and Retrofitter, 1997, p. 14.

19. Aerosol. *Aerosol Duct Sealing Technology* [online]. [Cited October 9, 2003]. The California Institute for Energy Efficiency, US Environmental Protection Agency, Electric Power Research Institute, US Department of Energy, February 2003. <http://epb1.lbl.gov/aerosol/body.html>.

20. Aerosol. *Aerosol Duct Sealing Technology* [online]. [Cited October 9, 2003]. The California Institute for Energy Efficiency, US Environmental Protection Agency, Electric Power Research Institute, US Department of Energy, February 2003. <http://epb1.lbl.gov/aerosol/body.html>.

21. Aerosol. *Aerosol Duct Sealing Technology* [online]. [Cited October 9, 2003]. The California Institute for Energy Efficiency, US Environmental Protection Agency, Electric Power Research Institute, US Department of Energy, February 2003. <http://epb1.lbl.gov/aerosol/body.html>.

22. Paul Hawkin, Amory B. Lovins, and L. Hunter Lovins. *Natural Capitalism: Creating the Next Industrial Revolution*. Little, Brown, 1999, p. 102.

23. Debra L. Dadd. *The Nontoxic Home and Office*. Jeremy P. Tarcher, 1992, p. 173.

24. US Environmental Protection Agency. *Sources of Indoor Air Pollution — Carbon Monoxide (CO)* [online]. [Cited October 13, 2003]. US Environmental Protection Agency, October 9, 2003. <www.epa.gov/iaq/co.html>.

25. American Lung Association Health House Program. *Carbon Monoxide: Tips for detecting the gas and preventing tragedy* [online]. [Cited October 13, 2003]. Clientek; American Lung Association of Minnesota, 1997 to 2003.<www.healthhouse.org/tipsheets/carbonmonoxide.asp>.

26. California Energy Commission. *Keeping Your Cool with Ceiling and Whole House Fans* [online]. [Cited October 13, 2003]. Energy Efficiency Division, Home Energy Guide Fact Sheets, n.d. <www.energy.ca.gov/efficiency/home_energy_guide/FANS.PDF>.

27. Thomas Klenck. *How It Works: Heat Recovery Ventilator* [online]. [Cited October 13, 2003]. Popular Mechanics, 2000.
<http://popularmechanics.com/home_improvement/how_it_works/2000/8/heat_recovery_ventilator/index.phtml>.

28 Canada Mortgage and Housing Corporation. *Healthy Housing Renovation Planner*. Canada Mortgage and Housing Corporation, 1999, p. 91.

29. The Air Filter Store. *Dust Fighter 95* [online]. The Air Filter Store, 2003. <www.theairfilterstore.com/>.

30. Trane. *Looking for a Healthier Home Environment? Then Look into Today's Electronic Air Cleaners* [online]. American Standard, Inc., 2003. <www.trane.com/Residential/NewSystem/Guides/ElectronicAirCleaners.asp>.

31. North Carolina State University. *Air Filters and Cleaners* [online]. [Cited December 2, 2003]. North Carolina Cooperative Extension Service, 1995. <www.ces.ncsu.edu/depts/fcs/housing/pubs/fcs3606.html>.

32. "Cleaning the Air: Portable room air cleaners, whole-house air cleaners." *Consumer Reports*, February, 2002.

33. Sofia Perez. *Air Conditioners* [online]. [Cited October 13, 2003]. The Green Guide Institute, 2003. <www.thegreenguide.com/reports/product.mhtml?id=9>.

34. Built Online. *General Questions and Answers on Energy Savings* [online]. [Cited October 13, 2003]. Environmental Energy Technologies Division, Lawrence Berkeley National Laboratory; MicroAssist, Inc., 1998. <www.builtonline.com/articles.cfm?P_ID=334#use>.

35. Alex Wilson and John Morrill. *Consumer Guide to Home Energy Savings*. American Council for an Energy-efficient Economy, 1991, p. 95.

36. Judy Bucher. "The Air Inside: You Are What You Breathe." *Natural Home*, July/August, 1999.

37. Alex Wilson. "Daylighting: Energy and Productivity Benefits." *Environmental Building News*, Vol. 8, no. 9 (September 1999), p. 12.

38. Judy Bucher. "The Air Inside: You Are What You Breathe." *Natural Home*, July/August, 1999.

39. Alex Wilson and John Morrill. *Consumer Guide to Home Energy Savings*. American Council for an Energy-efficient Economy, 1991. p. 95.

40. Alex Wilson and John Morrill. *Consumer Guide to Home Energy Savings*. American Council for an Energy-efficient Economy, 1991. p. 113.

Chapter 17 Water Heating

1. Alex Wilson and John Morrill. *Consumer Guide to Home Energy Savings*. American Council for an Energy-efficient Economy, 1991, p. 125.

2. Rob deKieffer. *Combustion Safety Checks: How Not to Kill Your Clients* [online]. [Cited October 13, 2003]. Home Energy Magazine Online, March/April 1995. <http://hem.dis.anl.gov/eehem/95/950308.html>

3. Alex Wilson and John Morrill. *Consumer Guide to Home Energy Savings*. American Council for an Energy-efficient Economy, 1991, p. 126.

4. Alex Wilson and John Morrill. *Consumer Guide to Home Energy Savings*. American Council for an Energy-efficient Economy, 1991, p. 127.

5. Home Energy Magazine. *No Regrets Remodeling*. Energy Auditor and Retrofitter, 1997, p. 115.

6. Iowa Energy Center. *Upgrading Your Current Water Heater* [online]. [Cited October 13, 2003]. Iowa Energy Center, 1997, 1998. <www.energy.iastate.edu/efficiency/residential/homeseries/waterheaters/upgrade.htm>.

7. "No Regrets Remodeling." *Home Energy Magazine.*. Energy Auditor and Retrofitter, 1997, p. 106.

8. Iowa Energy Center. *Upgrading Your Current Water Heater* [online]. [Cited October 13, 2003]. Iowa Energy Center, 1997, 1998. <www.energy.iastate.edu/efficiency/residential/homeseries/waterheaters/upgrade.htm>.

9. Iowa Energy Center. *Upgrading Your Current Water Heater* [online]. [Cited October 13, 2003]. Iowa Energy Center, 1997, 1998. <www.energy.iastate.edu/efficiency/residential/homeseries/waterheaters/upgrade.htm>.

Chapter 18 Appliances

1. Arizona Department of Commerce. *Energy-saving Tips: Appliances* [online]. [Cited October 13, 2003]. Arizona Department of Commerce, 2002. <www.commerce.state.az.us/Energy/tips1.htm>.

2. "New Generation of Horizontal-axis Washing Machines on the Way." *Environmental Building News*, Vol. 6, no. 4 (April 1997).

3. EnergyStar®. *Clothes Washers* [online]. [Cited October 13, 2003]. Environmental Protection Agency, Department of Energy, n.d. <www.EnergyStar.gov/index.cfm?c=clotheswash.pr_clothes_washers>.

4. EnergyStar®. *Dishwashers* [online]. [Cited October 13, 2003]. Environmental Protection Agency, Department of Energy, n.d. <www.EnergyStar.gov/index.cfm?c=dishwash.pr_dishwashers>.

5. EnergyStar®. *Dishwashers* [online]. [Cited October 13, 2003]. Environmental Protection Agency, Department of Energy, n.d. <www.EnergyStar.gov/index.cfm?c=dishwash.pr_dishwashers>.

6. Paul Hawkin, Amory B. Lovins, and L. Hunter Lovins. *Natural Capitalism: Creating the Next Industrial Revolution*. Little, Brown, 1999, pp. 104 to 105.

7. Paul Hawkin, Amory B. Lovins, and L. Hunter Lovins. *Natural Capitalism: Creating the Next Industrial Revolution*. Little, Brown, 1999, p. 102.

8. Alan Meier. *Performance Versus Projections: Does Your Refrigerator Measure Up?* [online]. [Cited October 13, 2003]. Home Energy Magazine Online, January/February 1993. <http://hem.dis.anl.gov/eehem/93/930113.html>.

9. EnergyStar®. *Refrigerators* [online]. [Cited October 13, 2003]. Environmental Protection Agency, Department of Energy, n.d. <www.EnergyStar.gov/index.cfm?c=refrig.pr_refrigerators>.

10. Western Energy Services. *How to Buy Energy Efficient Appliances* [online]. [Cited October 13, 2003]. Washington State University Cooperative Extension Energy Program. <www.es.wapa.gov/pubs/files/appliances.pdf>.

Chapter 19: Interior Materials/Finishes

1. Wayne R. Ott and John W. Roberts. "Everyday Exposure to Toxic Pollutants." *Scientific American*, February 1998.

2. Green Resource Center. *"Greener" Carpet* [online]. [Cited October 15, 2003]. Green Resource Center; Eric Spletzer, n.d. <www.greenresourcecenter.org/MaterialsSheets/GreenerCarpet.html>.

3. The American Lung Association of Michigan. *Traverse City* [online]. [Cited October 15, 2003]. Plante and Moran, 2002. <www.alam.org/programs_services/Air_Quality/traverse_city1.asp>.

4. Green Resource Center. *"Greener" Carpet* [online]. [Cited October 15, 2003]. Green Resource Center; Eric Spletzer, n.d. <www.greenresourcecenter.org/MaterialsSheets/GreenerCarpet.html>.

5. The Green Guide. *Carpets* [online]. [Cited October 15, 2003]. Green Guide Institute, 2003. <www.thegreenguide.com/reports/product.mhtml?id=35>.

6. The Green Guide. *Carpets* [online]. [Cited October 15, 2003]. Green Guide Institute, 2003. <www.thegreenguide.com/reports/product.mhtml?id=35>.

7. The Green Guide. *Carpets* [online]. [Cited October 15, 2003]. Green Guide Institute, 2003. <www.thegreenguide.com/reports/product.mhtml?id=35>.

8. Annie Berthold-Bond. *Better Basics for the Home*. Three Rivers Press, 1999, p. 291.

Glossary

A

Above-grade walls: Those walls on the exterior of the building and completely above grade, or the above grade portion of a basement or first-story wall that is more than 15 percent above grade.

Active solar: A system using mechanical devices (pumps, fans, etc.) that transfers collected heat to the storage medium and/or the end-use.

Addition: An extension or increase in the height, conditioned floor area, or conditioned volume of a building. Building codes apply to additions of existing buildings. Additions include new construction, such as a conditioned bedroom or sunspace. Additions also include existing spaces converted from unconditioned to conditioned spaces (converting an existing porch to a conditioned sun room).

Adhesive: Material that bonds surfaces of different materials. Adhesives may be liquid or tacky semisolids, natural or synthetic, organic or inorganic, waterborne, solvent-borne, or solventless. Solvent-based adhesives can be sources of VOCs (see VOCs definition).

Advanced framing techniques (AFT): Reduce the amount of lumber used to build a home while maintaining the structural integrity of the building. Using AFT results in lower material and labor costs and improved energy performance for the building.

Aggregate: Natural sands, gravels, and crushed stone used for mixing with cementing materials in making mortars and concretes. Can also be added to paint for texture or non-slip flooring surface.

Air barrier: A flexible or rigid membrane designed to reduce the movement of moisture-laden air between the interior and exterior of a building. See also 'air retarder.'

Air changes per Hour (ACH): The number of times in one hour the entire volume of air in a house or room is replaced with outdoor air.

Air handler: The fan on a forced-air heating or cooling system.

Air-to-air heat exchanger: Also known as a heat recovery ventilator, the air-to-air exchanger transfers heat directly from one air stream to another through either side of a metal transfer surface.

Ammoniacal copper arsenate (ACA): A variety of toxic pressure-treated wood.

Annual fuel utilization efficiency (AFUE): A measure of efficiency of a combustion heating appliance. AFUE differs from steady-state efficiency since it employs an empirical equation to deduct all operational losses, such as vent losses, cyclic effects, and part-load operation. New equipment typically ranges from about 78 to 96 percent AFUE. Higher AFUE ratings indicate more efficient equipment.

Arsenic: Poisonous metal that is identified by the Environmental Protection Agency as a persistent, bioaccumulative, toxic pollutant. Obtained as by-product in flue gases from smelting copper, lead, cobalt, and gold ores. May cause acute and chronic toxicity either by inhalation or ingestion, e.g., skin, liver, bladder, kidney, or lung cancer. Direct contact can cause local irritation and dermatitis. Used as a wood preservative, herbicide, pesticide, and in the manufacture of low melting-point glass.

Asbestos: A mineral fiber that has been commonly used in many building construction materials for insulation and as a fire-retardant. Invisible fibers of asbestos may be inhaled and result in a diffuse interstitial fibrosis of the lung tissue (asbestosis) or lung cancer.

Asphalt: Black, semi-soft mixture of hydrocarbons from animal origins. Also, residue left after removing tar tailings from petroleum distillation. Asphalt fumes are carcinogenic. Used in adhesives, coatings (roofs, floors, wood), sealants, rubber, and paint.

Autoclaved cellular concrete (ACC): Concrete made with an aluminum powder that causes the mass to expand. It is then steam-cured, which renders it inert. The resulting material has many benefits (non-combustible, easily worked, R-value of 1.25 per inch, etc.)

B

Backdrafting: The pressure-induced spillage of exhaust gases from combustion appliances into the living space.

Baffle: A single opaque or translucent element used to diffuse or shield a surface from direct or unwanted light.

Ballast: A device used to operate fluorescent lamps. The ballast provides the necessary starting voltage, while limiting and regulating the lamp current during operation.

Batt insulation: Glass or mineral wool, which may be faced with paper, aluminum, or other vapor retarder. Used in walls and ceiling cavities.

Benzene: Very flammable, volatile liquid. Long-term exposure has been associated with bone marrow depression, leukemia, and cancer. Direct contact may cause irritation of the eyes, nose, respiratory system, and skin, as well as narcotic effects. Widely used in the manufacture of polymers, detergents, pesticides, plastics, resins, dyes, and as a solvent in waxes, resins, oils, and natural rubber.

Benzo(a)pyrene: A toxic combustion by-product.

Biocide: Toxic chemical or physical agent capable of killing or inactivating one or more groups of microorganisms such as vegetative bacteria, mycobacteria, bacterial spores, fungi parasites, or viruses. Used in paint, preservatives, floor coverings, or fabrics. Biocides are safe only in low concentrations.

Blackwater: Water from the toilet, kitchen sink, or other sources, which may be contaminated with microorganisms or harmful bacteria, which should not be reused until such sources of contamination are removed.

Blown-in batt: Loose insulation that is installed in wall cavities using a powerful blower and a fabric containment screen.

Blower door: A variable speed fan used to pressurize or depressurize a house to measure air leakage. It is mounted in an adjustable frame that fits snugly in the doorway.

Blowing agent: Also known as foaming agent, this chemical agent is added to plastics and rubbers. When heated, it generates inert gases to give resins in the material cellular structure, thereby strengthening materials like carpet foam cushions. Traditionally, chlorofluorocarbons (CFCs) were used as blowing agents and contributed to the degradation of ozone. Although very few manufacturers use CFCs, some have substituted methylene chloride, a gas that also damages the upper ozone layer.

Borate: Salt or ester of boric acid; a mineral product derived from borax. Used for moisture and insect protection in cellulose insulation applications. Wood treated with borates is resistant to termites and moisture.

British thermal unit (Btu): The energy needed to raise the temperature of 1 pound (0.454 kg) of water 1°F (0.56°C). 1 Btu = 1,055 J

Building envelope: The elements of a building that enclose conditioned spaces through which thermal energy may be transferred to or from the exterior or to or from unconditioned spaces.

Building pressurization: The air pressure within a building relative to the air pressure outside. Positive building pressurization is usually desirable to avoid infiltration of unconditioned and unfiltered air. Positive pressurization is maintained by providing adequate outdoor makeup air to the HVAC system to compensate for exhaust and leakage.

C

Carbon dioxide (CO_2): A heavy, colorless gas that does not support combustion. Made of one carbon atom and two oxygen atoms, it is formed especially in animal respiration and in the decay or combustion of animal and vegetable matter. It is absorbed from the air by plants in photosynthesis, and is an atmospheric greenhouse gas. Significant overexposure from combustion processes in the home may cause headache, dizziness, restlessness, increased heart rate and pulse pressure, and elevated blood pressure.

Carbon monoxide (CO): A colorless, odorless, very toxic gas made up of carbon and oxygen that burns to carbon dioxide with a blue flame and is formed as a product of the incomplete combustion of carbon. Overexposure may result in headache, mental dullness, dizziness, weakness, nausea, vomiting, loss of muscular control, increased then decreased pulse and respiratory rate, and collapse.

Carpet backing: The supportive structure for carpet, providing a durable surface to bond the face fibers to each other and to the flooring system. Backing enhances the dimensional stability, resilience, and comfort; however, it may also be a source of VOC emissions.

Cathedral ceiling: A sloped or vaulted ceiling, usually with the rafters serving as the ceiling joists. There is generally no attic above a cathedral ceiling, although some cathedral ceilings have a short truss attic.

Caulk: Heavy paste incapable of significant expansion or contraction, used to make a seam airtight, watertight, or steamtight. Caulks fabricated with butyl rubber or polyurethane can be sources of VOC emissions indoors.

Cellulose insulation: Thermal insulation manufactured from recycled newspaper, typically treated with natural borates to provide vermin and fire protection. More recently, fungicides and chemical fire retardants have been used and may prove problematic for sensitive individuals.

Certified sustainably managed forest: Some certifying organizations have been established that oversee the harvesting of wood for lumber. The underlying guideline is preservation of a diverse forest that exhibits the same ecological characteristics as a healthy natural forest.

Charrette: An intensive design process that involves the collaboration of all project stakeholders at the beginning of a project to develop a comprehensive plan or design. Although it may only take place over a few short days, it establishes groundwork for communication and a team-oriented approach to be carried throughout the building process.

Chase: A vertical opening, often from the basement all the way to the attic, for ducts, plumbing, telephone, and electrical lines.

Chlorofluorocarbons (CFCs): Any of a group of compounds that contain carbon, chlorine, fluorine, and sometimes hydrogen and have been used as refrigerants, cleaning solvents, aerosol propellants, and in the manufacture of plastic foams. CFCs have been linked to the destruction of the ozone layer and the US Environmental Protection Agency banned their use in 1997.

Chromated copper arsenic (CCA): A variety of pressure treated wood classified as toxic waste by the Environmental Protection Agency.

Cistern: A tank to hold a supply of fresh water. May be above or below ground.

Clerestory: That part of a building rising above the roofs or other parts, whose walls contain windows specifically intended to provide lighting to the interior.

Coefficient of performance (COP): The ratio of the rate of heat delivery or removal to the energy input of the machine in consistent units. COP = energy output / energy input.

Coliform bacteria: These indicate the presence of dangerous microbes in drinking water such as Cryptosporidium, which can be life-threatening to people with weak immune systems.

Combined/Indirect water heater: A boiler made to heat domestic hot water as well as the house air.

Combustion air: The air that chemically combines with a fuel during combustion to produce heat and flue gases, mainly carbon dioxide and water vapor.

Combustion gases: The gases, such as carbon monoxide, that result from the process of burning. In a building, these are produced by gas appliances, such as furnaces and water heaters. Proper venting must be assured.

Compact fluorescent lamp (CFL): A lamp that produces visible light by emitting electromagnetic radiation. Commonly they consist of a glass tube filled with argon, along with krypton or other inert gas. When electrical current is applied, the resulting arc emits ultraviolet radiation that excites the phosphors inside the lamp wall, causing them to radiate visible light. Fluorescent lighting is significantly more efficient than incandescent lighting.

Composting toilet: A toilet that uses little or no water and in which the waste composts to a material that can be safely used as a soil amendment.

Conditioned space: Portion of a building where the air is conditioned to provide comfort to the occupants.

Cooling/heating load: A building's demand for heat/cool to offset a deficit/surplus of the opposite.

Cross ventilation: Passive building strategy which aids in cooling a building using outdoor breezes which enter the building by passing through an opening, then exit the building through another opening, extracting some of the heat from inside the building as it leaves.

Cubic foot of water: The volume of water contained in a cube that measures one foot on each side; equal to 7.48 gallons.

Cullet: Crushed, waste glass that is returned for recycling.

Curing: Conversion of raw products to finished material. In rubber technology, vulcanization. In thermosetting resins, the final cross-linking procedure. In concrete, the process of bringing freshly placed concrete to required strength and quality by maintaining humidity and temperature at specified levels for a given length of time. In textile manufacture, heat treatment with chemicals or resins to impart water or crease resistance.

D

Damper: A device in a duct or vent that restricts airflow.

Damp proofing: A treatment, such as a sealer or asphalt coating, that inhibits the transfer of moisture. Used in locations where higher-quality waterproofing is not needed.

Daylighting: The use of natural light to minimize the need for artificial lighting during the day. Common daylighting strategies include the proper orientation and placement of windows, use of light wells, light shafts or tubes, skylights, clerestory windows, light shelves, reflective surfaces, and shading, and the use of interior glazing to allow light into adjacent spaces.

Dense-pack insulation: Loose-fill insulation (like cellulose) that is blown into enclosed building cavities at a relatively high density so that it will stop air movement and insulate the cavity.

Dimensional lumber: Conventional lumber, such as two-by-fours and two-by-sixes, typically used for studs, joists and rafters.

Directional lighting: The distribution of all, or nearly all, of the light from a fixture in one direction.

Domestic hot water: The water that comes out of the taps or appliances, as distinguished from the hot water circulated through the radiators and pipes of a hydronic heating system for the house.

Double cropping: When a farmer can grow crops while leasing the wind rights for utility scale wind turbine(s). The income-generating wind turbine is essentially a second, or "double" crop.

Drip irrigation: Aboveground, low-pressure watering system with flexible tubing that releases small, steady amounts of water through emitters placed near individual plants.

Drywall: Also called gypsum board or Sheet rock. The most common interior wall and ceiling finish material. Can be part of the home's air barrier if it is sealed to make the barrier continuous.

Duct System: A continuous passageway for the transmission of air that, in addition to ducts, includes duct fittings, dampers, plenums, fans, and accessory air-handling equipment and appliances.

E

Egress window: A window large enough for a person to climb out in case of emergency. There are national egress requirements and egress requirements dictated by local building codes.

Electronic ballast: Type of ballast for a fluorescent light, which increases efficiency and reduces flicker and noise.

Embodied energy: The energy required to grow, harvest, extract, manufacture, refine, process, package, transport, install and dispose of a particular product or building material. A useful measure of ecological cost.

Emissivity or Emittance: The property of radiating heat. Values range from 0.05 for brightly polished metals to 0.96 for flat black surfaces. Most non-metals have high values of emittance.

Energy efficiency ratio (EER): The ratio of cooling capacity in BTU per hour divided by the electrical power input in watts under designated operating conditions.

Energy Heel: Elevating the heel of roof/ceiling construction to allow the insulation to achieve its full thickness over the plate line of exterior walls. Sometimes referred to as an energy truss, raised-heel truss, or Arkansas truss.

Energy factor (EF): The portion of the energy going into the water heater that gets turned into usable hot water under average conditions. Takes into account heat lost through the walls of the tank, up the flue, and in combustion.

Engineered wood: Reconstituted wood products that result in appropriate-to-use strength and consistent quality with less material. Engineered products use less wood or wood from smaller trees and are generally stronger, straighter, and of more consistent quality than dimensional lumber. Examples are engineered I-beams, laminated veneer lumber, laminated strand lumber, and finger-jointed studs.

Environmental cost: A quantitative assessment of impacts such as resource depletion; air, water, and solid waste pollution; and disturbance of habitats.

Environmental impact: The net change (positive or negative) in human health and the condition of the environment that results from human actions, activities, or development.

Environmental Protection Agency (EPA): An independent executive agency of the US federal government, established in 1970, responsible for the formulation and enforcement of regulations governing the release of pollutants, to protect public health and the environment.

Evaporative cooling: Passive building strategy employing the evaporation of water directly into hot, dry air streams to produce cooling; limited to arid climates. Also called a swamp cooler.

Expanded polystyrene (EPS): A rigid insulation material (also called bead board) made by heating pentane-saturated polystyrene pellets. (Pentane is used instead of the CFCs or HCFCs used to make extruded polystyrene, which cause damage to the ozone layer.) Frequently has a high recycled content. Comes in various densities for different purposes.

Exterior grade plywood: Uses phenol formaldehyde (a volatile organic compound) as an adhesive that is released in much smaller amounts compared to urea formaldehyde used in interior grade plywood and particle board.

Extruded polystyrene (XPS): Rigid insulation that has high insulation value and is largely impervious to water. XPS is manufactured using HCFCs as the blowing agent; HCFCs are strong greenhouse gases.

F

Fiberglass: Produced by spinning molten glass. Used for fireproof and acid-resistant textiles, in fiber-reinforced plastics, and in insulation. Man-made mineral fibers used to make fiberglass are suspected carcinogens.

Finger-jointed: High-quality lumber formed by joining small pieces of wood glued end to end, so named because the joint looks like interlocked fingers.

Fire retardant: Material that reduces or eliminates the tendency of flammable or combustible materials to burn. Fire retardants may be applied to the surface, impregnated, or incorporated during polymerization of plastics and rubbers.

Flat plate collector: An assembly containing a panel of metal or other suitable material that absorbs sunlight and converts it into heat. In the collector, the heat transfers to a circulating liquid or gas and is either utilized immediately or stored for later use.

Fly ash: Fine, non-combustible particulate suspended in flue gases during coal combustion, from which it is collected. Most commonly used to replace a portion of Portland cement. Fly ash is also used for fill material, soil stabilization and waste remediation. Also called bottom ash.

Forest Stewardship Council (FSC): An independent, non-profit organization that promotes responsible forest management by evaluating and accrediting certifiers, by encouraging the development of national and regional forest management standards, and by providing public education about independent, third-party certification.

Formaldehyde: Poisonous, reactive, flammable gas with pungent suffocating odor. Many cause irritation of the eyes, nose, throat, and respiratory system. Contact may result in sensitization. Carcinogen. Used in wood products, plastics, fertilizer, and foam insulation. Incorporated in synthetic resins by reaction with urea, phenols and melamine. Urea-formaldehyde (UF) resin is used in particleboard, hardwood plywood paneling, and medium-density fiberboard (MDF). MDF is generally the highest formaldehyde-emitting pressed wood product. Softwood plywood and oriented strand board (OSB) produced for exterior applications contain phenol-formaldehyde (PF) resin. Pressed wood products containing PF resin generally emit less formaldehyde than those containing UF resin.

Frost-protected shallow foundation: A foundation which is 12 to 16 inches below grade and has been insulated to protect against frost heave, rather than extending to below the local frost line to protect against frost heave.

Fungicide: An agent that destroys molds, mildews and yeasts. The destruction of fungi does not necessarily destroy its toxic or allergenic properties.

G

Gas-filled window: Double- or multiple-glazed window systems where the air space(s) is (are) filled with a low-conducting gas, like argon or krypton. The fill reduces the heat exchange rate associated with windows. Gas-filled windows provide the greatest benefit when used in conjunction with low-emissivity coatings.

Geothermal heat pump: In winter, geothermal heat pump technology utilizes heat from subsurface water and soils to heat buildings; in summer, this technology extracts heat from the building into subsurface water and soils for cooling.

Glazing: Any translucent or transparent material in exterior openings of buildings, including windows, skylights, sliding doors, the glass area of opaque doors, and glass block.

Greywater: Water that has been used for showering, clothes washing, and faucet uses. Kitchen sink and toilet water is excluded. This water can be reused in subsurface irrigation for yards.

Green roof: Vegetation cover on roof surfaces.

Greenhouse gas: Any of several dozen heat-trapping (radiatively active) trace gases in the earth's atmosphere that absorb infrared radiation. The two major greenhouse gases are water vapor and carbon dioxide; lesser greenhouse gases include methane, ozone, CFCs, and nitrogen oxides.

Grid-connected: Attached to the utility electric service, or grid.

Grout: Fluid mixture of cement, water, and possibly sand.

H

Hardwood: Deciduous trees with broader leaves and slower growth rates compared to conifers, or softwoods. Examples of hardwoods include oak, maple, cherry, walnut, beech, birch, cypress, elm, and hickory. Used in furniture and flooring, for appearance, excellent durability, and resistance to wear.

Harvested rainwater: The rain that falls on a roof or yard and is channeled to a storage tank (cistern). The first wash of water on a roof is usually discarded and the subsequent rainfall is captured for use if the system is being used for potable water. Good quality water is available by this method in most areas.

Hazardous chemical: Any hazardous material requiring a Material Safety Data Sheet under the Occupational Safety and Health Administration's hazard communication standard. Includes those associated with physical hazards such as fire and explosion, or health hazards such as cancer and dermatitis.

Hazardous waste: A class of waste materials that poses immediate or long-term risks to human health or the environment and requires special handling for detoxification or safe disposal. Both industrial and household wastes include hazardous materials.

Heat exchanger: A device that transfers heat from one medium to another, physically separated medium of different temperature.

Heat island effect: The rise in ambient temperature that occurs over large paved areas. Strategic placement of trees can reduce this effect and reduce energy consumption for cooling by 15 to 30 percent.

Heat pump: A mechanical device that removes heat from one medium, concentrates it, and distributes it in another. This device can be used to heat or cool indoor space. The heat source can be air or water.

Heat reclaimer: A device that takes waste heat from an air conditioner, heat pump, or ground source heat pump and diverts it into the domestic hot water.

Heat recovery ventilator (or air-to-air heat exchangers): Exhaust fans that warm the incoming air with the heat from the outgoing air, recovering about 50 to 70 percent of the energy. In hot climates the function is reversed so that the cooler inside air passes by the incoming hot air and reduces its temperature.

Heating seasonal performance factor (HSPF): A rating for heat pumps describing how many Btus they transfer per kilowatt-hour of electricity consumed.

Heating, ventilating and air-conditioning (HVAC) system: The equipment, distribution network, and terminals that provide either collectively or individually the processes of heating, ventilating, or air conditioning to a building.

Heavy metals: Elements such as copper, lead, cadmium, mercury, and other toxic metals used in industrial processes and often released as both air and water pollutants. They may accumulate to hazardous concentrations in sediments and sludge. Many are classified by EPA as persistent, bioaccumulative, toxic pollutants.

High-efficiency particulate air (HEPA) filter: A designation for very fine air filters (usually exceeding 98 percent atmospheric efficiency) typically used only in surgeries, clean rooms, or other specialized applications.

Horizontal axis (H-axis) washing machine: A washing machine designed to clean without an agitator. It uses much less water than vertical-axis models, reduces wear and tear on clothes, and result in drier clothes.

House wrap: Material, such as Tyvek, that acts as a barrier to protect insulation from wind. Can be an air barrier, but only if the seams are all sealed and the material is sealed to adjoining parts of the house barrier.

Humidistat: Device for measuring and controlling relative humidity.

Hydrochlorofluorocarbon (HCFC): Hydrogenated chlorofluorocarbons. Since HCFCs are generally less detrimental to ozone depletion (one-twentieth as potent as CFCs), they are a substitute for CFCs. A total ban on all CFCs and HCFCs is scheduled effective 2030.

Hydronic: A heating system using hot water or steam as the heat-transfer fluid; a hot-water heating system (common usage).

I

I-joist: A manufactured wood product so named because in cross-section it looks like an upper case 'I.' The top and bottom chord are lumber or laminated wood, and the vertical web is plywood or oriented strand board.

Ice dam: the bands of ice that form along the eaves of many roofs during cold winters, caused by heat leaking from the house into the attic.

Impervious surface: Area that has been sealed and does not allow water to infiltrate, such as roofs, plaza, streets, and other hard surfaces.

Incandescent bulb: An incandescent bulb is the most common and least energy-efficient lamp. Electricity runs through a tungsten filament that glows and produces a soft, warm light. Because so much of the energy used is lost as heat, these are highly inefficient sources of light.

Indoor air quality (IAQ): Air in which there are no known contaminants at harmful concentrations, as determined by cognizant authorities, and with which a substantial majority (80 percent or more) of the people exposed do not express dissatisfaction. IAQ is heavily influenced by both choice of building materials (and cleaning procedures) and ventilation rates.

Inert: Inert ingredients listed by a manufacturer are distinct from the product's active ingredients. However, these materials may not be inert from a health standpoint.

Infiltration: The uncontrolled inward air leakage through cracks and interstices. Infiltration occurs in any building element and around windows and doors of a building. It is caused by the pressure effects of wind or the effect of differences in the indoor and outdoor air density or both. In stormwater management, infiltration refers to the entry of runoff into the soil.

Insulating sheathing: An insulating board with a minimum thermal resistance of R-2 of the core material.

Insulation: Material that has a resistance to transfer of energy (e.g., acoustic, electric, thermal, vibrational, or chemical).

Inverter: Device that converts direct currents (DC) generated by photovoltaic modules to alternating current (AC), or energy currents used by utility companies.

J

Joists: The horizontal structural framing for floors and ceilings. The ceiling joists of the first story are also the floor joists of the second story, or of the attic.

Joist cavity: The space between joists.

K

Kelvin (K): The standard unit of temperature that is used in the Systeme Internationale d'Unites (SI) system of measurements. The Kelvin temperature scale is used to describe the correlated color temperature of a light source.

Kilowatt (kW): A unit of power equal to 1,000 watts. It is used as a measure of electrical power. On a hot summer afternoon a home with central air conditioning and other equipment in use might have a demand of 4 kilowatts each hour.

Kilowatt-hour (kWh): A measure of energy usage equal to the amount of power multiplied by the amount of time the power is used. A 100-watt light bulb burning for ten hours uses one kilowatt-hour of power.

L

Laminate: A thin layer of material (veneer) bonded to another surface. Wood and plastics are both commonly laminated.

Laminated veneer lumber: A manufactured wood product similar to plywood but made in thick sections with all the grain oriented one way for use as beams.

Latex: A naturally occurring, sticky resin from rubber tree sap used for rubber products, carpet backings, and paints.

Lead: Soft heavy metal. A harmful environmental pollutant that can be found in the home in lead-based paints and in lead solder used in plumbing before 1978. Lead is toxic to many organs and can cause serious damage to the brain, kidneys and nervous system.

Leadership in Energy and Environmental Design (LEED): A building rating and certification program developed and operated by the US Green Building Council.

Leak detection: Systematic method of using listening equipment to survey the water distribution system, identify leak sounds, and pinpoint the exact locations of hidden underground leaks.

Life cycle cost (LCC): The cost of a design feature that allows for production, sales, operation, maintenance, and demolition or recycling costs. The cost also encompasses all the environmental burdens of a product or process through its entire service life. This approach can often be used to justify more expensive and energy efficient systems, which save money over the life of the product.

Light shelf: Horizontal projections at the building interior that reflect direct sun rays onto the ceiling deep into a space and that shield direct sunlight from the area immediately adjacent to the window. The upper surface of the shelf may be specular or nonspecular but should be highly reflective (that is, having 80 percent or greater reflectance). Light shelves work best on facades that are generally south facing.

Linoleum: Natural linoleum is made from natural, minimally processed ingredients, including linseed oil, pine rosin, cork dust, wood flour, limestone, mineral pigments, and jute backing. Offgassing will depend primarily on the adhesive used for installation. Some long-term offgassing will occur with the linoleum due to oxidation of the linseed oil, resulting in emission of a mixture of aldehydes and carboxylic acids, which are not considered hazardous but may affect sensitive individuals.

Linseed oil: Derived from flaxseed by crushing and pressing. Used as a vehicle in oil paints and as a component of oil varnishes.

Loose-fill insulation: insulation made from vermiculite, perlite, glass, or mineral wool, shredded wood, or shredded paper, loosely packed to allow pockets of dead air space. For use in wall and ceiling cavities.

Louvers: A series of baffles used to shield a light source from view at certain angles, or to absorb unwanted light, or to allow selective ventilation.

Low-emissivity windows (low-e): Glazing that has special coatings to permit most of the sun's light radiation to enter the building, but prevents heat radiation from passing through.

Low-flush toilet: a toilet that requires 1.6 gallons of water per flush or less, as compared to the 3.5 to 5 gallons of water required to flush most older standard toilets.

M

Mastic: A thick, creamy substance used to seal ducts and cracks in buildings. It dries to form a permanent, flexible seal.

Mechanical system: The system and equipment used to provide heating, ventilating, and air conditioning functions as well as additional functions not related to space conditioning, such as, but not limited to, freeze protection in fire protection systems and water heating.

Medium Density Fiberboard (MDF): Composite panel product generally made from wood fibers combined with a synthetic resin or other bonding system and joined together under heat and pressure. Used in kitchen cabinets, paneling, doors, jambs and millwork, and laminate flooring.

Megawatt (MW): A million watts, or 1,000 kilowatts.

Methane (CH₄): An odorless, colorless, flammable gas that is a major component of natural gas; it is a more powerful global warming agent than carbon dioxide.

Mineral fibers: Very fine insulation fibers made from glassy minerals that have been melted and are hazardous to inhale.

Mineral wool: A fibrous glass made from molten slag, rock, and/or glass, produced by blowing or drawing, used for insulation and fireproofing. Also known as rock wool.

Moisture barrier: A material, membrane, or coating used on the inside of foundation, below grade, to keep moisture from penetrating interior wall or flooring materials.

N

Natural cooling: Use of environmental phenomena to cool buildings, e.g., natural ventilation, evaporative cooling, and radiative cooling.

Natural draft: A vent system that relies on the forces of gravity to draft exhaust gases out of the house.

Negative indoor pressure: When air is exhausted from a space faster than it is replaced.

Net metering: A program in which electric utilities buy power from customers with photovoltaics at the same rate as the utility charges the customer for electricity.

O

Occupancy sensor: A device that detects the presence or absence of people within an area and causes any combination of lighting, equipment, or appliances to be adjusted accordingly.

Offgassing: The release of gases or vapors into the air.

Old growth: Wood from trees found in mature forests. In many cases, the trees have never been exposed to logging operations. In the northwest United States, only about 10 percent of these biologically rich areas are left.

On center: The distance from the center of one stud or joist to the center of the adjacent stud or joist. Most homes are built with studs and joists 16 inches on center.

Organic compound: Chemical compound based on carbon chains or rings, and containing hydrogen with or without oxygen, nitrogen, or other elements. Organic compounds are the basis of all living things; they are also the foundation of modern polymer chemistry.

Orientation: The relation of a building and its associated fenestration and interior surfaces to compass direction and, therefore, to the location of the sun. It is usually given in terms of angular degrees away from south, i.e., a wall facing due southeast has an orientation of 45 degrees east of south.

Oriented strand board (OSB): A manufactured wood sheet product made from large flakes of wood pressed together with glue, usually a dry phenolic type. OSB is used for structural sheathing and subfloors.

Outgas: When a solid material releases volatile gases as it ages, decomposes, or cures.

Ozone (O₃): A molecule made of three oxygen atoms instead of the usual two. Ozone is a poisonous gas and an irritant at the earth's surface, capable of damaging lungs and eyes. But the ozone layer in the stratosphere shields life on earth from deadly ultraviolet radiation from space.

P

Particleboard: Sawdust and resin compressed into sheets. See formaldehyde.

Passive cooling: The building's structure (or element of it) is designed to permit increased ventilation and retention of coolness within the building components. The intention is to minimize or eliminate the need for mechanical means of cooling.

Passive solar: The building's structure (or an element of it) is designed to allow natural thermal energy flows such as radiation, conduction, and natural convection generated by the sun to create heat. The home relies solely or primarily on non-mechanical means of heating.

Permeable paving: Paving material that allows water to penetrate the soil below; this reduces the amount of water that needs to be treated by the water system and increases water in the aquifer.

Pesticide: Lethal chemical that destroys pests, e.g., insects, rodents, nematodes, fungi, seeds, viruses, or bacteria. Term includes insecticides, herbicide, rodenticide, and fungicide. Also known as biocide. Many pesticides have been identified by the Environmental Protection Agency as persistent, bioaccumulative toxic pollutants. Alternative, safe forms of controlling pests include use of sticky or mechanical traps, companion planting and natural deterents like tomato leaf spray.

Phenol-formaldehyde: Although both phenol and formaldehyde are toxic, phenolic resin is said to be inert as a fully cured polymer and a relatively low emitter of formaldehyde compared to urea formaldehyde resin. Used as adhesive for particleboard, decorative laminate backing, coatings, and insulation. See Formaldehyde.

Photocell: A light sensing device used to control luminaires and dimmers in response to detected light levels.

Photovoltaic (PV) cells: Thin silicone wafers that convert any light, not only sunlight, directly into electricity.

Plywood: Material composed of thin sheets of wood glued together, with the grains of adjacent sheets oriented at right angles to each other.

Polyethylene: A semitransparent plastic used in sheets as vapor barriers or for packaging and containers. Made in high density (HDPE) and low density (LDPE) varieties. Low in toxicity, it produces low risk vapors when burned.

Polyvinyl chloride (PVC): Thermoplastic polymer of vinyl chloride. Rigid material with good electrical properties and flame and chemical resistance. PVC is a known human carcinogen. Due to the environmental releases during manufacture, it is banned in many parts of Europe. Greenpeace has developed an online resource to PVC alternatives. Used in soft flexible films, including flooring, and in molded rigid products like pipes, fibers, upholstery, and siding. Identified by a "3" inside a recycling triangle found on packaging.

Positive indoor pressure: When air is delivered to a space faster than it is exhausted.

Pressure-treated wood: Wood that is chemically preserved to prevent moisture decay. The chemicals typically used are health hazards for workers. Such wood should not be burned because it produces toxic fumes, and must be treated as a hazardous waste when disposed of.

R

R-value: R-values are used to rate insulation and are a measurement of the insulation's resistance to heat flow. The higher the R-value, the better the insulation. R-value is the reciprocal of U-factor.

Radiant barrier: A material (typically an aluminum foil) that is good at blocking the transfer of radiant heat across a space. In a hot climate it is often installed in attics under the roof decking to keep the attic cooler.

Radon gas: A radioactive, colorless, odorless gas that occurs naturally in soil in many areas. When trapped in buildings, concentrations build up, and can cause health hazards.

Rapidly renewable wood: Quickly growing trees that offer the general benefits of trees (helping replenish oxygen in the air, and removing harmful CO_2), but also offer a consistent supply of material for construction. Bamboo and Aspen are two examples.

Recirculating system: A distribution system that keeps hot water running in the pipes to provide immediate hot water at the taps.

Reclaimed wood: Wood salvaged from buildings that are being remodeled or torn down, abandoned railroad trestles and sometimes logs that sank decades ago during river log drives. It is desirable from an environmental perspective because is not associated with recent timber harvesting, it reuses materials, and it can reduce the construction and demolition load on landfills. Additionally, reclaimed wood is often available in species, coloration and wood quality not found in today's forest.

Recycled plastic lumber: Structural components fabricated from recycled plastic as a replacement for lumber. Insect and water resistant.

Reflective glass: Glass coated with metallic oxide or other material to increase the amount of solar energy reflected. The result is a reduction in the cooling load within the building.

Reflector: Highly reflective backing inserted behind a light fixture, used to reclaim lighting and improve efficiency.

Register: A grille covering a duct outlet.

Relative humidity: The percent saturation of a moist air mixture at given conditions. Hot air can hold more moisture than cold air before reaching its dew point and condensing.

Renewables: Resources that are created or produced at least as fast as they are consumed, so that nothing is depleted. If properly managed, renewable energy resources (e.g., solar, hydro, wind power, biomass, and geothermal) should last as long as the sun shines, rivers flow, wind blows, and plants grow.

Resilient flooring: includes rubber, vinyl, or linoleum floor coverings. These floorings are able to return to their original forms after subjection to static or dynamic loads, or sudden impact.

Retrofit: The replacement, upgrade, or improvement of a piece of equipment or structure in an existing building or facility.

Return air: Air that has circulated through a building as supply air and has been returned to the HVAC system for additional conditioning or release from the building.

Return ducts: Ducts in a forced-air heating or cooling system that bring house air to the furnace or air conditioner to be heated or cooled.

Reverse osmosis: Common process used to produce de-ionized water from municipal water.

Ridge vent: A vent installed continuously along the ridge of the attic peak, to allow ventilation air from the soffit vents to flow out of the attic.

Rigid insulation: Insulation, such as foamed plastic, wood, cork, glass or mineral fibers, pressed into standard-sized boards for easy handling. Used as a surface insulation.

Rock wool: See Mineral wool.

Roof Assembly: Considered to be all roof/ceiling components of the building envelope through which heat flows, thus creating a building transmission heat loss or gain, where such assembly is exposed to outdoor air and encloses conditioned space.

Roof truss: A form of structural support of a building's roof.

S

Sealed combustion: A type of fossil-fuel heater that provides outside air directly to the burner rather than using household air.

Seasonal Energy Efficiency Ratio (SEER): The total cooling output of an air conditioner during its normal annual usage period for cooling, in Btu/h (W), divided by the total electric energy input during the same period, in watt-hours. New equipment ranges from about 10 to 16 SEER. Higher SEER ratings indicate more efficient equipment.

Sheathing: A structural component, such as plywood, installed as part of the wall assembly. Sometimes rigid insulation board is also called nonstructural wall sheathing.

Sick building syndrome (SBS): According to the Environmental Protection Agency and The National Institute for Occupational Safety and Health (NIOSH), Sick Building Syndrome is defined as "situations in which building occupants experience acute health and/or comfort effects that appear to be linked to time spent in a particular building, but where no specific illness or cause can be identified. The complaints may be localized in a particular room or zone, or may be spread throughout the building." Occupants experience relief of symptoms shortly after leaving the building.

Skylight: Glazing that is horizontal or sloped at an angle less than 60° (1.1 rad) from horizontal.

Slab-On-Grade Floor: A floor that is poured in direct contact with the earth.

Slag: Nonmetallic by-product of smelting and refining of metals. Used in cement manufacture, lightweight concrete, and rock wool.

Soffit: The exposed underside of the part of a roof that extends out beyond the top of the wall.

Soffit vent: A screened vent located in the soffit of the roof overhang that allows air to flow through the eaves into the attic.

Softwood: Wood from a coniferous tree, such as pine, fire, hemlock, spruce, or cedar. Softwoods are fast growing and primarily used for construction. See also Hardwood.

Solar collector: Device that uses the sun's energy to perform some kind of mechanical advantage which would normally be supplied by an non-renewable energy source. Photovoltaic (PV) panels which convert the sun's energy directly into electricity, and solar hot water panels, which preheat water before sending it into a hot water heater are two examples.

Solar heat gain coefficient (SHGC): Glazing's effectiveness in rejecting solar heat gain. SHGC is part of a system for rating window performance used by the National Fenestration Rating Council (NFRC). SHGC is gradually replacing the older index, shading coefficient (SC), in product literature and design standards. If you are using glass whose performance is listed in terms of SC, you may convert to SHGC by multiplying the SC value by 0.87.

Solar reflectance: The percentage of the sun's energy that is absorbed as heat by a roofing material.

Solar water heater: A water heating system in which heat from the sun is absorbed by collectors and transferred by pumps to a storage unit. The heated fluid in the storage unit conveys its heat to the domestic hot water of the building through a heat exchanger.

Sone: A sound rating. Fans rated 1.5 sones and below are considered very quiet.

Source reduction: Elimination of waste at the beginning of a process. Sometimes called "pre-cycling".

Stack effect: The upward movement of air in a building or chimney, due to the buoyancy of warm air. It is responsible for pushing warm house air out though leaks in the upper floors and drawing air in on the lower levels in winter.

Standby loss: The energy lost through the walls and up the flue of a hot water tank.

Storage water heater: A water heater that stores hot water in a tank; the most common type of water heater.

Structural insulated panels (SIPs): A type of building system combining exterior sheathing, structural support, and insulation, and interior sheathing into one modular factory-assembled unit, thus reducing the number of vertical joints, interior voids, and assembly time.

Studs: vertical framing in the walls of the house.

Superwindows: Double- or triple-glazed window sandwiches that contain a center sheet of coated mylar "low-emissivity" film and are filled with argon or krypton gas. This construction and the coating on the film allow short-wave radiation (visible light) to pass through, but reflects long-wavelength radiation (infrared or heat) so heat cannot pass through. R-values of 4.5 or more are achieved.

Supply air: The total quantity of air supplied to a space of a building for thermal conditioning and ventilation. Typically, supply air consists of a mixture of return air and outdoor air that is appropriately filtered and conditioned.

Supply ducts: The ducts in a forced-air heating or cooling system that supply heated or cooled air from the furnace or air conditioner to the house.

T

Tankless/Instantaneous water heater: A water heater that heats water as needed with no storage tank, using large electric elements or gas burners.

Task lighting: Lighting to provide illumination for a specific activity in a specific place.

Thermal boundary: The border between conditioned and unconditioned space, where insulation is placed. Should be lined up with the air barrier.

Thermal break: A material of low heat conductance used to reduce the flow of heat. For example, the vinyl separating the interior and exterior frames in some metal windows.

Thermal bridge: A component, or assembly of components, in a building envelope through which heat is transferred at a substantially higher rate than through the surrounding envelope area.

Thermal mass: Materials that have a high capacity for absorbing heat, and change temperature slowly. These materials are used to absorb and retain solar energy during the daytime for release at night or during cloudy periods. They include water, rocks, masonry, and earth.

Thermal resistance (R): See R-value.

Tight buildings: Buildings that are designed to let in minimal infiltration air in order to reduce heating and cooling energy costs. In actuality, buildings typically exhibit leakage that is on the same order as required ventilation; however, this leakage is not well distributed and cannot serve as a substitute for proper ventilation.

Tipping fees: Fees charged for dumping large quantities of trash into a landfill.

Tint: A mineral coloring incorporated into the glass pane of a window to reduce solar heat gain. Tints also reduce visible transmittance through the window.

Ton: A common unit of measure for central air conditioning systems. (Before mechanical refrigeration, block ice was used for cooling. A ton of ice – 2,000 pounds— absorbs 12,000 Btu of heat per hour.) The actual capacity of the air conditioner will depend on the temperature and relative humidity in your area.

Topsoil: The uppermost soil horizon (layer), containing the highest amounts of organic material; depth varies greatly from region to region.

Trihalomethanes (THMs): Chemicals formed when chlorine used to disinfect water reacts with organic matter, such as animal waste, treated sewage, or leaves and soil. They can increase the risk of cancer and may damage the liver, kidneys, and nervous system, and increase rates of miscarriage and birth defects.

Tropical hardwood: Wood products harvested from tropical rainforests. Tropical rainforests are not being harvested sustainably except in a few isolated cases.

Truss: Structural members of an engineering material (like wood) joined together at their ends with connectors, normally in a triangular pattern for a (stable) rigid framework.

Tubular skylights: Channels daylight into dimly lit areas for residential and commercial buildings. They spread natural light evenly throughout a desired area.

U

U-factor: A measure (British thermal units per hour square foot degrees Fahrenheit) of how well a material or series of materials conducts heat. U-factors for window and door assemblies are the reciprocal of the assembly R-value. The smaller the number, the less the heat flow.

Ultraviolet (UV) radiation: Electromagnetic radiation, usually from the sun, that consists of wavelengths in the range of 4 to 400 nanometers; shorter than the violet end of the visible spectrum. UV radiation is a health hazard that can lead to skin cancer or cataracts.

Unconditioned space: An enclosed space within a building that is not insulated and/or heated.

Underlayment: A sheet material laid under finish flooring material to minimize irregularities in the subfloor or to add acoustic separation.

Underwriters Laboratories (UL): the US testing agency responsible for verifying product or electrical safety, fire ratings, etc.

Urea formaldehyde: Synthetic resin made by the reaction of urea with formaldehyde. Used as binder for interior composite wood products. See Formaldehyde.

V

Vapor barrier: A material that prevents or drastically reduces the passage of water in vapor form. In cold climates, vapor barriers are typically installed on the inside of the wall frame. In hot humid climates, they are installed on the outside, or preferably omitted entirely. In hot dry climates they are not needed.

Varnish: Clear or pigmented surface coating that changes to a hard solid when dried from a liquid.

Vinyl chloride: Flammable, explosive gas. Highly toxic, damages the liver, affects the central nervous system, blood, respiratory system. Carcinogen. Used widely in plastics and adhesives.

Vinyl flooring: Flooring manufactured from polyvinyl chloride (PVC) resin over a backing material like paper or foamed plastic. See also Polyvinyl chloride.

Vinyl wall covering: Sheets of PVC colored and usually embossed, and laminated to a backing that imparts stability and adhering qualities. The backing is usually paper or a loosely woven fabric of polyester or cotton. See also Polyvinyl chloride.

Volatile organic compound (VOC): A class of chemical compounds that can cause nausea, tremors, headaches, and, some doctors believe, longer-lasting harm. VOCs can be emitted by oil-based paints, solvent-based finishes, and other products on/in construction materials.

W

Waste heat: Heat escaping from combustion that can be captured and used for other processes.

Waste water: Spent or used water from individual homes, a community, a farm, or an industry that contains dissolved or suspended matter.

Water-based paint: Paint in which the vehicle or binder is dissolved in water or in which the vehicle or binder is dispersed as an emulsion. Latex paint is a water-based paint.

Weatherstripping: Thin strips of metal, rubber, vinyl or foam around doors and windows that prevent infiltration of air or moisture.

Wet spray: A method of applying insulation by adding water or an adhesive and spraying material into open cavities, such as walls.

Whole-house fan: A fan typically centrally located in the ceiling of a house that draws fresh outside air into the living space, flushes hot air up to the attic and exhausts it to the outside. Windows must be open and adequate venting area must be present in the attic which must be well sealed and insulated during cold weather. Whole-house fans are an inexpensive way to cool a house when outside air is cooler than inside air, and not excessively humid.

Whole-systems thinking: A process through which the interconnections of systems are actively considered, and solutions are sought that address multiple problems at the same time.

Wind energy: Energy from moving air, which is converted to electricity, by using wind to turn electricity generators. Wind energy has a number of advantages over conventional forms of energy. It is pollution-free and renewable.

Wind farm: A cluster of wind turbines (up to several hundred) for generating electrical energy, erected in areas where there is a nearly steady prevalent wind; such areas generally occur near mountain passes.

Wind Turbine: A windmill that generates electricity.

Window: The terms "fenestration," "window," and "glazing" are often used interchangeably. However, window actually describes a system of several components. Window is the term given to an entire assembly comprised of the sash, glazing, and frame.

X

Xeriscaping: Landscaping design for conserving water that uses drought-resistant or drought-tolerant plants (trademarked name).

Z

Zone: A space or group of spaces within a building with any combination of heating, cooling, or lighting requirements sufficiently similar so that desired conditions can be maintained throughout by a single controlling device.

Index

Locators in boldface indicate tables and locators in italics indicate figures

About the Authors

DAVID JOHNSTON has been in the construction industry designing, building, and consulting on environmental construction for over 30 years. He founded *Lightworks Construction, Inc.* in Bethesda MD, was named Builder of the Year by the Washington, D.C. chapter of the National Association of the Remodeling Industry and has been inducted into the *Remodeling Magazine* Hall of Fame.

Johnston founded *What's Working, Inc.* in 1992 and became the original designer of the Denver Metro Home Builders Association's Built Green Program. David is also the main designer and consultant to the Alameda County Waste Management Authority's program, stimulating market transformation programs in the San Francisco Bay Area.

He presents his expertise across the country and internationally to groups including real estate professionals, builders, developers, city/state officials, civic organizations, utilities, and universities. Johnston has received numerous awards for his work, including the prestigious Corporate Excellence Award for Sustainable Development from the University of Colorado, and was named Environmental Hero for 2004 by *Interiors and Sources Magazine.* He has represented the United States at International Energy Agency meetings to develop international research agreements in the area of sustainable construction.

David lives and works in Boulder, Colorado.

KIM MASTER, LEED® AP has focused her education and work experience on national and international environmental and health issues for over 10 years. She holds BA in Human Biology, an MA in Anthropology, and a minor in photography from Stanford University, in addition to being a Leadership in Energy and Environmental Design Accredited Professional (LEED® AP). Since joining *What's Working* in April 2003, Kim has consulted on various green building and sustainable business initiatives throughout the country, including projects for Alameda County, CA, The University of Colorado, Future500, and Sustainable Travel International. She has also lead the "Sustainable Building Solutions" track for Sustainable Resources 2004 Conference; and successfully helped initiate, organize and manage the Boulder Green Building Guild, a group of building professionals working to make green building common practice. Kim lives, works and plays in Boulder, Colorado, with partner Noah and Ginger the dog.

If you have enjoyed *Green Remodeling,* you might also enjoy other

BOOKS TO BUILD A NEW SOCIETY

Our books provide positive solutions for people who want to make a difference. We specialize in:

**Sustainable Living • Ecological Design and Planning • Natural Building & Appropriate Technology
New Forestry • Environment and Justice • Conscientious Commerce • Progressive Leadership
Educational and Parenting Resources • Resistance and Community • Nonviolence**

For a full list of NSP's titles, please call 1-800-567-6772 or check out our web site at:
www.newsociety.com

New Society Publishers

ENVIRONMENTAL BENEFITS STATEMENT

New Society Publishers has chosen to produce this book on recycled paper made from
100% post consumer waste, processed chlorine free, and old growth free.

For every 5,000 books printed, New Society saves the following resources:[1]

67	Trees
6,033	Pounds of Solid Waste
6,638	Gallons of Water
8,658	Kilowatt Hours of Electricity
10,967	Pounds of Greenhouse Gases
47	Pounds of HAPs, VOCs, and AOX Combined
17	Cubic Yards of Landfill Space

---

[1]Environmental benefits are calculated based on research done by the Environmental Defense Fund and other members of the Paper Task Force who study the environmental impacts of the paper industry.

NEW SOCIETY PUBLISHERS